Livestock Farming

Livestock Farming

Editor: Elly Campbell

MURPHY & MOORE
www.murphy-moorepublishing.com

Murphy & Moore Publishing,
1 Rockefeller Plaza,
New York City, NY 10020, USA

Visit us on the World Wide Web at:
www.murphy-moorepublishing.com

ISBN: 978-1-63987-345-6 (Hardback)

Cataloging-in-Publication Data

Livestock farming / edited by Elly Campbell.
p. cm.
Includes bibliographical references and index.
ISBN 978-1-63987-345-6
1. Livestock. 2. Animal culture. 3. Agriculture. I. Campbell, Elly.
SF140.L65 L58 2022
636--dc23

Table of Contents

Preface

Over the recent decade, advancements and applications have progressed exponentially. This has led to the increased interest in this field and projects are being conducted to enhance knowledge. The main objective of this book is to present some of the critical challenges and provide insights into possible solutions. This book will answer the varied questions that arise in the field and also provide an increased scope for furthering studies.

The animals which are domesticated and raised in an agricultural setting are termed as livestock. Their key objective is the production of labor and commodities like leather, eggs, milk, fur and meat. Some of the common examples of livestock are sheep, goats and cattle. The practice related to the management and breeding of livestock are studied under livestock management. According to the production process involved, the discipline can be further classified into intensive livestock farming, semi-intensive livestock farming, nomadic livestock farming, extensive livestock farming, organic livestock farming, and transhumance livestock farming. It also involves the various aspects of veterinary medicine, transportation and logistics, marketing and economics. This book includes some of the vital pieces of work being conducted across the world, on various topics related to livestock farming. The various sub-fields of livestock farming along with the technological progress that have future implications are glanced at in it. This book is a complete source of knowledge on the present status of this important field.

I hope that this book, with its visionary approach, will be a valuable addition and will promote interest among readers. Each of the authors has provided their extraordinary competence in their specific fields by providing different perspectives as they come from diverse nations and regions. I thank them for their contributions.

Editor

1

RESPONSE TO THE MALE EFECT IN NUBIAN AND ALPINE GOATS MAINTAINED UNDER TROPICAL PHOTOPERIOD

Jorge Urrutia Morales[a*], José Francisco Cervantes Becerra[a], Héctor Raymundo Vera Ávila[b], Eugenio Villagómez-Amezcua Manjarrez[b] and María Teresa Rivera Lozano[a]

[a]San Luis Experimental Station, National Institute for Forestry, Agriculture and Livestock Research, San Luis Potosí, México.
E-mail: jorurrmo@hotmail.com
[b] National Center for Disciplinary Research on Animal Physiology and Breeding, National Institute of Forestry, Agriculture and Livestock Research, Querétaro, México.
**Corresponding Author.*

SUMMARY

In order to determine the extent of response to the male effect in two breed of goats, previously observed to express a similar pattern of reproductive seasonality under tropical photoperiod (22°N), 17 Nubian and 15 Alpine female goats were exposed to sexual active males at May 15th. A higher proportion of Nubian vs. Alpine does ovulated in response to the male effect (76.5 *vs* 33.3%; $P<0.02$). Two peaks of P_4 at 4 and 19 day after male introduction were observed, being more evident in Nubian does. Eight Nubian does showed P_4 increases >0.5 ng ml^{-1} within the first four days after male introduction, whilst Alpine does maintained basal P_4 levels during the same period (≤0.04 ng ml^{-1}). The higher proportion of Nubian does ovulating due to the male effect, indicate that female goats of this breed, maintained under tropical photoperiod, are more responsive than Alpine does to non-photoperiodic external stimuli, such as the presence of sexual active males.

Key words: Goats; male effect; Nubian; Alpine; tropical photoperiod.

INTRODUCTION

Under some circumstances, it might be necessary to mate female goats during any season of the year. However, reproductive seasonality which restricts continual ovulatory activity to the autumn and winter months (Duarte et al., 2008; De Santiago et al., 2009; Rivera et al., 2011), constitutes an impediment to obtain high conception rates during the spring and summer months (seasonal anestrous). Stimulation of ovulatory activity through the introduction of sexually active bucks has been used to counteract this impediment (Veliz et al., 2002; Luna-Orozco et al., 2012). With this method, the percentage of female goats showing estrus increases, however, responses are considerably variable (Delgadillo et al., 2002; Urrutia et al., 2003; Rivas-Muñoz et al., 2007). It has been proposed that this variability depends on the depth of anestrous experienced by the female goats when the male effect is attempted (Chemineau, 1987), condition which is related in part to the season of the year (proximity or remoteness to the reproductive season), and also apparently to the inherent seasonal reproductive characteristics of each particular goat breed. Recently, it has been reported that basal ovulatory activity in Nubian and Alpine does is much the same at a tropical latitude (Rivera et al., 2011), and limited to the months of September through February. Similar results were reported in temperate latitude conditions (Amoah et al., 1996). These findings, however, are in contrast with results reported by other authors who observed kidding throughout the whole year in Nubian (Mellado et al., 1991; 2006) and Alpine flocks (Silva et al., 1998), in which accordingly, a year round pattern of estrus and ovulatory activity was also probably expressed. These contrasting results, suggest that even if the basal seasonal ovulatory activity could be similar among Alpine and Nubian breeds, their responsiveness to non-photoperiodic modulatory effects as the male effect might defer significantly, possibly reflecting also differences in the depth of seasonal anestrous experienced by each breed (Chemineau, 1987). Mellado et al. (1991, 2006) and Silva et al. (1998) included only one of these breeds in their studies, thus no differences in patterns of reproductive seasonality among breeds could be inferred. Under the assumption of existing differences in the depth of seasonal anestrous between Alpine and Nubian female goats, the aim of this study was to evaluate their response to the male effect during the seasonal anovulatory period (Rivera et al., 2011) in the Mexican tropic (22° N).

MATERIALS AND METHODS

The study was conducted at the Goat Reproduction and Breeding Improvement Centre of San Luis Potosi State, México, located in the municipality of Soledad de Graciano Sanchez at 22° 11' N. Seventeen Nubian and fifteen Alpine female goats, 2 to 3 years old, free of clinical reproductive problems and of infectious diseases were used. Does were kept in confinement and fed with a diet based on alfalfa hay and cereal grains formulated to meet their maintenance requirements (NCR 1981). Isolation of buck contact was assured for the experimental does from August to initiation of experimental procedures on May 7th. In parallel, a group of 6 bucks (3 Nubian and 3 Alpine), were subjected to photoperiodic activation of sexual behavior (Delgadillo et al., 2002). Photoperiodic activation procedure was as follows: exposure to 17 h of daylight from January 15th to March 31st (complementary artificial light was supplied by incandescent bulbs which generated 300 lux light intensity at bucks eye height); return to natural photoperiod after March 31st (12 h light:12 h dark). Once on the reduced photoperiod, bucks were monitored 2 h a day for odor and sexual behavior. Five days after daylight length reduction, a characteristic odor of sexual active male started to be perceived, and after 2 to 7 days more, all the bucks began to show active sexual behavior. Two bucks in each breed, those with the highest grading on a sexual behavior test (10 min exposure to an estrus induced doe), were selected and used for doe bio-stimulation and mating in the next part of the experiment. On May 15th, each breed group of does was placed in contact with one of the two sexually activated bucks of its same breed, and thereafter, a 24 h rotation of the 2 bucks per breed was performed throughout a 21 day mating period. Body Condition Score (BCS) at initiation of mating was 3.18±0.43 for Nubian and 3.17±0.45 for Alpine female goats (scale 1 to 4; Honhold et al., 1991). During the mating period, breed groups were located in completely separated barns thus sensory contact was not allowed.

On days -7, -4, 0, +4, +7, +11, +14, +19 y +26 from buck introduction, blood samples were obtained in all the female goats by jugular venipuncture using vacuumed tubes with no anticoagulants. Blood samples were maintained at room temperature for 4 h to allow clotting and then centrifuged at 3000 rpm x 15 min for serum separation. Serum was frozen and kept at -20 °C until progesterone concentrations were determined through a solid phase RIA (Coat-a-Count; DPC, Los Angeles, CA, USA). RIA sensitivity was 0.05 ng ml-1 and the intra and inter-assay coefficients of variation were 8 and 12%, respectively. Serum progesterone concentrations were used to determine doe ovulatory activity before and after male bio-stimulation; it was considered that a female goat had ovulated when a serum sample showed progesterone concentration $\geq$ 1.0 ng mL-1. Supervision of experimental female goats was maintained until kidding and the number of kids born per pregnant doe was registered.

The response variables were the proportion of does that ovulated within the 0 to 7 and 0 to 26 day period from buck introduction, the proportion which get pregnant within the mating period, the proportion that kidded, and the percentage of kids born as related to female goat exposed to the male bio-stimulation. Statistical analyses were performed using Fisher and Yeates Exact Test for small samples or Chi Square Test, as appropriate (Castilla y Cravioto, 1991).

RESULTS

The percentage of goats showing progesterone concentrations indicative of luteal activity during the 26 d period after male bio-stimulation was higher

(P<0.02) in the Nubian (76.47%) as compared to the Alpine breed (33.33%). Serum progesterone concentrations, before and after introduction of sexual active bucks, are shown for both breeds in Figure 1. Serum progesterone concentrations remained below 0.2 ng ml-1 during the period prior to the introduction of males, indicating that irrespective of breed, all female goats were in anovulatory state during this period.

After the introduction of bucks, two peaks of serum progesterone concentration were observed; one occurred around day 4 and the other one around day 19, being more evident in the Nubian breed. Regarding the first rise in progesterone on the 4th day after male introduction (Figure 2), 8 Nubian does showed concentrations above 0.5 ng ml-1 (1.0±0.47 ng ml-1), while all the Alpine doe showed concentrations below 0.4 ng ml-1 (0.072±0.042 ng ml-1) (P<0.05). One Alpine doe presented a small rise in P4 on the day of male introduction (0.34 ng ml-1), but in the previous and posterior serum samples P4 levels were basal. After the first P4 peak, during the 7 to 26 d post male introduction period, 76.47% of Nubian females showed concentrations above 1 ng ml-1 in more than two consecutive samples, while only 33.33% of the Alpine females did.

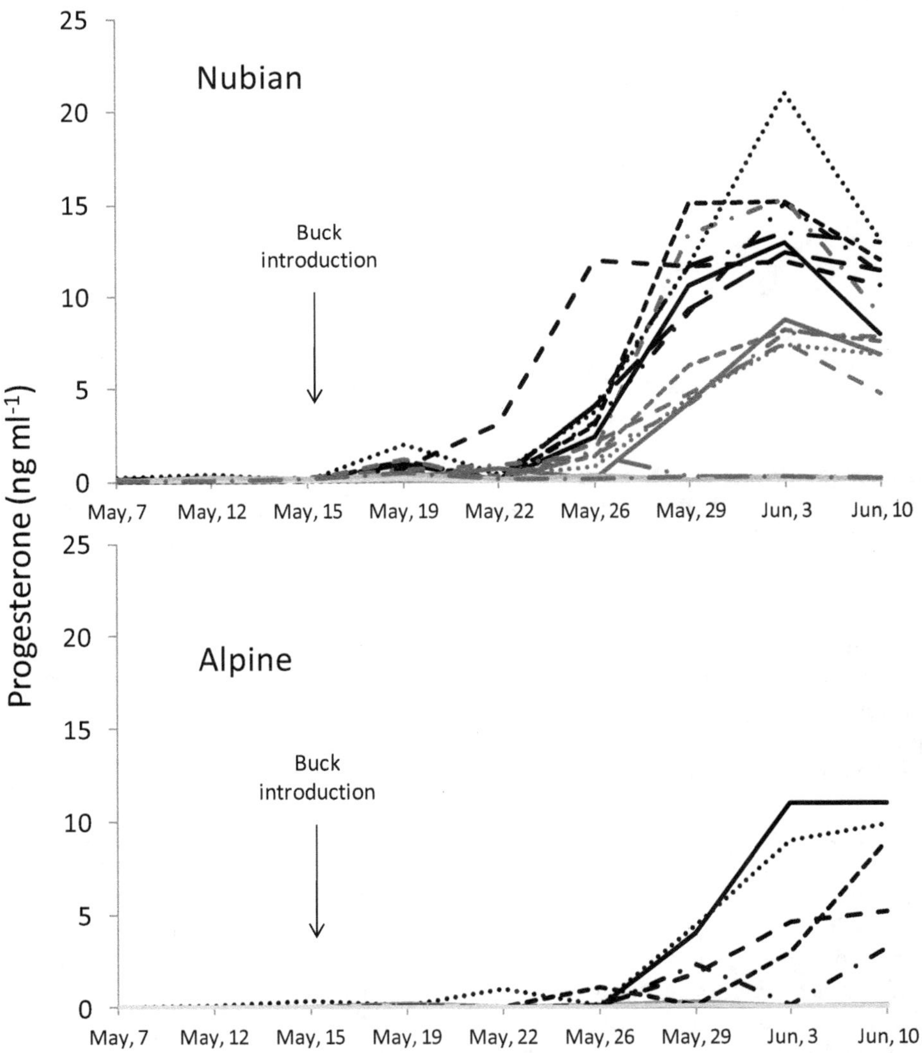

Figure 1.- Serum progesterone concentrations in Nubian and Alpine female goats exposed to sexual active bucks for 26 days, beginning on May 15th .The bucks were previously exposed to a light regime of 17 h d^{-1} from January 5th to March 15th and then returned to natural photoperiod (12 h light:12 h dark). The percentage of goats that showed a progesterone increase, indicative of luteal activity, was higher (P<0.02) in Nubian (76.5%) than in Alpine does (33.3%).

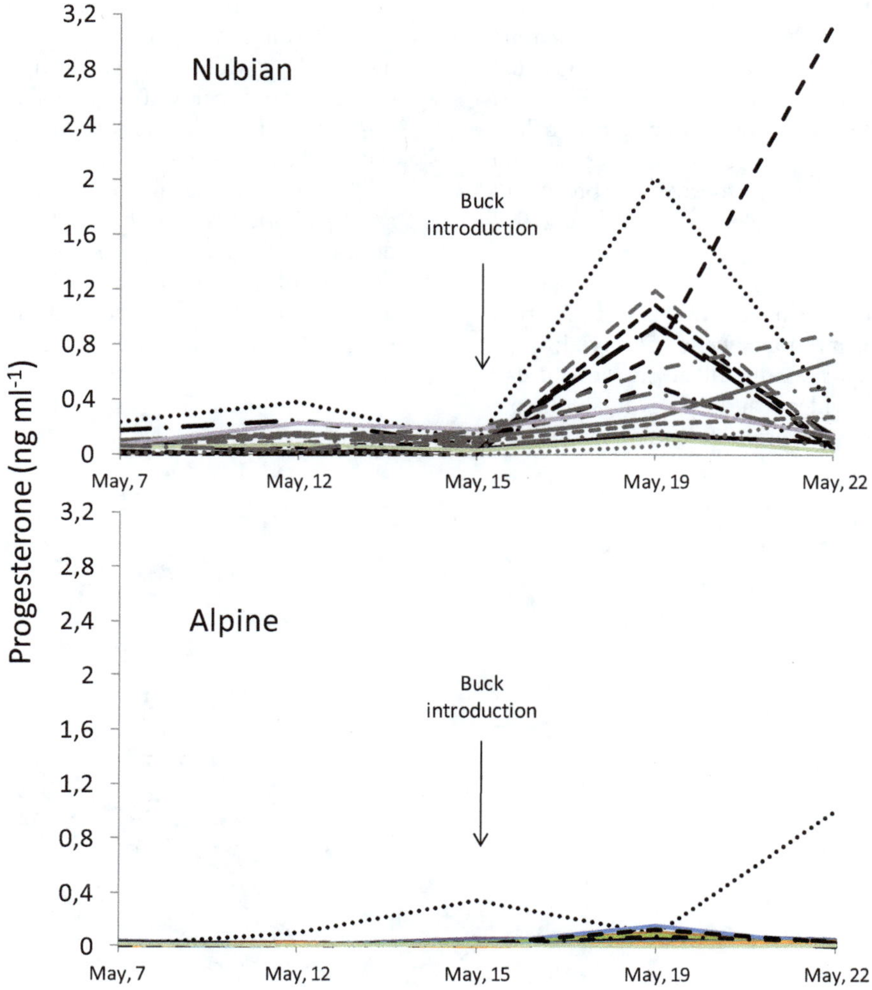

Figure 2. – Serum progesterone concentrations in Nubian and Alpine female goats during the first seven days after introduction of sexual active bucks, beginning on May 15th. The percentage of does that presented a progesterone rise above 0.4 ng ml^{-1} during this period was higher (P<0.01) in Nubian (47.06%) than in Alpine does (0%).

The higher response to male bio-stimulation observed in Nubian as compared to Alpine does was reflected as a greater percentage pregnant (58.8 vs 0%, P<0.001) and kidding (35.3 vs 0; P<0.02), and in the total number of kids born (52.9 vs 0; P<0.01).

DISCUSSION

The present study was designed with the aim to evaluate Nubian and Alpine doe response to the male effect, during the period considered to be the seasonal anovulatory period for these breeds in the Mexican Tropic. The higher response to sexually active buck bio-stimulation observed in the Nubian as compared to the Alpine female goats, could be a reflection of differences in the depth of anestrous experienced by does in each breed (Corteel, 1977; Restall, 1992).

Under photoperiodic conditions at subtropical latitudes, and in the presence of sexually active bucks, Nubian female goats present a wider active ovulatory period than Alpine does (Amoah et al., 1996). However, without external non-photoperiodic stimuli such as changes in feeding level, or presence of sexually active males, the patterns of seasonal reproductive activity are similar in both breeds (Rivera et al., 2011). This contrasting results, suggest that even if the basal seasonal ovulatory activity could be similar in both breeds, the responsiveness to external non-photoperiodic stimuli such as the buck presence might differ, probably due to breed differences in the depth of anestrous. Corteel (1977) introduced the concepts of "deep anestrous" and "transitional period" to name the stages of the anovulatory period in which female goats do not or do respond to the male effect, respectively. Afterwards, Restall (1992) suggested that the annual reproductive rhythm of the female goat could be divided in three phases; the active phase and the responsive phase which together make up the breeding season, and the phase of sexual rest which corresponds to the non-breeding or anovulatory season. During the responsive phase, does can ovulate or not, depending on the intensity of non-photoperiod external stimuli received, mainly those associated with the presence of sexually active bucks. Taken together, results obtained by Rivera et al (2011) and those in the present study, confirm the existence of breed differences in seasonal reproductive patterns between Nubian and Alpine does. In the present study, a difference in the depth of anestrous between these breeds is demonstrated.

It is known that a decreasing photoperiod pattern, which begins after Summer Solstice, constitutes the primary stimulus for initiation of ovulatory activity in the female goat, whilst an increasing photoperiod pattern, starting after the Winter Solstice, constitutes the stimulus to its ending (Mori et al., 1987; Chemineau et al., 1988). This suggests that the moment in which the ovulatory activity would be more reduced and the anestrous would be deeper, should be around the longest day of the year, during the Summer Solstice. In the current study, Nubian does responded to the male stimulus before the Summer Solstice, suggesting that during the anestrous period, females of this breed are sensitive to non-photoperiodic external stimuli such as the presence of sexually active males. In contrast, Alpine does showed a reduced responsiveness to the male presence, suggesting that female goats of this breed experiment a deeper anestrous condition as compared to Nubian does.

CONCLUSIONS

The higher response to the male effect showed by Nubian as compared to Alpine does, confirms the existence of breed differences in their seasonal reproductive pattern, particularly regarding to the depth of anestrous. Results also indicate that in Nubian does, manipulation of reproduction is possible by taking advantage of the male effect. This manipulation in turn, would allow attending market demands of goat meat as "cabrito" throughout the entire year in Mexico.

AKNOWLEDGEMENTS

This study was financed by the National Institute for Forestry Agricultural and Livestock Research, through Project 6467387 (Development of strategies to reduce reproductive seasonality in extensive goat production systems). We are grateful to the Goat Reproduction and Breeding Improvement Centre of San Luis Potosi State, for their support with animals, housing and management facilities.

REFERENCES

Amoah, E.A., Gelaye, S., Guthrie, P., Rexroad, C.E. 1996. Breeding season and aspects of reproduction in female goats. Journal of Animal Science. 74:723-728.

Castilla, S.L., Cravioto, J. 1991. Estadística simplificada para la investigación en ciencias de la salud. Primera Edición, Ed. Trillas. México.

Chemineau, P. 1987. Possibilities for using bucks to stimulate ovarian and oestrus cycles in anovulatory goats—a review. Livestock Production Science. 17:135–147.

Chemineau, P., Martin, G.B., Saumande, J., Normant, E. 1998. Seasonal and hormonal control of pulsatile LH secretion in the dairy goat (*Capra hircus*). Journal of Reproduction and Fertility. 83:91-98.

Corteel, J.M. 1977. Management of artificial insemination of dairy seasonal goats through oestrus synchronization and early pregnancy diagnosis. In : Proceedings of a Symposium on Management of Reproduction in Sheep and Goats, 24-25 July, Madison, WI, pp 1-20.

Delgadillo, J.A., Flores, J.A., Véliz, F.G., Hernández, H.F., Duarte, G., Vielma, J., Poindron, P., Chemineau, P., Malpaux, B. 2002. Induction of sexual activity in lactating anovulatory female goats using male goats treated only with artificially long days. Journal of Animal Science. 80:2780–2786.

De Santiago-Miramontes, M.A., Malpaux, B., Delgadillo, J.A. 2009. Body condition is associated with a shorter breeding season and reduced ovulation rate in subtropical goats. Animal Reproduction Science. 114:175–182.

Duarte, G., Flores, J.A., Malpaux, B., Delgadillo, J.A. 2008. Reproductive seasonality in female goats adapted to a subtropical environment persists independently of food availability. Domestic Animal Endocrinology. 35(4): 362-370.

Luna-Orozco, J.R., Guillet-Muñoz, J.M., De Santiago-Miramontes, M.A., García, J.E., Rodríguez-Martínez, R., Meza-Herrera, C.A., Mellado, M., Veliz, F.G. 2012. Influence of sexually inactive bucks subjected to long photopeeriod or testosterone on induction of estrus in anovulatory goats. Tropical Animal Health and Production. 44(1):71-75.

Mellado, M., Foote, R.H., Gomez, A. 1991. Reproductive efficiency of Nubian goats throughout the year in northern Mexico. Small Ruminant Research. 6:151-156.

Mellado, M., Valdéz, R., García, J.E., López, R., Rodríguez, A. 2006. Factors affecting the reproductive performance of goats under intensive conditions in a hot arid environment. Small Ruminant Research. 63:110–118.

Mori, Y., Tanaka, M., Maeda, K., Hoshino, K., Kano, Y. 1987. Photoperiodic modification of negative and positive feedback effects of oestradiol on LH secretion in ovariectomized goats. Journal of Reproduction and Fertility. 80:523-9.

N.R.C. Nutrient requirements of goats. National Academy Press. Washington. 1981; Number 15.

Restall, BJ. 1992. Seasonal variation in reproductive activity in Australian goat. Animal Reproduction Science. 27:305-318.

Silva, E., Galina, M.A., Palma, J.M., Valencia, J. 1998. Reproductive performance of Alpine dairy goats in a semi-arid environment of Mexico under continuous breeding system. Small Ruminant Research. 27:79-84.

Rivas-Muñoz, R., Fitz-Rodríguez, G., Poindron, P., Malpaux, B., Delgadillo, J.A. 2007. Stimulation of estrous behavior in grazing female goats by continuous or discontinuous exposure to males. Journal of Animal Science. 85:1257-1263.

Rivera Lozano, M.T., Díaz Gómez, M.O., Urrutia Morales, J.L., Vera Avila, H., Gámez Vázquez H., Villagómez-Amezcua Manjarrez, E., Aréchiga Flores, C.F., Escobar Medina, F. 2011. Seasonal variation in ovulatory activity of Nubian, Alpine and Creole x Nubian does under tropical photoperiod (22° N). Tropical and Subtropical Agroecosystems. 14:973-980.

Urrutia, M.J., Gámez, V.H.G., Ramírez, A.B.M. 2003. Influencia del pastoreo restringido en el efecto macho en cabras en baja condición corporal durante la estación de anestro. Técnica Pecuaria en México. 41: 251-260.

Véliz, F.G., Moreno, S., Duarte, G., Vielma, J., Chemineau, P., Poindron, P., Malpaux, B., Delgadillo, J.A. 2002. Male effect in seasonally anovulatory lactating goats depends on the presence of sexually active bucks, but not estrous females. Animal Reproduction Science. 72:197–207.

2

ENERGY AND PROTEIN REQUIREMENTS OF GROWING PELIBUEY SHEEP UNDER TROPICAL CONDITIONS ESTIMATED FROM A LITERATURE DATABASE ANALYSES

F. Duarte-Vera[1,2], C. A. Sandoval-Castro*[2], L. A. Sarmiento-Franco[2], L. O. Tedeschi[3], R. Santos-Ricalde[2]

[1] *Campo Experimental Mocochá, Centro de Investigación Regional del Sureste, INIFAP. Km 15 Carr. Mérida-Baca, Yucatán, México.*

[2] *Facultad de Medicina Veterinaria y Zootecnia, Universidad Autónoma de Yucatán, Mérida, Yucatán, México. Email: ccastro@uady.mx*

[3] *Department of Animal Science, Texas A&M University, College Station, TX 77845-2471, USA*

**Corresponding author*

SUMMARY

Data from previous studies were used to estimate the metabolizable energy and protein requirements for maintenance and growth and basal metabolism energy requirement of male Pelibuey sheep under tropical conditions were estimated. In addition, empty body weight and mature weight of males and female Pelibuey sheep were also estimated. Basal metabolism energy requirements were estimated with the Cornell Net Carbohydrate and Protein System – Sheep (CNCPS-S) model using the *a1* factor of the maintenance equation. Mature weight was estimated to be 69 kg for males and 45 kg for females. Empty body weight was estimated to be 81% of live weight. Metabolizable energy and protein requirements for growth were 0.106 Mcal ME_m/kg $LW^{0.75}$ and 2.4 g MP/kg $LW^{0.75}$ for males. The collected information did not allowed appropriate estimation of female requirements. The basal metabolism energy requirement was estimated to be 0.039 Mcal ME_m/kg $LW^{0.75}$. Energy requirements for basal metabolism were lower in Pelibuey sheep than those reported for wool breeds even though their total requirements were similar.

Key words: Energy and protein; Pelibuey sheep; CNCPS-S.

INTRODUCTION

Sheep production in Mexico has grown significantly, particularly in the Yucatan Peninsula in which the sheep population has more than doubled from 43,003 in 1990 to 87,508 in 2004 (SIAP, 2007). The Pelibuey and Black Belly breeds predominate in the Yucatan Peninsula as they do throughout the Mexican tropics, even though other breeds (e.g. Katahdín and Dorper) have been recently introduced through crossbreeding programs.

The main purpose of most sheep production systems in Yucatan is meat production (Torres et al., 2004) and these systems tend to involve a confined intensive growth stage after weaning. Under these conditions, energy is the first limiting nutrient for growth and rations are consequently formulated to meet energy requirements. These rations, however, are extrapolated from research on wool sheep in temperate climates. Very little research has been done on the nutritional needs of hair sheep in the tropics. Some authors have stated that Pelibuey breed has greater energy requirements than wool breeds, although this conclusion was based on limited data (Solís et al., 1991; Bores et al., 2001). The Pelibuey's productive potential has been considerably improved through selection and crossbreeding; therefore, its nutritional requirements should be revised and updated.

Nutritional models, such as the Cornell Net Carbohydrate and Protein System for sheep (CNCPS-Sheep; Cannas et al., 2004) and the Small Ruminant Nutrition System (SRNS, Cannas et al., 2007a; 2007b, 2007c) are continuously developed but updated data are not available for some of the parameters required by these models to predict the performance of Pelibuey sheep, including mature weight, empty body weight and maturity index. Given this, as well as the suggestion that the Pelibuey breed might have higher energy requirements for growth (Solís et al., 1991; Bores et al., 2001), the aim of the study was to estimate the protein and energy requirements for growth of Pelibuey sheep and to generate data on parameters than can be incorporated into nutrition models.

MATERIALS AND METHODS

Estimation of empty body weight

Adjusted empty body weight (EBW) was determined from data of a previous research studies on Pelibuey sheep that included carcass yield, organ weight and gastrointestinal fill (see Table 3 for references).

Estimation of maintenance and weight gain requirements

Data from 29 experiments with a total of 1304 sheep (1112 males and 192 females) and 93 diets during the growth-finishing stage were used in this study. Pelibuey sheep were used in 26 of these experiments and crossbreds between Pelibuey and paternal wool breeds (Dorset, Hampshire or Suffolk) in the remaining 3 experiments (Duarte, 2007) (Table 1). Animals were fed in growth corrals with complete diets formulated with agricultural by-products (e.g. corn stover), agroindustrial by-products (e.g. cane molasses), poultry manure, soybean or sunflower meal, grains (mainly corn and sorghum), grass hay, salt, minerals and additives, among other ingredients. Only 3% of the diets included a concentrate as a feed supplement using live weight (LW) as criteria for supplementation level and were fed in *ad libitum* regimen. The main characteristics of the diets are shown in Table 2. Values were obtained from the experimental data reported in the literature and used as input in the original CNCPS-S model (Cannas et al., 2004) to yield data on Rumen nitrogen balance (RNB) and total digestible nutrients (TDN) (Duarte, 2007).

Estimates of energy and protein requirements for maintenance and daily weight gain (DWG) were calculated using multiple linear regression analyses including metabolizable energy intake (MEI Mcal/kg $LW^{0.75}$) obtained from the literature reported values collected in database), daily metabolizable protein intake (MPI g/kg $LW^{0.75}$) calculated from the CNCPS-S, and weight gain in males and females. Regression analyses were performed using Minitab 15 statistical software procedures (Minitab, 2007). Equations are shown below.

$$NI = a\ LW^{0.75} + b\ DWG \quad \text{(eq. 1)}$$

$$NI = a\ LW^{0.75} + b\ DWG + c\ LW^{0.75} \quad \text{(eq. 2)}$$

Where:

NI = nutrient intake, i.e. metabolizable energy (Mcal/d) or metabolizable protein (g/d)
a = metabolizable energy or protein requirements for maintenance (Mcal/kg $LW^{0.75}$ or g/kg $LW^{0.75}$)
b = metabolizable energy or protein requirements for growth (Mcal/g DWG or g/g DWG)
c = metabolizable energy (Mcal/kg $LW^{0.75}$) or protein (g/kg $LW^{0.75}$) requirements for growth (when maintenance energy requirement was estimated parameter 'c' stands for protein and vice-versa).

Estimation of energy requirement for basal metabolism

Basal metabolism (*a1*) was estimated using the CNCPS-S model with the adjustments proposed by Cannas et al. (2004) and based on the experimental diets in the database. Preliminary assessment showed that nutrient requirement estimations for females were not reliable due to the number of data and were not further assessed. Given that the DWG and nutrient intake variables were known, the performance prediction equation (eq. 3) was rearranged to estimate basal metabolism (*a1*).

Table 1. General characteristics of sheep in database (males n=1112, females n=192) [1].

	Initial weight (kg)	Final weight (kg)	Average weight (kg)	Empty Weight (kg)	Metabolic Weight ($kg^{0,75}$)	Age (years)
Males						
Mean	21.6	34.9	28.2	22.9	12.2	0.5
Maximum	34.0	47.6	37.9	30.8	15.3	2.0
Minimum	12.0	26.8	19.3	15.7	9.2	0.3
SD	5.4	4.6	4.4	3.6	1.4	0.3
Females						
Mean	27.7	34.0	30.6	25.2	13.0	2.5
Maximum	32	42.1	36.7	30.3	14.96	3.0
Minimum	10.4	23.3	16.9	13.9	8.3	0.3
SD	6.4	4.8	5.1	4.2	1.7	1.1

[1] 1 From Duarte Vera (2007)

Table 2. Main characteristics of diets (males n=77, females n=16)[1]

	DMI	MEI	eNDF	CP	RNB*	Forage	TDN*
Males							
Mean	1.3	3.2	13.83	13.5	8.10	34.1	51.1
Maximum	2.0	5.6	32.91	20.0	16.85	73.4	79.5
Minimum	0.6	1.6	0.13	5.1	-3.67	6.0	29.9
SD	0.3	0.83	6.90	2.8	5.70	17.4	10.6
Females							
Mean	1.2	2.9	17.38	12.0	4.63	45.8	52.0
Maximum	1.7	4.8	22.95	16.9	14.46	68.7	62.3
Minimum	0.6	1.3	9.86	8.9	-0.05	13.2	32.6
SD	0.33	1.1	4.30	2.7	4.41	18.2	9.0

1 From Duarte Vera (2007)
DMI = dry matter intake (kg/animal/day);
MEI = metabolizable energy intake (Mcal ME/animal/day)
eNDF = effective neutral detergent fiber as a percentage of ration NDF
CP = crude protein, as a percentage of dry matter
RNB = rumen nitrogen balance
Forage = forage content as a percentage of diet dry matter
TDN = total digestible nutrients, as a percentage of dry matter
SD = standard deviation of the mean
*Values calculated with the CNCPS-S model (Cannas et al., 2004)

Table 3. Estimation of empty body weight (EBW) in Pelibuey sheep.

Sex	Age (months)	Initial weight (kg)	N	System	Weight at death (kg)	EBW * (%)	Reference
Males	3	16	15	Grazing	22.5	85.3	García et al., (1998)
Males	6	15.6	48	Supplementation	20-30	70.7	Romano et al., (1983)
Males	14	20.9	36	Feedlot	36	81.9	Martínez et al., (1990)
--	--	--	77	--	--	81.0	Cantón et al., (1992)
Males	5.5	28	4	Feedlot	42.3	83.3	Carvalho et al., (2005)
Males	5.5	30.4	4	Supplementation	43.3	81.5	Carvalho et al., (2005)
Males	5.5	29.5	4	Creep feeding	38.4	84.3	Carvalho et al., (2005)
Females	65	31.7	61	Supplementation	31.6	82.2	Martínez et al., (1987)
Females	--	--	16	Feedlot		82.7	Cantón et al., (1992)
Average		25.3			35.1	81.4	
SD		± 6			± 6.8	± 4	

*EBW as percentage of Body weight

$$EM_m = \{[SBW^{0.75} * a1 * a2 + e^{(-0.03*AGE)}] + (-0.09 * MEI * k_m) + ACT + NE_{mes} + UREA\}/k_m \quad \text{(eq.3)}$$

Where:
$SBW^{0.75}$ = shrunk body weight in Kg $LW^{0.75}$
a1 = energy requirement for thermoregulation in fasting metabolism (Mcal ME/ Kg $LW^{0.75}$).
a2 = adjustment factor for effect of environmental temperature (1 + 0.0091 x C); C = 20 – average temperature in previous month (ºC).
$e^{(-0.03*AGE)}$ = Adjustment for age in years.
MEI = ME intake in Mcal/day (0.106 Mcal ME/kg$^{0.75}$ obtained from equation 1)
k_m = energy utilization efficiency coefficient, fixed at 0.644.
ME_m = metabolizable energy requirement for basal metabolism in Mcal/day
ACT = effect of activity on maintenance requirements in Mcal/day, according to the equation of Cannas et al. (2004), considering animals walked 50 m on flat surfaces only.
NEmes = additional energy to compensate cold stress (this figure was considered to be 0 due to the predominant tropical climate)
UREA = energy cost in Mcal ME_m for elimination of excess N as urea.

The energy adjustment for cold stress equilibrium was not included in this equation due to climate differences between the region where the equation was developed and the Mexican tropics. MEI and $SBW^{0.75}$ were taken from the original data base; environmental data (i.e. average temperature in previous month in ºC) were acquired from the National Water Commission (Comisión Nacional del Agua -CNA) data base for the Yucatan Peninsula from four points (Merida Meteorological Observatory, Tizimín, Valladolid and Mocochá meteorological stations) nearest where most of the experiments in the data sample were done. For those cases in which the season and year of the beginning of the experiment were not stated, or in those done outside the Yucatan Peninsula but in tropical conditions, the average monthly temperature at the Merida Meteorological Observatory for the last 45 years (26.1 °C) was used in the CNCPS-S.

Estimation of mature weight

Age and weight data from birth to approximately 10 years of age were collected for Pelibuey sheep (n = 3,320 male; 1,857 female) from ranches and institutions in the region and arranged in separates datasets. These data were used to estimate growth curves for males and females (Brody, 1945) and then to estimate maturity rate (k) and mature weight (A) (Jenkins and Leymaster, 1993) (Equation 4):

$$Y_t = A(1 - Be^{-kt}) \quad \text{(eq. 4)}$$

Where:
Y_t = LW at time *t*
A = mature weight
B = relative weight change from birth to mature weight
k = maturity rate
t = age in days
The data were estimated and compared with a t test using the GraphPad Prism 4 (Graph Pad Prism 2003) program. The non-linear regression parameters were estimated by minimizing the sum of the squares of each value's relative distances from the predicted curve (Motulsky and Christopoulos, 2003) (equation 5), that is:

$$\Sigma[Y_{observed} - Y_{predicted}/Y_{observed}]^2 \quad \text{(eq. 5)}$$

RESULTS AND DISCUSSION

Database profile

A total of 83% of the experiments were done using growth-stage males, 7% with growth-stage females and 10% using initial gestation stage females. LW and initial age were highly variable between experiments, with an average of 21.6 kg and 0.5 years in males and 27.7 kg and 2.5 years in females (Table 1). Data for crude protein content, effective neutral detergent fiber, total digestible nutrients, dry matter intake (DMI), metabolizable energy, forage content and animal rumen nitrogen balance are shown in Table 2. Most of the animals had a positive rumen N balance, with only 6% of males and 25% of females with a negative balance as calculated with the CNCPS-S model.

Empty body weight

Estimation of basal metabolism nutritional needs is influenced by the amount of body tissue versus LW, which in turn depends on the gastrointestinal fill. Data on gastrointestinal fill is limited for Pelibuey sheep, the collected dataset showed that on average it was 81.4% ± 4 of the LW percentage a value 18% greater than in wool sheep (Table 3). Empty body weight (EBW) is affected by feed intake level and in turn by other factors such as diet type (NRC, 1985). However, the experimental data included in the database did not include enough information to generate a model that included these effects. An empirical relationship was generated between EBW and DMI percentage versus live weight (LW), assuming linearity and using relative values of 80 % minimum and 96% maximum for EBW and 3 to 7 % DMI versus LW:

$$EBW = LW * 0.96 \quad \text{if DMI} =< 3\% \text{ LW} \quad \text{(eq. 6)}$$

$EBW = LW * (1.08 - 0.04*DMI)$ if DMI =>3% LW (eq. 7)

The AFRC (1993) provides a requirement of 0.0687 Mcal ME/kg $LW^{0.75}$ for basal metabolism. In contrast, the NRC (1985) suggested that maintenance energy starts from 0.063 Mcal NE/kg $LW^{0.75}$ and an adjustment to EBW is necessary, assuming 6.1 kg gastrointestinal fill for a 40 kg animal (equivalent to 15.2 % of gastrointestinal fill). This adjustment would produce a value of 0.056 Mcal NE_m/kg $LW^{0.75}$ which is equivalent to a difference of 11.1 % in the requirement. However, the difference in gastrointestinal fill in Pelibuey sheep is 18%. These effects are quite remarkable and accurate data from Pelibuey animals are needed to determine EBW, and therefore better estimate SBW, and/or directly calculate SBW.

Energy and protein requirements

The linear regressions for MPI and MEI on DWG (Table 4) indicated that protein and energy requirements had linear pattern for both sexes. For males, most of the data is within 100 to 300 g/d of DWG. However, the data had not enough points for females.Therefore, the results for females were no longer included and are not reported. The r^2 improved considerably and the equation variation coefficient decreased when ME and MP were estimated with multiple regressions (Table 3). Using DWG and ME as predictors of the MP requirements yields a negative value indicating a catabolic and salvage/recycling process for protein when DWG and energy intake is equal to zero. Similarly, the protein requirements in the multiple regression for ME estimation results in positive values indicating the endogenous MP loss when animal are fed maintenance energy levels.

Estimated energy required for maintenance, by males, expressed as Mcal/ kg $LW^{0.75}$, energy requirements was 0.143 Mcal ME/kg $LW^{0.75}$ (Table 4). One of the main factors involved in calculating maintenance energy requirements is basal metabolism and Cannas et al. (2006) proposed an adjustment to this factor, decreasing it from 0.062 to 0.052 Mcal ME/kg $LW^{0.75}$, to improve predicted LW gain in hair sheep using the CNCPS-S. However, the average value generated using the present data when calculating the *a1* factor for each diet used in the experiments with males (n = 77) was 0.039 Mcal ME_m/kg $LW^{0.75}$, lower than the 0.062 considered in the original CNCPS-S model (Cannas et al., 2004) and the 0.052 suggested by Cannas et al. (2006). Using the 0.039 value improved predicted LW gain in Pelibuey sheep with the CNCPS-S (Duarte, 2007), supporting the hypothesis that basal metabolism energy requirements in this breed are lower than in wool breeds. Maintenance energy requirements would consequently be less than previously reported (Solis et al., 1991; Bores et al., 2001). A possible explanation for this may be higher gastrointestinal fill in the Pelibuey breed compared to wool breeds, which, together with its higher bone proportion of LW (30.5 % in Pelibuey vs. 16 to 24 % in wool breeds; Martínez et al., 1987), provide this breed with a smaller amount of metabolically active tissue.

Mature weight

Average mature weight based on the collected data was 69.8 ± 1.98 kg for males and 44.6 ± 1.276 kg for females (Figure 1). Growth equations for males (eq. 8) and females (eq. 9) respectively were:

$$LW_t(kg) = 69.8 (\pm 1.98) * (1 - 0.963 (\pm 0.0011) * e^{-0.0033 (\pm 0.00012) * t}) \quad r^2 = 0.904 \quad \text{(eq. 8)}$$

$$LW_t(kg) = 44.6(\pm 1.27) * (1 - 0.943 (\pm 0.0017) * e^{-0.0031 (\pm 0.0001) * t}) \quad r^2 = 0.923 \quad \text{(eq. 9)}$$

It is noteworthy that in some animals aprox. 4-5 years of age weights up to 100 kg for males and 70 kg for females were observed.

Table 4. Multiple regressions between daily weight gain (DWG), metabolizable protein intake and metabolizable energy intake over DWG in male Pelibuey sheep.

Equation	R^2	P	CV
MP = 2.40 (± 0.6185) + 0.0285(±0.003328)* DWG	0.49	<0.01	28.5
ME = 0.143 (±0.01324) + 0.000686 (±0.00007124)*DWG	0.55	<0.01	15.9
MP = -2.41 (±0.6891) + 0.00538 (±0.003471) *DWG + 33.7 (±3.761)* ME	0.76	<0.01	1.4
ME = 0.106 (±0.01012) + 0.000246 (±0.00006986)* DWG + 0.1154 (±0.001723) *MP	0.78	<0.01	11.5

MP = metabolizable protein (g/kg $LW^{0.75}$)
ME = metabolizable energy (Mcal/kg $LW^{0.75}$)
DWG = daily weight gain (g/d)
±SE = standard error
r^2 = determination coefficient
P = statistical probability
CV = coefficient of variation

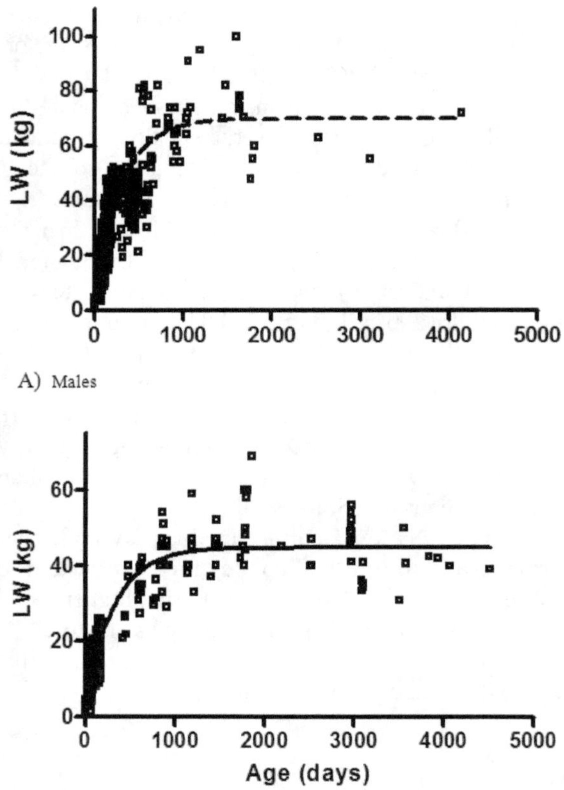

Figure 1. Growth curve for Pelibuey sheep a) males (n=3320), b) females (n=1857)

To calculate net energy requirements for weight gain (NE_g), the NRC (1985) uses the equation: $NE_g = 276$ LWG x $W^{0.75}$, where LWG is expected LW gain and $W^{0.75}$ is mature weight (assumed to be 115 kg), with a positive adjustment of 21 Kcal per 10 kg mature weight below the assumed weight. However, the mature weight of Pelibuey sheep has increased over 40% versus the breed standard (Berruecos, 1975). Determination of a more accurate mature weight for Pelibuey sheep allowed better estimation of the corresponding maintenance and weight gain energy needs, and could be incorporated into any other predictive model that uses this variable.

When using proper mature weight and SBW values, the energy requirements of growing Pelibuey sheep were similar to those of wool breed sheep. Genetic improvement programs may have increased this breed's productive potential to the point where it exhibits weight gains similar to wool breeds. This would allow establishment of more objective criteria which will in turn produce similar energy requirements.

CONCLUSIONS

This study indicated that basal metabolism energy requirements are lower in Pelibuey sheep than in wool breeds. However, the total energy requirements for weight gain in Pelibuey sheep are greater than in wool breeds, although the difference is less than previously reported. The mature weight in Pelibuey sheep is about 69 kg for males and 45 kg for females. The present database was insufficient to generate trustworthy energy and protein requirements results for female Pelibuey.

REFERENCES

AFRC, 1993. Necesidades energéticas y proteicas de los rumiantes. CAB Internacional Editorial Acribia, Zaragoza (España). Cap. 2 Necesidades de Energía Metabolizable, pp. 25-37.

Berruecos, J.M.V., Valencia, Z.M., Castillo, R.H., 1975. Genética del Borrego Tabasco o Peligüey. Técnica Pecuaria México 29: 59-56.

Bores, R.Q., Velásquez, M.A., Heredia, M.A., 2001. Evaluación de niveles energéticos y pesos de sacrificios de corderos terminales Suffolk X Pb. Procceedings Segundo Congreso Latinoamericano de especialistas de pequeños rumiantes y camélidos sudamericanos Mérida Yucatán México.

Brody S., 1945. Bioenergetics and Growth. Reinhold Publishing, NY.

Cannas, A., Tedeschi, L.O., Fox D.G., Pell, A.N., Van Soest, P.J., 2004. A mechanistic model for predicting the nutrient requirements and feed biological values for sheep. Journal Animal Science 82: 149-169.

Cannas, A; Tedeschi, LO; Atzori, AS; Fox, DG., 2006. Prediction of Energy Requirement for Growing Sheep with the Cornell Net Carbohydrate and Protein System. In: Kebreab, E., Dijkstra, J., Bannink, A., Gerrits, W.J.J. and France, J. (eds), Nutrient Digestion and Utilization in Farm Animals. Modelling Approaches. CABI Publishing, Wallingford, UK. Pp. 99-113.

Cannas, A., Tedeschi, L.O., Atzori, A.S., Fox, D.G., 2007a. Prediction of the growth rate of kids with the Small Ruminant Nutrition System. In: Annual Meeting of the Brazilian Society of Animal Science, 44, July 27, 2007, Jaboticabal, SP, Brazil. Pp- 1-3-

Cannas, A., Tedeschi, L.O., Fox, D.G., 2007b. Prediction of metabolizable energy intake and energy balance of goats with the Small Ruminant nutrition System. In: 2nd

International Symposium of Energy and Protein Metabolism and Nutrition, September 9-13, 2007, Vichy, France. Pp. 569-570.

Cannas, A., Tedeschi, L.O., Fox, D.G., 2007c. The Small Ruminant Nutrition System: development of a goat submodel. Proceedings American Society of Animal Science, San Antonio, TX. Pp. 505.

Carvalho, S., Da Silva, M. de F., Cerutti, R., Kieling, R., De Oliveira, A., Deleastre, M., 2005. Performance and live weight components of lambs submitted to different feeding systems. Ciencia Rural. 35:650-655.

Duarte Vera, F., 2007. Evaluación del CNCPS-S para estimar los requerimientos de energía durante el crecimiento de borregos Pelibuey. Tesis de doctorado Universidad Autónoma de Yucatán, México. 98 pp

Graph Pad Prism 4., 1994-2003. Graph Pad Software, San Diego California, USA.

García, M.J.A., Núñez, G.F.A., Rodríguez, A.F.A., Prieto, C., Molina, D.N.I., 1998. Calidad de la canal y de la carne de borregos Pelibuey castrados. Técnica Pecuaria México 36: 225-232.

Canton, J.G.C., Velazquez, A.M., Castellanos, A.R., 1992. Body Composition of pure and crossed blackbelly sheep. Small Ruminant Reseach, 7:61-66.

Jenkins, T.G. and Leymaster, K.A., 1993. Estimates of Maturing Rates and Masses at Maturity for Body Components of Sheep. Journal Animal Science 71: 2952-2957.

Martínez Avalos., A.M.M., Bores Quintero, R.F., Castellanos Ruelas, A.F., 1987. Zoometría y predicción corporal de la borrega Pelibuey. Técnica Pecuaria México 25: 72-84.

Martinez Avalos, A.M.M., Bores Quintero, R., Velazquez Madrazo, P.A., Castellanos Ruelas, A.F., 1990. Influencia de la castración y del nivel energético de la dieta sobre el crecimiento y composición corporal del borrego pelibuey. Tecnica Pecuaria Mexico, 28:125-132.

Minitab, 2007. Minitab Release 15. Minitab Reference Manual. Minitab, State College, Philadelphia, USA.

Motulsky, H., Christopoulos, A., 2003. Fitting models to biological data using linear and nonlinear regression. A practical guide to curve fitting. Graph Pad Software, San Diego California, USA.

NRC 1985. Nutrient Requirements of Sheep. 5th Revised Edition, National Academy Press Washington, D.C.

SIAP, 2007. Secretaría de Agricultura Ganadería Desarrollo Rural Pesca y Alimentación. Coordinación General de Ganadería. Servicio de Información Estadística Agroalimentaria y Pesquera, México, febrero de 2007. http//:www.siap.sagarpa.gob.mx/ar_comcada gral.html.

Romano, M.J.L., Hernandez, G.J., Castellanos, R.A., 1983. Repercusion del valor nutritivo de la dieta sobre el crecimiento del borrego pelibuey. Tecnica Pecuaria Mexico. 45:67-79.

Solís, R.G.E., Castellanos, A.F., Velázquez, M., Rodríguez, G.F., 1991. Determination of nutritional requirements of growing hair sheep. Small Ruminant Research 4: 115-125.

Torres, A.J.F., Cámara, S.R., Aguilar, C.A.J., López, C.C., Ruíz, P.C.H., Vargas, M.J.J., Calderón, Q.J.A., Cánto D.J.H., Canul, K.H.L., 2004. Transferencia de Tecnología a Productores de Ovinos del Estado de Yucatán. Informe Final. FMVZ-UADY. Fundación Produce Yucatán A. C. Mérida Yucatán México. 123 pp.

3

BIOSTIMULATION IN CATTLE: STIMULATION PATHWAYS AND MECHANISMS OF RESPONSE

Carolina Fiol[1]* and Rodolfo Ungerfeld[2]

[1]*Departamento de Bovinos;*
[2]*Departamento de Fisiología; Facultad de Veterinaria, Universidad de la República,*
Lasplaces 1550, Montevideo, Uruguay
E-mail address: cfiol@yahoo.com
** Corresponding author*

SUMMARY

The stimulatory effects of males on female cyclic activity through genital stimulation, pheromones or other less defined external cues, is known as biostimulation. In cattle, biostimulation may be provoked with vasectomized bulls, testosterone-treated cows, or testosterone-treated steers. Several studies reported that exposure of cows to bulls decreases postpartum anestrous interval in cows, and advances puberty in heifers. Moreover, male exposure before the breeding period or estrous synchronization treatment, can improve breeding performance. However, stimulation pathways and mechanisms of response to male effect in cattle are not fully understood. The objective of this review is to summarize and discuss the principal advances in our understanding of biostimulation in cattle.

Key words: Anestrus; male effect; puberty; postpartum.

INTRODUCTION

Anestrus is defined as the absence of estrous cycles. In livestock industry, one of the main objectives is to decrease the duration of anestrus periods in order to obtain a more sustainable production. In beef and dairy cattle, this imply: 1) reducing the age at first service (Patterson *et al.,* 1992) and 2) decreasing the interval to resumption of cyclic activity during postpartum (Yavas and Walton, 2000; Peter *et al.,* 2009). The first one determines a reduction in the age at first calving, and so it is necessary to achieve optimal lifetime productivity (Lesmeister *et al.,* 1973; Day and Grum, 2005). The second allows early postpartum re-breeding (80-85 d postpartum) in order to maintain a 365 d calving interval (review: Crowe, 2008). Many alternatives are available to achieve those goals (e.g. nutritional managements; review: Hess *et al.,* 2005). Although hormonal treatments appear to be highly effectives (Diskin *et al.,* 2002; Patterson *et al.,* 2003), growing concern in the development of *"clean, green and ethical"* techniques, in particular those linked to *"reduced usage, if not elimination, of practices that depend on drugs, chemicals and exogenous hormones"* (Martin *et al.,* 2004; Martin, 2009), have determined that the evaluation of other management tools, like socio-sexual stimuli, became more important during recent years.

Biostimulation (male or bull effect) can be defined as the stimulus provoked by the presence of males, which induce estrous and ovulation through genital stimulation, pheromones, or other external cues

(Chenoweth, 1983). In 1901, Heape proposed that the presence of males stimulates the onset of puberty in several mammals (cited by Izard, 1983). In cattle, Neresjan in 1959 reported that postpartum anestrus interval is reduced if cows are exposed to vasectomized bulls. Since then, many studies reported that male exposure stimulates cyclic activity in postpartum cows (Zalesky *et al.*, 1984; Alberio *et al.*, 1987; Rekwot *et al.*, 2000a; Landaeta-Hernández *et al.*, 2004, 2006; Berardinelli and Joshi, 2005a; Tauck, 2008). However, there is scarce information on the stimulant effect of males on the advancement of puberty in heifers (Izard and Vandenbergh, 1982a; Roberson *et al.*, 1991; Rekwot *et al.*, 2000b; Lima *et al.*, 2008; Oliveira *et al.*, 2009). The objective of this review is to summarize and discuss the principal advances on the understanding of biostimulation in cattle.

STIMULATION PATHWAYS

The stimulus provoked by the introduction of the males can act through different pathways, including olfactory, visual and auditory signals (Ungerfeld, 2007). In small ruminants, olfactory cues are undoubtedly major components in the response of females to male effect, but recent work have determined that maximum ovulatory response is obtain even when females are in full contact with males, which indicates that non-olfactory signals are complementary or synergistic with chemical cues (Delgadillo *et al.*, 2009). A complementary explanation may be that chemical signals act –at least partially– as an individual signature (Wyatt, 2010). When it comes to olfactory cues, pheromones were the first ones considered as male effect mediators in ewes and pigs (Rekwot *et al.*, 2001), and, in several species, it has been documented the influence of male signals on puberty onset in prepubertal females (mouse: "Vandenbergh effect": Vandenbergh, 1983; ewes: O'Riordan and Hanrahan, 1989; gilts: Thompson and Savage, 1978). Pheromones refer to airborne chemical substances released into the urine, feces or cutaneous glands, which cause a behavioural and/or specific endocrine reaction when they are perceived by a recipient of the same specie (Izard, 1983; Rekwot *et al.*, 2000a). In most mammals, pheromones are behavioral modulating substances (signals), and affect both maternal and reproductive behaviors, as the alarm and aggression responses (Swaney and Keverne, 2009). However, as discussed below, at least in small ruminants the type of signal exerted by the male does not strictly adjust to the pheromone definition (Delgadillo *et al.*, 2009), which determined that the term "olfactory cue" (Chanvallon and Fabre-Nys, 2009) or "chemical signal" (Ungerfeld *et al.*, 2008) instead of "pheromone" is sometimes prefered.

The detection and integration of olfactory signals in mammals occurs through the main olfactory system (MOS) and accessory olfactory system (AOS) pathways. Grus and Zhang (2008) developed the hypothesis that environmental cues are perceived by MOS, while specie-specific signals (e.g. pheromones) may be perceived by AOS. However, inactivation of the projections from the accessory olfactory bulb in ewes did not affect the response to ram odor (Cohen-Tannoudji *et al.*, 1989). In contrast, destruction of the main olfactory epithelium and inactivation of the cortical amygdale blocked the endocrine response to male odor, but did not affect the response to male presence (Gelez and Fabre-Nys, 2004). In turn, AOS neural pathways are activated during the response to ram odor in ewes (Gelez and Fabre-Nys, 2004). Thus, chemical signals associated to male effect may mainly act through MOS and not AOS, and, as mentioned before, sensory signals other than olfaction appear to be involved in the response to males (review: Gelez and Fabre-Nys, 2006).

Learning and effects of previous experience in the response to the male effect in small ruminants are other characteristics that differentiate these signals from the classic definition of pheromone (Delgadillo *et al.*, 2009). Unlike in adults and experienced ewes, the exposure to a ram or to ram odor in sexually naive ewes did not determined activation of AOS (Chanvallon and Fabre-Nys, 2009). Young and sexually naive females generally have poor responses to male effect, and those responses are not improved by prior sexual experience (Chanvallon *et al.*, 2010a). However, in adult and sexually inexperienced ewes, pre-exposure to rams influenced positively the response to the ram effect (Chanvallon *et al.*, 2010b). Endocrine response to ram odor in young and adult sexually naive ewes was reduced in contrast to the experienced ewes, but the response to the further exposure to males did not differ between naive and experienced females (Gelez *et al.*, 2004). Similarly, prior exposure to rams is not a pre-requisite for the endocrine response to ram exposure (Hawken *et al.*, 2008).

In cattle, males' excretory products and cervical mucus from estrous females enhance ovarian function, both in postpartum cows (Berardinelli and Joshi, 2005a; Wright *et al.*, 1994) and prepubertal heifers (Izard and Vandenbergh, 1982a). Moreover, vaginal products can advance the luteolysis induced by the administration of PGF2α (Izard and Vandenbergh, 1982b). This demonstrates the existence of some sort of "pheromone" on females' body fluids, which may exert a female-female effect, even though it remains unknown if stimulus provoked by females act similarly as products secreted by males (Wright *et al.*, 1994).

Berardinelli and Joshi (2005a) evaluated resumption of cyclic activity in postpartum, anovular, primiparous cows exposed to bulls or to excretory products of bulls. They introduced cows for 12 h daily during 70 d in pens in where males were kept during the other 12 h of the day. Anestrus postpartum length did not differ between females exposed to males or to excretory products of males, and was shorter than in non-exposed females. On the other hand, continuous exposure (24 h/d) of primiparous cows to male urine using a urine delivery system during 64 d did not affect postpartum anestrus interval compared to cows not exposed to urine (Tauck and Berardinelli, 2005; Tauck *et al.*, 2006). Therefore, authors concluded that in cattle urine may not be the biological agent by which pheromones are delivered, or that continuous urine exposure may "overstimulate" females. However, in contrast with these results, in prepuberal beef heifers, oronasal treatment with bull urine during 7 wk determined an increase of 35% in females that began cyclic activity in contrast to heifers treated with water (Izard and Vandenbergh, 1982a). Moreover, a 7 d length oronasal treatment with bull urine induced an increase in LH mean concentrations in postpartum dairy cows (Baruah and Kanchev, 1993), but these authors did not include information on resumption of cyclic activity. Therefore, although Tauck *et al.* (2006) explanation cannot be discarded, according to our knowledge there is no previous information on possible "overstimulation" of females by chemical signals action. Thus, even though positive effect of biostimulation in postpartum cows and prepubertal heifers may be at least partially mediated by olfactory cues, intensity of that stimulus and other communication pathways (visual and auditory) should affect cattle response.

MECHANISMS OF RESPONSE

In small ruminants, the first response to male exposure is an increase in LH pulse frequency that can be observed even a few minutes after males are introduced (reviews: Martin *et al.*, 1986; Walkden-Brown *et al.*, 1999; Ungerfeld, 2007). The increase in LH pulse frequency is also observed in ovariectomized ewes, which means that the change on steroid negative to positive feedback is not the only mechanism involved in the response (Martin *et al.*, 1983). Although the first ovulation occurs approximately 48 h after the initial exposure to males, at least in ewes this ovulation is not associated with estrous behavior (Signoret, 1991), and in approximately half of the females the resulting corpus luteum is abnormal and regresses 6-7 d later (Gelez and Fabre-Nys, 2006; Ungerfeld, 2003). Estrus is normally displayed 17 to 25 d after the introduction of males.

In cattle, unlike in small ruminants, physiological mechanisms involved in the response to biostimulation are not well understood, and most studies have been performed with postpartum cows. In this section we aim to summarize the scarce information available on the responding mechanism of cows to biostimulation.

LH and ovarian response

Custer *et al.* (1990) did not observe any change in LH secretion pattern in primiparous beef cows exposed to bulls during postpartum period, although biostimulation reduced the interval to first estrus. In contrast, Fernandez *et al.* (1996) observed an increase in LH mean concentration and pulse frequency in postpartum cows during continuously or intermittently exposure to epididectomized bulls since 30 d postpartum. In dairy cows, Roelofs *et al.* (2007) found increased mean and basal LH concentration, and LH pulse frequency during 8 h fenceline contact with males. Similarly, as mentioned before, oronasal treatment with bull urine during 7 d enhanced LH mean concentrations in postpartum dairy cows (Baruah and Kanchev, 1993). In more recent studies, acute (5 h daily for 9 d; Tauck *et al.,* 2010a) exposure to males of primiparous, suckled cows since 67 d postpartum determined an increase in LH pulse frequency compared to females exposed to steers, but no differences were found on mean, baseline, amplitude and duration of LH pulses between treatments. Moreover, none of the females resumed cyclic activity during the experiment (Tauck *et al.,* 2010a). It should be noticed that sampling scheme was different between the above mentioned experiments: while Custer *et al.* (1990) began blood samples 10 d after males were introduced, and sampled at weekly intervals, on the other studies the first blood sample was taken at the moment (Baruah and Kanchev, 1993; Fernandez *et al.,* 1996), 1 d (Roelofs *et al.,* 2007) or 1 h (Tauck, 2008; Tauck *et al.,* 2010a) before male introduction, and were repeated daily (Tauck *et al.,* 2010a). On the other hand, Fernandez *et al.* (1996) took blood samples every three days. Those differences in sampling scheme may explain why Fernandez *et al.* (1996) observed an LH response that was undetected by other researchers. In that sense, unlike in small ruminants, the ovulating response of cows is widespread on time. Therefore, endocrine changes induced by males may also differ between animals, making difficult to detect changes due to the different times in which those may be observed. In summary, most of the information indicates that positive effects of biostimulation in cattle would be mediated by an activation of the hypothalamic-pituitary-ovarian axis, but the moment in which the response appears may differ from that observed in small ruminants.

The ovarian response of ewes to the introduction of rams is well documented (Atkinson and Williamson, 1985; Martin *et al.,* 1986; Ungerfeld *et al.,* 2002). In cattle, Bastidas *et al.* (1997) performed weekly ovarian activity ultrasound observations in prepubertal heifers exposed or isolated from males. Although there were no differences on the onset of cyclic activity, the number of small and large follicles increased after biostimulation. In turn, Berardinelli *et al.* (2009) reported that the interval between follicular waves is shorter, and the size of the dominant follicle is greater in females exposed continuously or every 6 or 12 h to males during the first 45 d postpartum. We evaluated daily follicular size development in anestrus beef heifers exposed to androgenized steers (AS) during 30 d, and – in accordance with previous results– observed that the follicular diameter was greater after exposing females during 15 d (Fiol *et al.,* 2010a). In summary, in cattle male effect appears to induce an increase of LH secretion which is associated with a greater follicular development rate.

Cortisol and stress

Stress and linked hormones, in particular cortisol, had been generally associated with negative effects on reproduction. High cortisol concentrations affect LH secretion through an inhibition of GnRH pulsatility (Tilbrook *et al.,* 2000; Breen and Karsch, 2004), or through the increase in the sensibility to estradiol negative feedback (Oakley *et al.,* 2009). In ewes exposed to different stressors on day -1, 0 and +1 of male exposure, greater cortisol concentrations were associated with a lower response to male effect in contrast to no stressed ewes (Chanvallon *et al.,* 2010a). However, in sexually naive ewes, females selected for "nervous" temperament, which were expected to be inhibited by some kind of stress, responded better to male effect than "calm" ewes (Chanvallon *et al.,* 2010b).

In rodents there is evidence that adrenal activation is linked to female response to the pheromonal stimulus of the males (Mora and Sanchez-Criado, 2004). However, in ewes there were no changes observed in cortisol concentrations after the introduction of rams (Ungerfeld, 2003), as happens in rams stimulated by the presence of estrual ewes (Gonzalez *et al.,* 1988). Tauck *et al.* (2007) evaluated the effects of close physical contact of primiparous, postpartum, anestrous cows to males during 30 d, or to bull, or steer urine during 64 d, on cyclic activity and plasma cortisol concentrations. Interval to resumption of ovarian activity was shortened in exposed cows compared to isolated ones, with no differences between females exposed to bulls or steer urine. Cortisol concentrations raised and remained high during exposure to the males, but decreased in cows exposed to urine. No differences were found between cows exposed to bulls or steers: in both groups cortisol gradually decreased during the experimental period (Tauck *et al.,* 2007). More recent studies reported no differences in mean cortisol concentrations between postpartum cows exposed acutely (5 h/d during 9 d; Tauck *et al.,* 2010a) or continuously and chronically (24 h/d during 44 d; Tauck, 2008) to males, with those isolated from males. However, cows exposed acutely to bulls had less cortisol pulse frequency and tended to have greater pulse duration than non-exposed females (Tauck *et al.,* 2010a). It should be considered that as mentioned before, in this experiment none of the females resumed ovarian cyclic activity (Tauck *et al.,* 2010a). Similarly, continuous exposure to males decreased cortisol pulse frequency, but differences were only seen before cows resumed cyclic activity (Tauck, 2008). In turn, cortisol and LH concentrations were negatively correlated in exposed groups: while cortisol pulse frequency decreased, LH pulse frequency increased (Tauck *et al.,* 2010a), which also agrees with results found in other species (Breen and Karsch, 2004). Considering all those findings, Tauck (2008) suggested that one component of male stimulus may be an alteration of cortisol secretion pattern before resumption of cyclic activity, which might determine an activation of hypothalamus-pituitary-ovarian axis.

Leptin

Leptin is a metabolic hormone (Clarke and Henry, 1999) that has also been evaluated in relation to biostimulation. Although the proportion of primiparous estrous cows increased with biostimulation during 40 d postpartum, leptin mean concentrations were not affected (Olsen *et al.,* 2009). Leptin concentrations increased in cows that resumed cyclic activity 15, 18 and 27 d after biostimulation began, but while in exposed females levels remained high during that period, in the cows that remained isolated leptin concentrations decreased earlier, by day 21. These authors considered that biostimulation may only have an indirect effect on leptin levels, as it advances postpartum rebreeding (Olsen *et al.,* 2009).

FACTORS THAT AFFECT THE RESPONSE TO BIOSTIMULATION

The main factors that affect the response to biostimulation in cattle may be classified as those linked to stimulus characteristics and management (type, intensity, males' characteristics, stimulus length), and those linked to females receptivity (breed, age and parity, time of the year, reproductive status, body development and nutritional status). Most of the information of factors affecting the response has been generated in postpartum cows.

Stimulus characteristics and management

Animals used for biostimulation

Positive effects of biostimulation in cyclic activity in cows have been demonstrated with bulls (Fernandez *et al.,* 1993; Berardinelli and Tauck, 2007), AS (Ungerfeld, 2009; Fiol *et al.,* 2010b) and androgenized cows (Burns and Spitzer, 1992). Burns and Spitzer (1992) observed a similar reduction in postpartum interval to first estrus in cows exposed to bulls or to androgenized females. Similarly, Ungerfeld (2009) reported an increase of pregnancy rates in heifers exposed to AS before the breeding period. Eventhough, others observed that the use of bulls was more effective than exposure to androgenized females (De Souza, 2002). According to our knowledge, there are no published articles comparing the effectiveness of AS and bulls. Despite the effectiveness, considering the positive results of the use of testosterone-treated animals, it may be affirmed that at least part of the biostimulatory signals on females' cyclic activity are androgen-dependent.

Stimulus intensity or proximity

The percentage of females' that respond to biostimulation is related to the intensity of the stimulus. Fenceline exposure to males (visual, chemical and auditory stimulus) stimulated resumption of postpartum cyclic activity in primiparous cows (Fike *et al.*, 1996). Similarly, Berardinelli and Tauck (2007) reported that fenceline contact of primiparous cows with bulls increased the proportion of cyclic females after 14 d of contact. However, a greater proportion of cyclic females was obtained with direct physical contact compared to fenceline exposure to males (Berardinelli and Tauck, 2007). In agreement, Sato *et al.* (1994) reported better results when cows were physically in contact with bulls than when they could only watch and smell them. Physical proximity also influences the response, as the distance between AS and females was positively associated with the probability of ovulating in prepubertal heifers (Fiol *et al.,* 2010b). Therefore, signals associated with courtship (tactile, physical contact) are also important, and stimulus is not only dependent upon chemical signals.

In dairy cows, response appears to be different. In high producing dairy cows, Shipka and Ellis (1999) did not find positive effects of direct physical contact with bulls during 120 d, nor of twice daily fenceline male exposure on resumption of postpartum cyclic activity. Eventhough, surprisingly postpartum anestrus was reduced in non exposed females. In turn, fenceline exposure of postpartum dairy cows to bulls did not affect estrus behavioral response (Shipka and Ellis, 1998; Roelofs *et al.*, 2008). Roelofs *et al.* (2007) found positive effects of fenceline male exposure on LH mean concentration in postpartum dairy cows, but cyclic activity was not determined in that study. It may be possible to speculate that the great negative energy balance that occurs during early postpartum in dairy cows may explain the different results reported in dairy and beef cows (Shipka and Ellis, 1998; Roelofs *et al.,* 2007).

Therefore, considering the relationship between intensity of male-female link and the response to biostimulation, the final response appears to be linked to a summatory of pathways by which females received the stimulus, and it is not possible to simplify the mechanism to the effects of olfactory signals only.

Males' characteristics and management

Males' sexual behavior and "novelty" effect influence the response to male exposure in sheep (Perkins and Fitzgerald, 1994) and goats (Delgadillo *et al.,* 2006). The alternate exposure to different males determined a greater response in females as a consequence of intensifying the stimulus with new males (Cushwa *et al.,* 1992; Hawken and Beard, 2009). In the same sense, Miller and Ungerfeld (2008) reported that weekly bull exchange during the breeding period of postpartum cows decreased the anestrus interval and increased early pregnancy rates. In contrast, Berardinelli *et al.* (2005) did not find any differences on resumption of cyclic activity in postpartum cows exposed to the same bulls during 95 d postpartum or after exchanging bulls once 35 d after parturition. Therefore, the effectiveness of bull exchange may be related to the frequency with which it is performed: positive results were obtained with weekly exchange (Miller and Ungerfeld, 2008), but not with a single exchange (Berardinelli *et al.,* 2005).

In sheep (Ungerfeld *et al.,* 2008) and swine (Kirkwood and Hughes, 1981) the response is greater when adult males rather than young males are used. However, no differences were found in either testosterone concentrations reached by adult or young rams used to stimulate anestrous ewes (Ungerfeld, 2003), or in sexual behavior between them (Ungerfeld *et al.,* 2008). However, more ewes ovulated and came into estrus after being in contact with wool from adult rams than with wool from young males. Thus, these authors proposed that chemical signals contained in wool may be responsible for that difference (Ungerfeld *et al.,* 2008). In cattle there is only one trial that compared the use of adult and young bulls. Cupp *et al.* (1993) observed a reduction in anestrus length in females exposed to males during 63 d postpartum, but of similar magnitude in cows exposed to young (1 yr old) or adult (> 3 yr old) bulls. Therefore, according to the scarce information available, bulls' age- at least in the evaluated ranges- appears not to affect females' response.

Duration of stimulation periods

The duration of the males' stimulation influences the response in sheep (Hawken and Beard, 2009) and goats (Rivas-Muñoz *et al.,* 2007). Intermittent exposure of postpartum primiparous suckled beef cows to males (2 h every 3 d, during 18 d) since 30 d postpartum had no effects on resumption of cyclic activity compared to isolated females (Fernandez *et al.,* 1996). Moreover, in those groups, postpartum anestrus was longer than in females continuously exposed to males since early postpartum or since 30 d postpartum. Berardinelli and Joshi (2005a) reported shorter anestrus intervals in females exposed to excretory products of bulls 12 h daily than in non exposed females. Recently, Tauck *et al.* (2010b) evaluated resumption of cyclic activity in postpartum cows exposed to males during 6 or 12 h daily, or not exposed for 45 d. Exposure during 6 or 12 h reduced anestrus postpartum compared to isolated females, but the cumulative proportion of cyclic females at 10 d intervals was greater for females exposed during 12 than during 6 h. Therefore, daily and continuous stimulus from the males appears to be necessary to obtain a positive response on cyclic activity, and it seems that there is a negative correlation between exposure periods length and time needed for resumption of cyclic activity. Eventhough, as daily exposure during 24 h to bull urine had no effects on cyclic activity (Tauck and Berardinelli, 2005; Tauck *et al.,* 2006), Tauck *et al.* (2010b) proposed that the alternation of "stimulation" and "relaxation" periods may be necessary to obtain a positive response to biostimulation.

In prepubertal heifers, most of the studies reported longer periods of exposure to males than those reported in postpartum cows to obtain a positive response (175 d, Roberson *et al.*, 1991; 450 d, Rekwot *et al.,* 2000b; 180 d, Lima *et al.*, 2008; 210 d, Oliveira *et al.,* 2009). However, it seems that the length may be related to the time close to spontaneous puberty achievement that heifers are. In that sense, others found positive effects with shorter periods of exposure (75 d, Assis *et al.,* 2000; 50 d, Quadros and Lobato, 2004; 15 d, Ungerfeld, 2009; 30 d, Fiol *et al*., 2010b) (Table 1). In addition, there are some studies in which no effects were observed on the onset of puberty between heifers exposed during short (21 d, Macmillan *et al.*, 1979; 30 d, Berardinelli *et al.,* 1978) or long (152 d, Roberson *et al.,* 1987) periods. The inconsistency of the results obtained in prepubertal heifers may be due to the great influence of body development and nutritional status at the beginning of the exposure and during biostimulation on the response to male exposure.

Females receptivity

Breed, age and parity

Although effects of male exposure on cyclic activity have been evaluated mainly in *Bos taurus taurus*, others reported positive effects of biostimulation in zebu (*Bos taurus indicus)*, both in postpartum cows (Rekwot *et al.,* 2000a; Soto Belloso *et al.,* 1997; Bolaños *et al.,* 1998) and prepubertal heifers (Rekwot *et al.,* 2000b). In turn, biostimulation was also effective to reduce anestrus periods in buffaloes females (*Bubalus bubalis*; Ingawale and Dhoble, 2004; Gokuldas *et al.,* 2010).

In sheep, age and female parity at the beginning of the exposure affect the response to biostimulation. As it was mentioned before, young and sexually naive females generally have poor responses as a consequence of sexual inexperience and the need of some "learning" related to the male effect (Gelez *et al.,* 2004; Chanvallon *et al.,* 2010a). In postpartum cows, biostimulation has been studied especially in primiparous females as this category presents long anestrus intervals compared to multiparous cows (Short *et al.,* 1990). In some studies, exposure to males was effective in reducing anestrus periods in primiparous but not in multiparous cows (Gifford *et al.,* 1989; Fike *et al.,* 1996), while others found positive effects in both categories (Soto-Belloso *et al.,* 1997). The greater duration of postpartum anestrus in primiparous cows could determine that multiparous females are less dependent on external stimuli. Thus, in those cows the effect of male exposure may be less effective in reducing anestrus length (Gifford *et al.,* 1989; Fike *et al.,* 1996).

Macmillan *et al.* (1979) reported positive effects of male exposure during 20 d on cyclic activity in postpartum cows but not in prepubertal heifers. In turn, Small *et al.* (2000) found positive effects of biostimulation in puberty onset in winter-born heifers (puberal females), but negative effects in spring-born heifers (prepuberal females). Those authors concluded that response to male effect may be related to the time close to spontaneous puberty achievement that heifers are at the beginning of the exposure. Sensitivity to stress in young female sheep exposed to males for first time, affected negatively the response to male exposure (Chanvallon, 2009). In cattle, regrouping with strange animals determines a stress response (Veissier *et al.,* 2001), similar to what may be happening when heifers were exposed to males for first time. Therefore, nutritional status and stress may be main factors that determined differences in the response to biostimulation between prepubertal heifers and postpartum cows.

Table 1. Summary of studies that reported positive effects on the onset of cyclic activity in heifers exposed to males, in relation to the length of the exposure period, age and body weight of the heifers, and the interval from the beginning of the exposure to the response (IETR).

Exposure length (d)	Age (mo-old)	Body weight (kg)	IETR (d)	Reference
30	22	282	15	Fiol and Ungerfeld, 2011b
35	12	226	21	Fiol *et al.,* 2010b
50	24.5	278	35	Quadros and Lobato, 2004
75	24	269	< 75	Assis *et al.,* 2000
154	9	300	14	Small *et al.,* 2000
210	12	180	210	Oliveira *et al.,* 2009
460	14.8	148	189	Rekwot *et al.,* 2000b

Reproductive status at the beginning of the exposure period

Physiological status of females exposed to males has a strong influence on their response. Moreover, it has been proposed that the wide variation between animals' physiological status in postpartum cattle at the time of male exposure may partially explain the high variability in the response to biostimulation (Ungerfeld, 2007). The exposure of multiparous cows from the first week postpartum during 50 (Gokuldas *et al.,* 2010) or 90 (Landaeta-Hernández *et al.,* 2008) d, decreased anestrus length compared to isolated females, but no effects on cyclic activity were found when exposure began 33 d postpartum (Bonavera *et al.,* 1990). The exposure of multiparous cows to bulls during early postpartum (day 3 to 85) decreased anestrus length compared to those females exposed only from day 53 to 85 postpartum (Zalesky *et al.,* 1984). In contrast, Gifford *et al.* (1989) using the same category did not find any difference between early (day 3 to 85 postpartum) or late (from day 55 to 85 postpartum) male exposure on duration of postpartum anestrus. The differences could be explained by postpartum anestrus length in cows exposed to males during late postpartum between experiments (62 and 31 d, Zalesky *et al.,* 1984 and Gifford *et al.,* 1989, respectively). According to those findings, it is interesting to speculate that positive response to male exposure during early postpartum in multiparous cows may be only observed when females present long anestrus intervals- e.g. 50 or more days.

It appears that the response of postpartum primiparous cows is more homogeneous than that of multiparous cows. Early postpartum exposure to males decreased anestrus interval more dramatically than late exposure (Gifford *et al.,* 1989), but Fernandez *et al.* (1993) reported no differences in anestrus length between primiparous cows exposed continuously to males since calving, exposed during the first 30 d postpartum, or only after the first 30 d postpartum. Moreover, in the three groups of females, anestrus was shorter than in cows isolated from males during all the postpartum period. Similarly, there were no differences on anestrus length between primiparous cows exposed to males since 15, 35 or 55 d postpartum (Berardinelli and Joshi, 2005b). However, cows exposed to males since 55 d postpartum had shorter intervals from the beginning of the exposure to resumption of cyclic activity than females exposed since 15 d, but the final interval between partum and resumption of cyclic activity was similar. Therefore, in primiparous cows the positive response to biostimulation appears to be faster in late postpartum. This might be related with a greater sensitivity to male stimulus in late postpartum cows.

The high variability of results could be due to one or more physiological events linked to ovarian cyclicity resumption during postpartum (Rekwot *et al.,* 2001; Berardinelli and Joshi, 2005b). During early postpartum, females' metabolism undergoes a period of negative energy balance (NEB), caused by high nutritional requirements and a decrease in food intake (Short *et al.,* 1990; Montiel and Ahuja, 2005). As reproduction is one of the more affected functions, there is a direct relation between resumption of postpartum cyclic activity and duration and severity of NEB (Short *et al.,* 1990). In turn, duration and severity of the NEB is determined by the energy stores present at partum, lactation and animal parity (Short and Adams, 1988; Montiel and Ahuja, 2005). In general, when age at first partum occurs at 2 years-old or earlier, growth requirements are added to lactation requirements, so severity and duration of postpartum anestrus is greater than in multiparous cows (Short *et al.,* 1990).

Summarizing all the above information, while primiparous cows appear to respond to male exposure during the entire postpartum period, in multiparous cows positive effects may be obtained during a shorter and more variable period of time. In multiparous cows with long postpartum anestrus intervals (more than 50 d), early exposure to males has positive effects (Zalesky *et al.,* 1984; Landaeta-Hernandez *et al.,* 2008). In turn, when multiparous cows presented short anestrus periods, or when exposure to males begins late in the postpartum, biostimulation may have no effect (Gifford *et al.,* 1989; Bonavera *et al.,* 1990). In contrast, in primiparous cows with extended postpartum anestrus, response to male exposure appears to be more homogeneous along the postpartum period, but it may be faster as postpartum time increases (Gifford *et al.,* 1989).

Time of the year

MacMillan *et al.* (1979) found positive effects of male exposure in postpartum cows during spring but not during winter. Similarly, anestrus postpartum decreased when multiparous cows were exposed to bulls during spring-summer, but not in winter (Alberio *et al.,* 1987; Agabriel *et al.,* 2004). Differences were attributed to the differences in food availability (Agabriel *et al.,* 2004), which determined that exposed females increased body weight during spring but lost body weight during winter (Alberio *et al.,* 1987). However, exposure to males during spring began 60 d postpartum, while winter exposure began 10 d postpartum (Alberio *et al.,* 1987). According to the authors, early exposure during winter could be the main factor contributing to the absence of effects of biostimulation. Moreover, regardless males' presence, winter females had short anestrus intervals (53.2 vs 36.6 d in exposed and isolated females, respectively; Alberio *et al.,* 1987). Overall, considering that reproduction in domestic cattle are not strongly influenced by photoperiodic changes, it seems that differences in responses in relation to the period of the year are more related to changes in food consumption, and thus in metabolic status, than to direct effects of season.

Heifers body development and nutritional status

Nutritional status and body weight gain are related to age at puberty in beef heifers (Wiltbank *et al.,* 1969; Short and Bellows, 1971; Quintans *et al.,* 2004), and both appear to affect the response to biostimulation. Period of time necessary to obtain a positive response might be related to heifers' body development at the beginning of male exposure (Table 1). More 25.5 mo-old heifers began cyclic activity after 50 d of bull exposure than 23.5 mo-old old heifers (Quadros and Lobato, 2004). Moreover, in heifers that were younger at the beginning of biostimulation, longer periods of exposure were necessary to obtain a positive response in cyclic activity (Roberson *et al.,* 1991; Oliveira *et al.,* 2009). Rekwot *et al.* (2000b) reported that heifers reared in mixed groups attained puberty at lower ages and body weights compared to those isolated from males.

In turn, Roberson *et al.* (1991) found that heifers with high body weight gains during male exposure began cyclic activity earlier than those with medium body weight gains. Oliveira *et al.* (2009) obtained positive effects of long exposure (210 d) on the onset of cyclic activity in 12 mo-old prepuberal heifers, but they did not obtained positive results of supplementing those heifers. However, exposed females had greater final body weight, which was positively correlated to age at puberty. Body weight at the beginning of the exposure period affected the response to male exposure: positive effects were found in heavier heifers (Quadros and Lobato, 2004; Fiol *et al.,* 2010b), which agree with previous studies performed in goats (Véliz *et al.,* 2006). Moreover, heavier heifers had closer proximity with the AS than lighter ones (Fiol *et al.,* 2010b). In order to isolate the attractiveness related to body weight from the competitive effects of having heifers from different body weights stimulated together, we evaluated sexual behavior of AS toward heavier or lighter heifers in competitive and non-competitive sexual tests (Fiol and Ungerfeld, 2011a). We found that males prefer heavier heifers only when they were in competitive situations with lighter heifers, but courtship was similar in individual tests despite heifers' body weight. We concluded that factors associated with females' competition and not the body weight of the heifers itself, may be determinant of the greater stimulus received by the heavier heifers. Body weight is one of the main factors implicated in the determination of social rank in cattle (Bouissou, 1972). In turn, social rank in goats affected the response to male exposure: dominant females maintained a closer contact with the male, and had a higher endocrine and ovulatory response to male exposure (Alvarez *et al.,* 2003, 2007, 2009). Therefore, it is possible to speculate that dominance-submission relationships when cattle of different body weights are mixed are major determinants in the different response to male exposure due to competition in access to males.

Nutritional status in postpartum cows

In postpartum cows, nutritional status influence resumption of cyclic activity (Montiel and Ahuja, 2005) and response to biostimulation, as reported in small ruminants (Wright *et al.,* 1990). Monje *et al.* (1992) evaluated the effects of two energy levels since 12 d postpartum (30% more and 30% less of energy requirements) and male exposure during 80 d (from 30 d postpartum). Those authors reported a positive relationship between nutritional status and response to biostimulation: cows with high energy level and exposed to males had more synchronized ovulations and less variability on resumption of cyclic activity than exposed cows that were submitted to low energy levels, and to non-exposed females. Moreover, between low energy groups, non-exposed females began cyclic activity earlier than exposed ones. The authors attributed that difference to the fact that cows in the low energy level, and those not exposed to males improved their nutritional status since the moment in which all the groups were joined together (110 d postpartum), and responded positively to male exposure later. In contrast, low energy cows exposed to males from 30 d postpartum might have been used to male presence, losing the novel effect when both groups were joined (Monje *et al.,* 1992).

Energy supplementation and male exposure during 180 d on multiparous zebu cows had a positive synergistic effect on resumption of postpartum cyclic activity (Rekwot *et al.,* 2004). Stumpf *et al.* (1992) evaluated the effects of body condition score (BCS) at partum on the response to biostimulation in postpartum multiparous cows, and found that cows with moderate BCS (4.9 units in a 1 to 10 scale) responded better to male exposure than cows with high BCS (5.9 units). Similarly, Madrigal *et al.* (2001) reported an interaction between BCS and biostimulation on resumption of cyclic activity and pregnancy rates, which was more pronounced in low BCS cows. In turn, Landaeta-Hernández *et al.* (2008) found that high BCS multiparous cows (4.8 units) exposed to bulls since the first week postpartum began cyclic activity earlier than non exposed females (38 vs 51 d, respectively).

Considering all the above information, nutritional status and body development may be main factors affecting the response to male exposure in beef cattle. In prepubertal heifers, body development at the beginning of the exposure period and body weight gain during biostimulation, influence the response to male exposure (Roberson *et al.,* 1991; Rekwot *et al.,* 2000b; Quadros and Lobato, 2004; Fiol *et al.,* 2010b). In postpartum cows, high nutritional status at the beginning of the exposure period determined a low response to biostimulation because females in those situations have short anestrus postpartum regardless males' presence (Stumpf *et al.,* 1992; Madrigal *et al.,* 2001). On the other hand, cows with very low BCS are unable to respond to male exposure due to the low nutritional status (Monje *et al.,* 1992). Therefore, moderate nutritional status appears to be necessary to obtain positive response to male exposure. This should be specifically considered under grazing conditions in which impact of nutritional status is more difficult to be controlled.

PRE-BIOSTIMULATION AND BREEDING PERFORMANCE

In Tables 2 and 3 we summarize the results of different studies that evaluated the effects of biostimulation before a bull or AI breeding period, both in prepubertal heifers (Table 2) and postpartum cows (Table 3). Exposure of beef heifers to AS (Ungerfeld, 2009) or to vasectomized bulls (Oliveira *et al.,* 2009) before the breeding period increased natural-service pregnancy rates. Moreover, male exposure improved AI pregnancy rates in beef heifers (Roberson *et al.,* 1991; Quadros and Lobato, 2004) and in primiparous (Fernandez *et al.,* 1993) and multiparous (Zicarelli *et al.,* 1997; Gokuldas *et al.,* 2010) cows. However, some of those studies did not find differences in overall pregnancy rates between exposed and isolated females (Roberson *et al.,* 1991; Fernandez *et al.,* 1993). As pregnancy rates are improved when beef heifers begin cyclic activity before the onset of the breeding season (Byerley *et al.,* 1987), the greater pregnancy rates reported in some of the aforementioned studies may be a consequence of a greater proportion of exposed females cycling at the beginning of the natural service or AI breeding periods. In turn, age and body weight of the heifers at the beginning of the exposure period influenced the response to males: positive and significant differences on pregnancy rates were limited to older (Quadros and Lobato, 2004) and heavier (Ungerfeld, 2009) females.

Table 2. Summary of studies that evaluated breeding performance in terms of onset of cyclic activity and pregnancy rates in heifers exposed to males (EXP) or isolated from males (ISO)

Exposure type and length (d)	Cyclic activity (%)		Pregnancy (%)		Reference
	EXP	ISO	EXP	ISO	
ASE/DPC (30 d)	55a	32b	83	86.5	Fiol and Ungerfeld, 2011b
TBU (56 d)	67a	32b	79	85	Izard and Vandenbergh, 1982a
BE/DPC (175 d)	**	---	50.5a	16.5b	Roberson *et al.,* 1991
BE/DPC (75 d)	86.9c	71.1d	89.1	88.9	Assis *et al.,* 2000
BE/DPC (200 d)	**	---	58.9a	32.5b	Small *et al.,* 2000
BE/DPC (50 d)	76c	56d	90c	73d	Quadros and Lobato, 2004
ASE/DPC (15 d)	---	---	61a	32.3b	Ungerfeld, 2009
BE/DPC (210 d)	**	---	63a	21.5b	Oliveira *et al.,* 2009

ASE: androgenized steers exposure; BE: bull exposure
DPC: direct physical contact; TBU: treatment with bull urine
**Decrease age at puberty in EXP heifers
Different letters in the same row and for each experiment: a vs b: $P < 0.05$; c vs d: $P=0.06$

Table 3. Summary of studies that evaluated breeding performance in terms of resumption of cyclic activity and pregnancy rates in postpartum cows exposed to males (EXP) or isolated from males (ISO).

Exposure type and length (d)	Cyclic activity (%)		Pregnancy (%)		Reference
	EXP	ISO	EXP	ISO	
ASE/DPC (20 d)	---	---	58.5	50.0	Ungerfeld, 2010
BE/DPC (60 d)	81a	41b	67	63	Berardinelli *et al.*, 2001
BE/DPC-EPB (63 d)	87a	19b	87a	56b	Anderson *et al.*,2002
BE/DPC-EPB (60 d)	85.1a	31.3b	66.3a	51.5b	Berardinelli *et al.*, 2007
BE/DPC (35 d)	100a	70.4b	85a	60b	Tauck and Berardinelli, 2007
BE/DPC (50 d)	82a	38.5b	54.5a	15.4b	Gokuldas *et al.*, 2010
BE/FCB (42 d)	86a	73b	58	77	Tauck and Berardinelli, 2007
TBU (64 d)	15	33	89.5a	55b	Tauck and Berardinelli, 2007

ASE: androgenized steers exposure; BE: bull exposure

DPC: direct physical contact; EPB: excretory products of bulls; FCB: fence-line contact with bulls; TBU: treatment with bull urine

Different letters in the same row and for each experiment: a vs b: $P < 0.05$

Different results have been obtained when cows were exposed to males before or during an estrous synchronization treatment (EST). Exposure of postpartum beef cows to bulls or excretory products of bulls before or during an EST increased fixed-timed AI pregnancy (Anderson *et al.,* 2002) and conception (Berardinelli *et al.,* 2007) rates, but no differences were found on cows bred by AI 12 h after estrus (Anderson *et al.,* 2002; Berardinelli *et al.,* 2007). Similarly, Small *et al.* (2000), using an EST that included two doses of PGF2α, reported a 30% increase in timed AI pregnancy rates in winter-born, but not in spring-born heifers, exposed to bulls before the beginning of the EST compared to isolated heifers. In contrast, others did not find any positive effect on pregnancy or conception rates with direct (Ungerfeld, 2010; Berardinelli *et al.,* 2001) or fence-line (Fike *et al.,* 1996; Tauck and Berardinelli, 2007) exposure to males before the beginning of an EST in postpartum cows and heifers. Recently, we evaluated the onset of cyclic activity and breeding performance of heifers exposed to AS for 30 d before the beginning of an EST with double injection of PGF2α (Fiol and Ungerfeld, 2011b). We found a positive but transitory effect of biostimulation on ovarian cyclic activity, which determined that similar proportion of exposed and isolated females was cycling at the beginning of the EST. However, no effects of biostimulation were found on estrous response and AI pregnancy rates (Fiol and Ungerfeld, 2011b). According to those results, the induction of differences in cyclic activity at the beginning of the EST appears as necessary to obtain positive effects of male exposure on breeding performance.

Results were different when exposure to male urine through a delivery device was evaluated. Conception rate increased in postpartum cows exposed to bull urine before an EST, but not in females exposed to steer urine (Tauck and Berardinelli, 2007). At the same time, no effects of exposure to bull or steer urine on resumption of ovarian cyclic activity were observed in those experiments, which agree with previous studies (Tauck *et al.,* 2006). Thus, proportion of cyclic females at the beginning of the EST was similar between females exposed to bull or steer urine. Based on those results, Tauck and Berardinelli (2007) hypothesized the existence of two components of the biostimulatory effect of bulls: one that accelerates the resumption of cyclic activity and another that improves breeding performance. According to the authors, the last may only be effective with very close proximity to bull urine.

Summarizing all the above information, in order to have positive effects of biostimulation on breeding performance in cattle, exposure length and type (physical presence vs male urine or excretory products of males), and natural vs artificial breeding and subsequent use of an EST, should be considered.

FINAL CONSIDERATIONS AND FUTURE PERSPECTIVES

Livestock production is challenged worldwide, especially because of the high competition for natural resources, particularly land and water (review: Thornton, 2010). Improving reproductive efficiency in beef and dairy cattle should be one of the main objectives to obtain a more sustainable production. In this context, socio-sexual stimulus, like biostimulation, represents low cost and hormone-free alternatives than can be used alone or in conjunction with other techniques to increase reproductive results. The information collected in this review regarding the effects of biostimulation in cattle, demonstrates the need of further studies to understand the mechanisms involved in this phenomenon. Differences in reproductive physiology in cattle determined that many of the characteristics of the male effect in small ruminants cannot be fully extrapolated to the first ones. In particular, some important questions remain to be answered, as: Why results in cattle are more unpredictable than those obtained in small ruminants? Which factors disperse the response in time? What is the role of non-olfactory cues in the response? What is the importance of sexual experience and learning in the response? The possibility to answer these questions and more will determine the most efficient use of biostimulation in cattle.

REFERENCES

Agabriel, J., Blanc, F., Egal, D., Dhour, P. 2004. Combined effects of male and calving period on post partum anoestrus duration of Charolais cows. Rencontres Recherches Ruminants. 11, 398.

Alberio, R.H., Schiersmann, G., Carou, N., Mestre, J. 1987. Effect of teaser bull on ovarian and behavioural activity of suckling beef cows. Animal Reproduction Science. 14, 263-272.

Alvarez, L., Martin, G.B., Galindo, F., Zarco, L.A. 2003. Social dominance of female goats affect their response to the male effect. Applied Animal Behavior Science. 84, 119-126.

Alvarez, L., Zarco, L., Galindo, F., Blache, D, Martin, G.B. 2007. Social rank and response to the "male effect" in the Australian Cashmere goat. Animal Reproduction Science. 102, 258-266.

Alvarez, L., Ramos, A.L., Zarco, L. 2009. The ovulatory and LH responses to the male effect in dominant and subordinate goats. Small Ruminant Research. 83, 29-33.

Anderson, K.A., Berardinelli, J.G., Joshi, P.S., Robinson, B. 2002. Effects of exposure to bull or excretory products of bulls on the breeding performance of first-calf restricted suckled beef cows using a modified Co-Synch protocol. Proceeding Western Section American Society Animal Science 53, 457-460.

Assis, R.R., Pimentel, M.A., Jardim, P.O., Osorio, J.C., Machado, J.P. 2000. Influência da bioestimulacâo com machos vasectomizados na eficiência reproductiva de novilhas Aberdeen Angus. Revista Brasilera de Agrociência. 6 (3), 226-231.

Atkinson S., Williamson, P. 1985. Ram-induced growth of ovarian follicles and gonadotrophin inhibition in anoestrous ewes. Journal of Reproduction & Fertility. 73, 185-189.

Baruah, K.K., Kanchev, L.N. 1993. Hormonal responses to olfactory stimulation with bull urine in postpartum dairy cows. Proceeding VII World Conference Animal Production 3,356–359.

Bastidas, P., Ruiz, J., Manzo, M., Silva, O., Guerrero, N., Trocóniz, J. 1997. Effect of bull exposure on corpus luteum function and ovarian activity in prepuberal Brahman heifers. Archivos Latinoamericanos de Producción Animal. 5 (1), 390-392.

Berardinelli, J.G, Fogwell, R.L., Inskeep, E.K. 1978. Effect of electrical stimulation or presence of a bull on puberty in beef heifers. Theriogenology. 9, 133-138.

Berardinelli, J.G., Anderson, K., Robinson, B., Adair, R. 2001. Effect of presence of bulls and restricted suckling on breeding performance of first-calf suckled beef cows using the Ovsynch protocol. Proceeding Western Section American Society Animal Science 52.

Berardinelli, J.G., Joshi, P.S. 2005a. Initiation of postpartum luteal function in primiparous restricted-suckled beef cows exposed to a bull or excretory products of bulls or cows. Journal of Animal Science. 83, 2495-2500.

Berardinelli, J.G., Joshi, P.S. 2005b. Introduction of bulls at different days postpartum on resumption of ovarian cycling activity in primiparous beef cows. Journal of Animal Science. 83, 2106-2110.

Berardinelli, J.G., Joshi, P.S., Tauck, S.A. 2005. Postpartum resumption of ovarian cycling activity in first-calf suckled beef cows exposed to familiar or unfamiliar bulls. Animal Reproduction Science. 90, 201-209.

Berardinelli, J.G., Joshi, P.S., Tauck, S.A. 2007. Conception rates to artificial insemination in primiparous, suckled cows exposed to the biostimulatory effect of bulls before and during a gonadotropin-releasing hormone-based estrus synchronization protocol. Journal of Animal Science. 85, 848-852.

Berardinelli, J.G., Tauck, S.A. 2007. Intensity of the biostimulatory effect of bulls on resumption of ovulatory activity in primiparous, suckled, beef cows. Animal Reproduction Science. 99, 24-33.

Berardinelli, J.G., Wilkinson, J.R.C., Tauck, S.A., Olsen, J.R., Wedlake, R.J. 2009. Duration of daily bull exposure on follicular wave dynamics of postartum, anovular, suckled cows. Biology of Reproduction. 81, 544.

Bolaños, J.M., Forsberg, M., Kindahl, H., Rodriguez-Martinez, H. 1998. Biostimulatory effects of estrous cows and bulls on resumption of ovarian activity in postpartum anestrous zebu (*Bos Indicus)* cows in the humid tropics. Theriogenology. 49, 629-636.

Bonavera, J.J., Schiersmann, G.C.S., Alberio, R.H., Mestre, J. 1990. A note on the effects of 72-hour calf removal and/or bull exposure upon postpartum reproductive performance of Angus cows. Animal Production. 50, 202-206.

Bouissou, M.F. 1972. Influence of body weight and presence of horns on social rank in domestic cattle. Animal Behaviour. 20, 472-477.

Breen, K.M., Karsch, F.J. 2004. Does cortisol inhibit pulsatile luteinizing hormone secretion at the hypothalamic or pituitary level? Endocrinology. 145, 692-698.

Burns, P.D., Spitzer, J.C. 1992. Influence of biostimulation on reproduction in postpartum beef cows. Journal of Animal Science. 70, 358-362.

Byerley, D.J., Staigmiller, R.B., Berardinelli, J.G., Short, R.E. 1987. Pregnancy rates of beef heifers bred either on puberal or third estrus. Journal of Animal Science. 65, 645-650.

Chanvallon, A. 2009. Origine de la variabilité de résponse à l´effect mâle. PhD Thesis. University Françoise-Rabelais de Tours, Nouzilly, Francia.

Chanvallon, A., Fabre-Nys, C. 2009. In sexually naive anestrous ewes, male odor is unable to induce a complete activation of olfactory systems. Behavioral Brain Research. 205, 272-279.

Chanvallon, A., Scaramuzzi, R.J., Fabre-Nys, C. 2010a. Early sexual experience and stressful conditions affect the response of young ewes to the male. Physiology & Behavior. 99, 457-465.

Chanvallon, A., Blache, D., Chadwick, A., Esmaili, T., Hawken, P.A.R., Martin, G.B., Viñoles, C., Fabre-Nys, C. 2010b. Sexual experience and temperament affect the response of Merino ewes to the ram effect during the anoestrous season. Animal Reproduction Science. 119, 205-211.

Chenoweth, P.J. 1983. Reproductive management procedures in control of breeding. Animal Production in Australia. 15, 28-31.

Clarke, I.J., Henry, B.A. 1999. Leptin and reproduction. Reviews of Reproduction. 4, 48-55.

Cohen-Tannoudji, J., Lavenet, C., Locatelli, A., Tillet, Y., Signoret, J.P. 1989. Non-involvement of the accessory olfactory system in the LH response of anoestrous ewes to male odour. Journal of Reproduction and Fertility. 86, 135-144.

Crowe, M. 2008. Resumption of ovarian cyclicity in postpartum beef and dairy cows. Reproduction in Domestic Animals. 43, 20-28.

Cupp, A.S., Roberson, M.S., Stumpf, T.T., Wolfe, M.W., Werth, L.A., Kojima, N., Kittok, R.J., Kinder, J.E. 1993. Yearling bulls shorten the duration of postpartum anestrus in beef cows to the same extent as do mature bulls. Journal of Animal Science. 71, 306-309.

Cushwa, W.T., Bradford, G.E., Stabenfeldtt, G.H., Berger, Y.M., Dally, M.R. 1992. Ram influence on ovarian and sexual activity in anestrous ewes: effects of isolation of ewes from rams before joining and date of ram introduction. Journal of Animal Science. 70, 1195-2000.

Custer, E.E., Berardinelli, J.G., Short, R.E., Wehrman, M., Adair, R. 1990. Postpartum interval to estrus and patterns of LH and progesterone in first-calf suckled beef cows exposed to mature bulls. Journal of Animal Science. 68, 1370-1377.

Day, M.L., Grum, D.E. 2005. Breeding strategies to optimize reproductive efficiency in beef herds. Veterinary Clinics Food Animal. 21, 367-381.

De Souza, F.G. 2002. Efeito da presença de rufiões sobre a primeira temporada reprodutiva de novilhas de corte. Acta Scientifica Veterinaria. 30, 72-73.

Delgadillo, J.A., Flores, J.A., Véliz, F.G., Duarte, G., Vielma, J., Hernandez, H., Fernandez, I.G. 2006. Importance of the signals provided by the buck for the success of the male effect in goats. Reproduction, Nutrition & Development. 46, 391-400.

Delgadillo, J.A., Gelez, H., Ungerfeld, R., Hawken, P.A.R., Martin, G.B. 2009. The 'male effect in sheep and goats-Revisiting the dogmas. Behavior Brain Research. 200, 304-314.

Diskin, M.G., Austin EJ, Roche JF. 2002. Exogenous hormonal manipulation of ovarían activity in cattle. Domestic Animal Endocrinology. 23, 211-228.

Fernandez, D., Berardinelli, J.G., Short, R.E., Adair, R. 1993. The time required for the presence of bulls to alter the interval from parturition to resumption of ovarian activity and reproductive performance in first-calf suckled beef cows. Theriogenology. 39, 411-419.

Fernandez, D.L., Berardinelli, J.G., Short, R.E., Adair, R. 1996. Acute and chronic changes in luteinizing hormone secretion and postpartum interval to estrus in first-calf suckled beef cows exposed continuously or intermittently to mature bulls. Journal of Anim Science. 74, 1098-1103.

Fike, K.E., Bergfeld, E.G., Cupp, A.S., Kojima, F.N., Mariscal, V., Sanchez, T.S., Wehrman, J.E., Kinder, J.E. 1996. Influence of fenceline bull exposure on duration of postpartum anoestrous and pregnancy rate in beef cows. Animal Reproduction Science. 41, 161-167.

Fiol, C., Curbelo, N., Larraz, G., Ungerfeld, R. 2010a. Ovarian response in anestrous beef heifers exposed to androgenised steers. *In:* Reproduction in Domestic Ruminants VII. Nottingham University Press. pp. 601.

Fiol, C., Quintans, G., Ungerfeld, R. 2010b. Response to biostimulation in peri-puberal beef heifers: influence of male-female proximity and heifer's initial body weight. Theriogenology. 74, 569-575.

Fiol, C., Ungerfeld, R. 2011a. Males′ sexual preference toward heavier heifers is only observed in competitive situations with lighter heifers. Livestock Science. 139, 281-284.

Fiol, C., Ungerfeld, R. 2011b. Estrous response and AI pregnancy rates of 2-year-old beef heifers exposed to androgenized steers before an estrous synchronization protocol with double injection of PGF2alpha. Animal Reproduction 8, 9-13.

Gelez, H., Fabre-Nyz, C. 2004. The "male effect" in sheep and goats: a review of the respective roles of the two olfactory systems. Hormones & Behaviour. 46, 257-271.

Gelez, H., Archer, E., Chesneau, D., Campan, R., Fabre-Nys, C. 2004. Importance of learning in the response of ewes to male odor. Chemical Senses. 29, 555-563.

Gelez, H., Fabre-Nyz, C. 2006. Role of the olfactory systems and importance of learning in the ewes' response to rams or their odors. Reproduction Nutrition Development. 46, 401-415.

Gifford, D.R., D′Occhio, M.J., Sharpe, P.H., Weatherly, T., Pittar, R.Y., Reeve, D.V. 1989. Return to cyclic ovarian activity following parturition in mature cows and first-calf beef heifers exposed to bulls. Animal Reproduction Science. 19, 209-216.

Gokuldas, P.P., Yadav, M.C., Kumar, H., Singh, G., Mahmood, S., Tomar, A.K.S. 2010. Resumption of ovarian cyclicity and fertility response in bull exposed postpartum buffaloes. Animal Reproduction Science. 121, 236-241.

Gonzalez, R., Orgeur, P., Signoret, J.P. 1988. Luteinizing hormone, testosterone and cortisol responses in rams upon presentation of estrous females in the nonbreeding season. Theriogenology. 30, 1075-1086.

Grus, W.E., Zhang, J. 2008. Distinct evolutionary patterns between chemoreceptors of 2 vertebrate olfactory systems and the differential tuning hypothesis. Molecular Biology Evolution. 25 (8), 1593-1601.

Hawken, P.A.R., Evans, A.C.O., Beard, A.P. 2008. Prior exposure of maiden ewes to rams enhances their behavioural interactions with rams but is not a pre-requisite to their endocrine response to the ram effect. Animal Reproduction Science. 108, 13-21.

Hawken, P.A.R., Beard, A.P. 2009. Ram novelty and the duration of ram exposure affects the distribution of mating in ewes exposed to rams during the transition into the breeding season. Animal Reproduction Science. 111, 249-260.

Hess, B.W., Lake, S.L., Scholljegerdes, E.J., Weston, T.R., Nayigihugu, V., Molle, J.D.C., Moss, G.E. 2005. Nutritional controls of beef cow reproduction. Journal of Animal Science. 83 (E. Suppl.), E90-E106.

Ingawale, M.V., Dhoble, R.L. 2004. Buffalo reproduction in India: an overview. Buffalo Bulletin. 23 (1), 4-9.

Izard, M.K., Vandenbergh, J.G. 1982a. The effects of bull urine on puberty and calving date in crossbred beef heifers. Journal of Animal Science. 55, 1160-1168.

Izard, M.K., Vandenbergh, J.G. 1982b. Priming pheromones from oestrous cows increase synchronization of oestrus in dairy heifers after PGF-2α injection. Journal of Reproduction & Fertility. 66, 189-196

Izard, M.K. 1983. Pheromones and reproduction in domestic animals. *In:* Pheromones and Reproduction in Mammals. Vandenbergh JG (Ed.). New York Academic Press, pp. 253-285.

Kirkwood, R.N., Hughes, P.E. 1981. A note on the influence of boar age on its ability to advance puberty in gilts. Animal Production. 31, 211.

Landaeta-Hernández, A.J., Giangreco, M., Meléndez, P., Bartolomé, J., Bennet, F., Rae, D.O., Hernández, J., Archbald, L.F. 2004. Effect of biostimulation on uterine involution, early ovarian activity and first postpartum estrous cycle in beef cows. Theriogenology. 61, 1521-1532.

Landaeta-Hernández, A.J., Meléndez, P., Bartolomé, J., Rae, D.O., Archbald, L.F. 2006. Effect of biostimulation on the expression of estrus in postpartum Angus cows. Theriogenology. 66, 710-716.

Landaeta-Hernández, A.J., Meléndez, P., Bartolomé, J., Rae, O., Archbald, L.F. 2008. The effect of bull exposure on the early postpartum reproductive performance of suckling Angus cows. Revista científica FCV-LUZ XVIII. 682-691.

Lesmeister, J.L., Burfening, P.J., Blackwell, R.L. 1973. Date of first calving in beef cows and subsequent calf production. Journal of Animal Science. 36, 1-6.

Lima, C.S., Gambarini, M.L., Viu, M.A.O., Filho, B.D.O., Santos, F.C., Caixeta, L.S. 2008. Effect of biostimulation, monensin and recombinant bovine somatotropin on the daily weight gain and onset of puberty of Girolando heifers raised in pastures. Archives of Veterinary Science. 13, 93-97.

Macmillan, K.L., Allison, A.J., Struthers, G.A. 1979. Some effects of running bulls with suckling cows or heifers during the premating period. New Zeland Journal of Experimental Agriculture. 7, 121-124.

Madrigal, M.A., Colin, J., Hallford, D.M. 2001. Influencia de la condición corporal y la bioestimulación sobre la eficiencia reproductiva en vacas de raza Simmental en agostadero. Veterinaria México. 32 (2), 87-92.

Martin, G.B., Scaramuzzi, R.J., Lindsay, D.R. 1983. Effect of the introduction of rams during the anoestrous season on the pulsatile secretion of LH in ovariectomized ewes. Journal of Reproduction & Fertility. 67, 47-55.

Martin, G.B., Oldham, C.M., Cognié, Y., Pearce, D.T. 1986. The physiological responses of anovulatory ewes to the introduction of rams- A review. Livestock Production Science. 15, 219-247.

Martin, G.B., Milton, J.T.B., Davidson, R.H., Banchero Hunzicker, G.E., Lindsay, D.R., Blache, D. 2004. Natural methods for increasing reproductive efficiency in small ruminants. Animal Reproduction Science. 82-83, 231-245.

Martin, G.B. 2009. The "Cleen, Green and Ethical" Concept in Animal Production. Agrociencia. XIII, Nº 3, 1-7.

Miller, V., Ungerfeld, R. 2008. Weekly bull exchange shortens postpartum anestrus in suckled beef cows. Theriogenology. 69, 913-917.

Monje, A.R., Alberio, R., Schiersmann, G., Chedrese, J., Caroú, N., Callejas, S.S. 1992. Male effect on the post-partum sexual activity of cows maintained on two nutritional levels. Animal Reproduction Science. 29, 145-156.

Montiel, F., Ahuja, C. 2005. Body condition and suckling as factors influencing the duration of postpartum anestrus in cattle: a review. Animal Reproduction Science. 85, 1-26.

Mora, O.A., Sánchez-Criado, J.E. 2004. Involvement of the corticoadrenal hormones in the pheromonal restoration of reproductive activity in aging rats. Life Science. 74, 3285-3290.

Oakley, A.E., Breen, K.M., Tilbrook, A.J., Wagenmaker, E.R., Karsch, F.J. 2009. Role of estradiol in cortisol-induced reduction of luteinizing hormone pulse frequency. Endocrinology. 150, 2775-2782.

Oliveira, C.M.G., Oliveira Filho, B.D., Gambarini, M.L., Viu, M.A.O., Lopes, D.T., Sousa, A.P.F. 2009. Effect of biostimulation and nutritional supplementation on pubertal age and pregnancy rates of Nelore heifers (*Bos indicus*) in a tropical environment. Animal Reproduction Science. 113, 38-43.

Olsen, J.R., Tauck, S.A., Wilkinson, J.R.C., Keisler, D.H., Berardinelli, J.G. 2009. Biostimulatory effect of bulls on temporal patterns of leptin concentrations and resumption of luteal activity in primiparous, postpartum, anestrous, beef cows. Biology of Reproduction. 81, 261.

O´Riordan, E.G., Hanrahan, J.P. 1989. Advancing first estrus in ewe lambs. Farm and Food Research 20 (6), 25-27.

Patterson, D.J., Perry, R.C., Kiracofe, G.H., Bellows, R.A., Staigmiller, R.B., Corah, L.R. 1992. Management considerations in heifer development and puberty. Journal of Animal Science. 70, 4018-4035.

Patterson, D.J., Kojima, F.N., Smith, M.F. 2003. A review of methods to synchronize estrus in replacement beef heifers and postpartum cows. Journal of Animal Science. 81, E166-177.

Perkins, A., Fitzgerald, J.A. 1994. The behavioral component of the ram effect: the influence of ram sexual behavior on the induction of estrus in anovulatory ewes. Journal of Animal Science. 72, 51-55.

Peter, A.T., Vos, P.L.A.M., Ambrose, D.J. 2009. Postpartum anestrus in dairy cattle. Theriogenology. 71, 1333-1342.

Quadros, S.A.F., Lobato, J.F.P. 2004. Biostimulation and reproductive performance of beef heifers. Revista Brasileira de Zootecnia. 33, 679-683.

Quintans, G., Straumann, J.M., Ayala, W., Vázquez, A. 2004. Effect of winter management on the onset of puberty in beef heifers under grazing conditions. 15th International Congress of Animal Reproduction, 8-12 August, Porto Seguro, Brazil, Abstract No. 22.

Rekwot, P.I., Ogwy, D., Oyedipe, E. 2000a. Influence of bull stimulation, season and parity on resumption of ovarian activity of zebu (*Bos indicus*) cattle following parturition. Animal Reproduction Science. 63, 1-11.

Rekwot, P.I., Ogwy, D., Oyedipe, E., Sekoni, V. 2000b. Effects of bull exposure and body growth on onset of puberty in Bunajii and Friesian x Bunajii heifers. Reproduction Nutrition Development. 40, 359-367.

Rekwot, P.I., Ogwy, D., Oyedipe, E., Sekoni, V. 2001. The role of pheromones and biostimulation in animal reproduction. Animal Reproduction Science. 65, 157-170.

Rekwot, P.I., Akinpelumi, O.P., Sekoni, V., Eduvie, L.O., Oyedipe, E. 2004. Effects of nutritional supplementation and exposure to bulls on resumption of post-partum ovarian activity in Bunaji (*Bos indicus*) cattle. The Veterinary Journal. 167, 67-71.

Rivas-Muñoz, R., Fitz-Rodríguez, G., Poindron, P., Malpaux, B., Delgadillo, J.A. 2007. Stimulation of estrous behavior in grazing female goats by continuous or discontinuous exposure to males. Journal of Animal Science. 85, 1257-1263.

Roberson, M.S., Ansotegui, R.P., Berardinelli, J.G., Whitman, R.W., McInerney, M.J. 1987. Influence of biostimulation by mature bulls on occurrence of puberty in beef heifers. Journal of Animal Science. 64, 1601-1605.

Roberson, M.S., Wolfe, M.W., Stumpf, T.T., Werth, L.A., Cupp, A.S., Kojima, N., Wolfe, P.L., Kittok, P.J. 1991. Influence of growth rate and exposure to bulls on age at puberty in beef heifers. Journal of Animal Science. 69, 2092-2098.

Roelofs, J.B., Soede, N.M., Dieleman, S.J., Voskamp-Harkema, W., Kemp, B. 2007. The acute effect of bull presence on plasma profiles of luteinizing hormone in postpartum, anoestrous dairy cows. Theriogenology. 68, 902-907.

Roelofs, J.B., Soede, N.M., Voskamp-Harkema, W., Kemp, B. 2008. The effect of fenceline bull exposure on expression of oestrus in dairy cows. Animal Reproduction Science. 108, 226-235.

Sato, S., Honda, Y., Ohta, M. 1994. Behavioural interactions of a bull with cows and the resumption of estrous. Animal Science & Technology (Japan). 65, 538-546.

Shipka, M.P., Ellis, L.C. 1998. No effects of bull exposure on expression of estrous behavior in high-producing dairy cows. Applied Animal Behavior Science. 57, 1-7.

Shipka, M.P., Ellis, L.C. 1999. Effects of bull exposure on postpartum ovarian activity of dairy cows. Animal Reproduction Science. 54, 237-244.

Short, R.E., Bellows, R.A. 1971. Relationships among weight gains, age at puberty and reproductive performance in heifers. Journal of Animal Science. 32, 127-131.

Short, R.E., Adams, D.C. 1988. Nutritional and hormonal interrelationships in beef cattle reproduction. Canadian Journal of Animal Science. 68, 29-39.

Short, R.E., Bellows, R.A., Staigmiller, R.E., Berardinelli, J.G., Custer, E.E. 1990. Physiological mechanisms controlling anestrus and infertility in postpartum beef cattle. Journal of Animal Science. 68,799-816.

Signoret, J.P. 1991. Sexual pheromones in the domestic sheep: importance and limits in the regulation of reproductive physiology. Journal of Steroid Biochemistry and Molecular Biology. 39, 639-645.

Small, J.A., Del Vecchio, R.P., McCaughey, W.P., Ward, D.R., Sutherland, W.P. 2000. The effects of bull exposure and lasalocid on the development of replacement beef heifers. Canadian Journal of Animal Science. 80, 615-624.

Soto Belloso, E., Ramírez Iglesia, E., Guevara, L., Soto Castillo, G. 1997. Bull effect on the reproductive performance of mature and first calf-suckled zebu cows in the tropics. Theriogenology. 48, 1185-1190.

Stumpf, T.T., Wolfe, M.W., Wolfe, P.L., Day, M.L., Kittok, R.J., Kinder, J.E. 1992. Weight changes prepartum and presence of bulls postpartum interact to affect duration of postpartum anestrus in cows. Journal of Animal Science. 70, 3133-3137.

Swaney, W.T., Keverne, E.B. 2009. The evolution of pheromonal communication. Behavioral Brain Research. 200, 239-247.

Tauck, S.A., Berardinelli, J.G., Geary, T.W., Johnson, N.J. 2006. Resumption of postpartum luteal function of primiparous, suckled beef cows exposed continuously to bull urine. Journal of Animal Science. 84, 2708-2713.

Tauck, S.A., Berardinelli, J.G. 2007. Putative urinary pheromone of bulls involved with breeding performance of primiparous beef cows in a progestin-based estrous synchronization protocol. Journal of Animal Science. 85, 1669-1674.

Tauck, S.A., Olsen, J.R., Berardinelli, J.G. 2007. Adrenal involvement in the biostimulatory effect of bulls. Reproduction Biology Endocrinology. 5, 33-39.

Tauck, S.A. 2008. The biostimulatory effect of bulls on the hypothalamic-pituitary-adrenal and-ovarian axes and on temporal aspects of resumption of ovarian cycling activity in primiparous, postpartum, anestrous, suckled, beef cows. PhD Thesis. Montana State University, Bozeman, Montana, USA.

Tauck, S.A., Olsen, J.R., Wilkinson, J.R.C., Wedlake, R.J., Davis, K.C., Berardinelli, J.G. 2010a. Characteristics of temporal patterns of cortisol and luteinizing hormone in primiparous, postpartum, anovular, suckled, beef cows exposed acutely to bulls. Reproductive Biology & Endochrinology. 8, 89.

Tauck, S.A., Olsen, J.R., Wilkinson, J.R.C., Berardinelli, J.G. 2010b. Duration of daily bull exposure on resumption of ovulatory activity in postpartum, primiparous, suckled, beef cows. Animal Reproduction Science. 118, 13-18.

Thompson, L.H., Savage, J.S. 1978. Age at puberty and ovulation rate in gilts in confinement as influenced by exposure to a boar. Journal of Animal Science. 47, 1141-1144.

Thornton, P.K. 2010. Livestock production: recent trends, future prospects. Philosophical Transactions of the Royal Society B: Biological Sciences. 365, 2853-2867.

Tilbrook, A.J., Turner, A.I., Clarke, I.J. 2000. Effects of stress on reproduction in non-rodent mammals: the role of glucocorticoids and sex differences. Reviews of reproduction. 5, 105-113.

Ungerfeld, R., Pinczak, A., Forsberg, M., Rubianes, E. 2002. Ovarian and endocrine responses of Corriedale ewes to "ram effect" in the non-breeding season. Canadian Journal of Animal Science. 82, 599-602.

Ungerfeld, R. 2003. The reproductive response of anestrous ewes to the introduction of rams. PhD Thesis. Swedish University of Agricultural Sciences, Uppsala, Sweden. Acta Universitatis Agriculturae Sueciae. Veterinaria 163, pp 66.

Ungerfeld, R. 2007. Socio-sexual signaling and gonadal function: Opportunities for reproductive management in domestic ruminants. *In:* Reproduction in Domestic Ruminants VI. Juengel JI, Murray JF, Smith MF (Eds,). Nottingham University Press, Nottingham, UK. pp. 207-221.

Ungerfeld, R., Ramos, M.A., González-Pensado, S.P. 2008. Ram effect: Adults rams induce a greater reproductive response in anoestrous ewes than yearling rams. Animal Reproduction Science. 103, 271-277.

Ungerfeld, R. 2009. Short-term exposure of high body weight heifers to testosterone-treated steers increases pregnancy rate during early winter bull breeding. Animal Reproduction. 6 (3), 446-449.

Ungerfeld, R. 2010. Exposure to androgenized steers did not improve the fertility obtained in progesterone-based fixed-timed artificial insemination programs in extensively managed cows and heifers. Animal Production Science. 50, 68-71.

Vandenbergh, J.G. 1983. Pheromonal regulation of puberty. *In:* Vandenbergh JG (Ed.). Pheromones and reproduction in mammals. New York Academic Press, pp. 95-112.

Veissier, I., Boissy, A., dePassille, A.M., Rushen, J., vanReenen, C.G., Roussel, S., Andanson, S., Pradel, P. 2001. Calves' responses to repeated social regrouping and relocation. Journal of Animal Science. 79, 2580-2593.

Véliz, F,G., Poindron, P., Malpaux, B., Delgadillo, J.A. 2006. Positive correlation between the body weight of anestrous goats and their response to the male effect with sexually active bucks. Reproduction Nutrition & Development. 46, 657-661.

Walkden-Brown, S.W., Martin, G.B., Restall, B.J. 1999. Role of male-female interaction in regulating reproduction in sheep and goats. Journal of Reproduction & Fertility (Suppl). 54, 243-257.

Wiltbank, J.N., Kasson, C.W., Ingalls, J.E. 1969. Puberty in crossbred and straightbred beef heifers on two levels of feed. Journal of Animal Science. 29, 602-605.

Wright, P.J., Geytenbeek, P.E., Clarke, I.J. 1990. The influence of nutrient status of post-partum ewes on ovarian cyclicity and on the oestrous and ovulatory responses to ram introduction. Animal Reproduction Science. 23, 293-303.

Wright, I.A., Rhind, S.M., Smith, A.J., Whyte, T.K. 1994. Female-female influences in the duration of the post-partum anoestrous period in beef heifers. Animal Production. 59, 49-53.

Wyatt, T.D. 2010. Pheromones and signature mixtures: defining species-wide signals and variable cues for identity in both invertebrates and vertebrates. Journal of Comp Physiology A. 196, 685-700.

Yavas, Y., Walton, J.S. 2000. Postpartum acyclicity in suckled beef cows: a review. Theriogenology. 54, 25-55.

Zalesky, D.D., Day, M.L., Garcia Winder, M., Imakawa, K., Kittok, R.J., D'Occhio, M.J., Kinder, J.E. 1984. Influence of exposure to bulls on resumption of estrous cycles following parturition in beef cows. Journal of Animal Science. 59, 1135-1139.

Zicarelli, L., Espósito, L., Campanile, G., Di Palo, R., Armostrong, D.T. 1997. Effects of using vasectomized bulls in artificial insemination practice on the reproductive efficiency of Italian buffalo cows. Animal Reproduction Science. 47, 171-180.

4

REPRODUCTIVE SEASONALITY AND ITS CONTROL IN SPANISH SHEEP AND GOATS

A. Gómez-Brunet, J. Santiago-Moreno, A. Toledano-Diaz and A. López-Sebastián

Departamento de Reproducción Animal, INIA, Madrid, Spain.
E-mail: gomez@inia.es
** Corresponding author*

SUMMARY

Sheep and goat breeds from subtropical, middle and high latitudes show seasonal changes in reproductive activity. In general, the breeding season starts in autumn and ends in winter, with anoestrus in spring/summer. An endogenous circannual rhythm driven and synchronised by the annual photoperiod cycle regulates the onset and offset of the breeding season. However, the timing and duration of the breeding season can be affected by interactions between the photoperiod and factors such as breed, geographical origin, nutritional and lactational status, social interactions, and the season of parturition. Seasonality in reproduction is naturally accompanied by variation in the availability and price of meat, milk and cheese over the year, affecting the economy of farmers, consumers and the food industry alike. The control of reproduction outside the normal breeding season by inducing and synchronizing oestrus and ovulation plus the use of artificial insemination and/or natural mating would help ensure the year-round availability of products. This review describes the seasonal variation in the sexual activity of ovine and caprine species with special regard to local Spanish sheep and goats breeds, examines how the photoperiod regulates their annual reproductive cycle, and discusses a number of strategies that can be used to induce and synchronise ovulation outside the natural breeding season.

Keywords: sheep; goats; reproductive seasonality; photoperiod; melatonin; male effect; progesterone; $PGF_{2\alpha}$

INTRODUCTION

Many types of production systems are used around the world to raise sheep and goats. In Spain, there are some 19.8 million sheep and 2.9 million goats. Once raised by families for their own use, these animals now lie at the centre of a meat and milk industry of great economic importance to many parts of the country (Perez *et al.*, 2007; Castel *et al.*, 2010). Most Spanish goat farming is orientated towards milk production.

This is a particularly important activity in the regions of Andalusia, the Canary Islands and Castilla-La Mancha, where 72% of the country's dairy goats are concentrated. Sheep farming may be orientated towards meat or milk production, though some breeds are raised to provide both commodities. Spain's sheep flocks are concentrated in five regions: Castilla y Leon (21%), Extremadura (19%), Castilla-la Mancha (16%), Aragón (11.4%) and Andalusia (12.1%) (Spanish Ministry for Agriculture, Food and Fisheries, 2009). At Spanish Mediterranean latitudes (36-42ºN), the local sheep and goats are seasonal breeders (Forcada *et al.*, 1992; Gómez-Brunet *et al.*, 2003, 2008; Zarazaga *et al.*, 2005). The reproductive season generally extends from summer/autumn to the end of winter, with the anoestrous period in the spring months. This leads to marked seasonal variation in the availability of meat, milk and cheese, as well as fluctuations in the price of these products, a problem that affects the economies of farmers, the food industry and consumers alike. Inducing and synchronizing ovulation outside of the natural breeding season, in conjunction with natural mounting or the use of artificial insemination (AI), might allow farmers to stabilise production year round. Knowledge of the reproductive characteristics of local breeds and the identification of the environmental factors that influence the timing of their annual breeding season, would allow breeding strategies to be adapted to local environmental, economic and social needs.

This review describes the seasonal variation seen in the sexual activity of sheep and goats with special regard to local Spanish breeds, taking into account the environmental factors - especially the photoperiod - that influence this. In addition, three approaches used to control reproductive activity and to optimise fertility in local Spanish breeds are examined: 1) the use of melatonin implants combined with the male effect, 2) the use of the male effect in combination with progesterone (P4) and $PGF_{2\alpha}$ administration, and 3) the use of the male effect in combination with $PGF_{2\alpha}$ administration.

REPRODUCTIVE SEASONALITY

In 1784, the Italian Biologist Lazzaro Spallanzani wrote: *It is well known that almost all animals, except man, have a stated season for the propagation of their species*. The restriction of breeding activity to a certain time of year is a common feature in mammalian species living outside of the tropics. This is commonly viewed as having a genetic basis and to reflect the optimisation of different phenotypes to their environments (Bronson, 1989). Seasonal reproductive rhythms in wild mammalian species reflect adaptations to changes in temperature and food availability that ensures birthing occurs at the time of year optimal for the survival of newborns. Wild ruminant species have a short annual period of sexual activity (breeding season) followed by a long period of inactivity (anoestrous season) (Santiago-Moreno *et al.*, 2006; Gómez-Brunet *et al.*, 2010b). In domestic species, however, normal environmental pressures have become much reduced and the duration of the period of sexual activity has expanded (Hafez, 1952).

Sheep and goats were first domesticated some 10,000 years ago. Originating from a few wild goat and sheep ancestor varieties in the region of today's Iraq, Iran, Syria and eastern Turky (Zeder and Hesse, 2000), hundreds of different breeds have been developed that differ from their wild cousins in many aspects of their reproductive physiology (Shechel, 1992). Breeds originating from high latitudes and colder climates have a shorter breeding season than those from nearer the tropics. Although the photoperiod is the main environmental cue influencing this seasonality, a genetic basis for photoresponsiveness has been demonstrated (Poulton and Robinson, 1987). Thus, a wide range of periods of breeding activity may be observed between and even within breeds.

Reproductive seasonality in sheep

Ewes are seasonally polyoestrous, and their reproductive activity is characterised by the alternation between breeding and anoestrous periods. The breeding season is characterised by a succession, at regular intervals of 16-17 days, of oestrus behaviour and ovulation, while an arrest of cyclic ovulatory activity occurs during the anoestrus period (Rosa and Bryant, 2003). The change in day length is the main environmental factor regulating seasonal breeding (Karsch *et al*, 1984; Malpaux *et al.*, 2001). Maximum reproductive activity is associated with short days, with the highest percentage of ewes displaying oestrus and ovulatory activity during the late summer, autumn and early winter.

Breeds of tropical and subtropical origin, those from the Mediterranean area, and those with Merino ancestry, have longer breeding seasons than those from higher temperate latitudes (Hafez, 1952). For breeds originating at latitudes of >40ºN, the breeding season starts in the early autumn when days are becoming shorter, and ends in mid-winter when day length is again increasing. Ovulation then ceases, and the animal remains anovulatory during the long days of spring-summer (Ortavant *et al.*, 1985). In the UK (at 56ºN), the ewes of British sheep breeds, such as the Finnish Landrace and the Scottish Blackface, show marked seasonality, with anoestrus lasting from March/April to October, a period during which no ewe ovulates spontaneously (Wheeler and Lands, 1977). In France (at 45ºN), the breeding season of Île de France ewes starts in August and last until February, although some females show isolated ovulation events during the anoestrus season (Thimonier and Mauleon, 1969). In contrast, sheep breeds raised in subtropical (around 24°N-34°N; 24°S-34°S) or tropical areas (23°N-23°S) perceive no strong seasonal changes in photoperiod cues, and generally have a long breeding season: in fact they may cycle all year around and show no anovulatory period (Yenikoye, 1984; Gonzalez *et al.*, 1992; Arroyo *et al.*, 2007). In these breeds, seasonal reproduction may have evolved in response to seasonal rainfall patterns or nutritional conditions (Thimonier *et al.*, 1986).

Breeds of sheep from middle latitudes (35-40°N), such as the Australian Merino and the Mediterranean breeds, have a short anoestrus period during which a variable proportion of ewes ovulate spontaneously. In Spain, most of the local sheep breeds show reduced seasonality, displaying a long breeding season (summer to winter) and a short anoestrus period of some 3-4 months, generally in the spring, during which some 15-30% of ewes remain cycling (Valls Ortiz, 1983). Measuring progesterone (P4) concentrations by radioimnmunoanalysis (RIA) in peripheral blood samples collected twice per week over a year, allowed characterizing the extension of the breeding season and the length of postpartum anoestrus in a number of local Spanish sheep breeds. For example, under natural environmental conditions at 40°N, Spanish Manchega ewes experience a long breeding season with cyclical ovulatory activity from early July to mid-March. Thereafter, they experience an anoestrus period of about 111 days from March to June, with some ewes showing isolated ovulations during the anoestrus period (Figure 1A). (Gómez-Brunet and López Sebastián, 1991; Santiago-Moreno *et al.*, 2000c; Gómez-Brunet *et al.*, 2008). Reduced seasonality is also apparent in the Rasa Aragonesa (at 41°N) and Merino (at 39°N) breeds, with 3-4 months anoestrus from March to May/June (Forcada *et al.*, 1992; Gonzalez-López, 1993; Folch and Alabart, 2000).

Although seasonal reproduction is mainly regulated by photoperiod, many other factors including breed, nutritional and lactational status, social interactions and the season of parturition, may interact with it to modulate its effect and thus influence the timing and length of the breeding season (Pope *et al.*, 1989; Ungerfeld *et al.*, 2004; Forcada *et al.*, 2006). For example, in Manchega ewes, the onset of the natural breeding season is affected by the lambing season (Gómez-Brunet *et al.*, 1992; Santiago-Moreno *et al.*, 2000b). When lambing occurs in June or November, Manchega ewes experience their first postpartum ovulation within the same breeding season, about 30 days after parturition, and daily or weekly weaning does not advance the beginning of the ovulatory activity (Gómez-Brunet *et al.*, 1992). In contrast, in ewes that lamb in April, postpartum anoestrus extends into the long days of spring, overlapping with seasonal anoestrus. Thus, the lambing-to-ovulation interval lasts about 112 days. All the ewes initiate cyclicity from 30 June to 17 August, coinciding with the beginning of the breeding season for this breed. Further, in a study performed under field conditions involving Manchega, Churra and Latxa ewes from commercial flocks in central and northern Spain, it was reported that the longest intervals between parturition and the onset of ovulation occurred at 150 days, when ewes lambed at the end of the breeding season (January-February) (López-Sebastián *et al.*, 1991). In contrast, when lambing occurred between June and November, the first ovulations started just 24-36 days after lambing.

It has been shown that the length of seasonal anoestrus can be reduced by approximately two months in mature Rasa Aragonesa ewes when these are maintained in good body condition (body score condition [BCS] 2.8 rather than 2.3) (Forcada *et al.*, 1992). In spring-born Manchega and Merino lambs, the failure to grow to appropriate body weight leads to the maintenance of the prepuberal anovulatory condition until the beginning of the next breeding season. Only lambs reaching 69.6 % and 60.1% respectively of their mean adult weight start cyclical activity in their first breeding season (López-Sebastián *et al.*, 1985b: Santiago-Moreno *et al.*, 2000a; Gómez-Brunet *et al.*, 2010a).

In temperate regions, rams also show seasonal changes in their sexual behaviour, testosterone secretion, testicular weight, and sperm quantity and quality (Ortavant *et al* 1985), although these changes are less pronounced than those experienced by ewes. Values for the above variables are high at the end of the summer through autumn, falling greatly by the end of winter and remaining low during spring (Ortavant *et al.*, 1985). The testicular volume of local Greek breeds follows a clear seasonal trend, with maximum and minimum volumes in July-August and February-April respectively (Avdi *et al.*, 2004). The males of Spanish breeds (e.g.,Merino, Churra, Latxa) also show reproductive seasonality in testosterone secretion and sperm quantity and quality (Beltran de Heredia and Gabiña, 2004; Santiago-Moreno *et al.*, 2005; Mazariegos *et al.*, 20110; Figure 2). In the Merino (a not very seasonal breed) rams, the highest levels of testosterone and maximum testicular diameter occur during summer (from June to August). However, these rams show limited annual fluctuations in the size of their accessory glands (vesicular and bulbourethral glands) (Santiago-Moreno *et al.*, 2005), in agreement with the relatively constant ejaculate volumes seen year round (Gonzalez-López, 1995). In this breed, high levels of testosterone coincide with the onset of cyclical ovulatory activity in the ewes, helping to ensure the success of sexual activity.

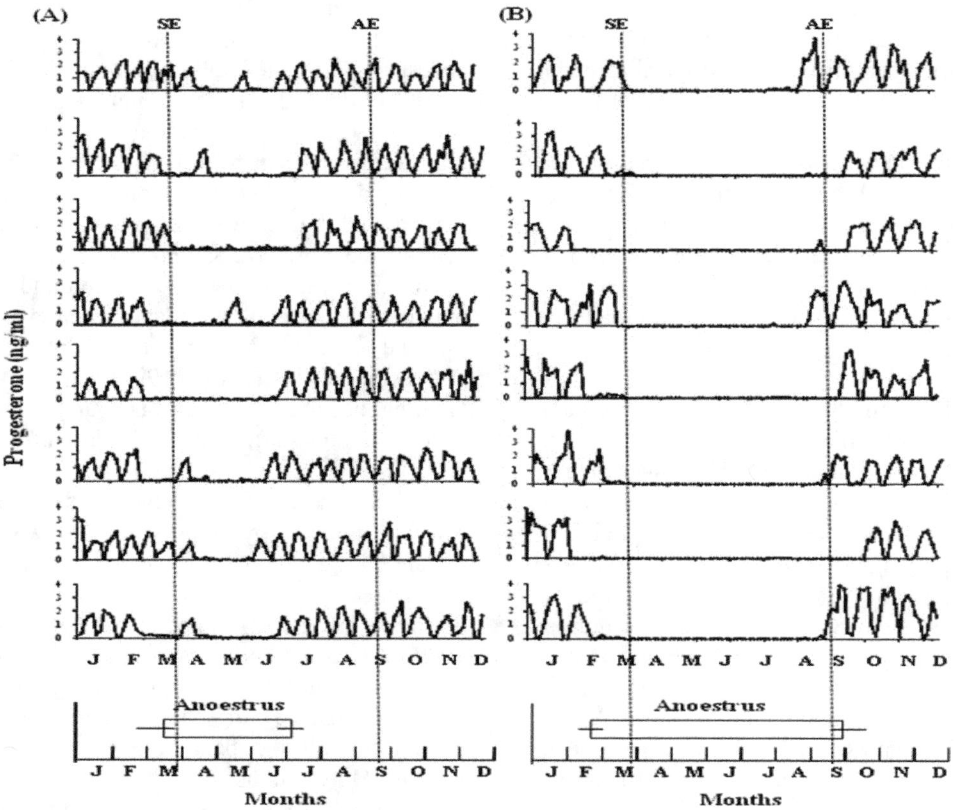

Figure 1. Annual changes in ovulatory activity in Manchega ewes (A) and Malagueña goats (B) raised at Mediterranean latitude (36-40ºN) and maintained under natural photoperiod conditions (Adapted from Gómez-Brunet et al., 2003; 2008).

Reproductive seasonality in goats

Reproductive seasonality is observed in all temperate goat breeds (Shelton, 1978; Chemineau *et al.*, 1992a), and in some local breeds adapted to or originating from subtropical latitudes (Rivera *et al.*, 2003; Duarte *et al.*, 2008). Goats alternate annually between two distinct seasons: the sexual season characterised by a regular succession of oestrus activity and ovulation about every 21-22 days, and an anoestrus season characterised by the total absence of sexual activity. In most goat breeds, the breeding season occurs in the autumn/winter, and the anoestrus period in the spring/summer. However, great variability exists between and within breeds in terms of the timing and duration of the seasonal reproductive cycle, depending on geographic origin. For example, in France (at 46ºN), the frequency of ovulation in Alpine dairy goats varies from 0% between March and September to 100% between October and January, reflecting a long period of anoestrus (Chemineau *et al.*, 1992a). In Australia (at 39ºS), Cashmere goats have a short period of ovulatory activity in autumn and winter (April to August) (Restall, 1992). In subtropical Creole goats in Argentina (at 30ºS) the breeding season extends from February/March to September, while anoestrus occurs during the months of October to January (Rivera *et al.*, 2003). In goats adapted to the subtropical conditions of northern Mexico (at 26ºN), the breeding season lasts from September to February and the anoestrus season from March to August, with some goats having one or two isolated ovulations in June or July (Duarte *et al.*, 2008). In other goats originating from tropical and subtropical environments, reproductive seasonality is less marked and some local breeds have just a short anoestrous period or breed all year around (Chemineau, 1986). In Chile, local goats experience only three months of anoestrus in late spring and early summer (Santa Maria *et al.*, 1990), while Boer goats in South Africa experience oestrous activity all year round, with the highest percentage in oestrous in the autumn (Greyling, 2000).

In the Mediterranean regions of Spain (36-40ºN), ovulatory activity over the year has only been accurately studied in Malagueña and Payoya breeds (Gómez-Brunet *et al.*, 2003, 2010c; Zarazaga *et al.*, 2005). The measurement of plasma progesterone concentrations in blood samples collected twice a week showed the existence of a marked seasonality in cyclical ovulatory activity over the year. In Malagueña goats, onset and offset occur in late September/early October and February respectively (Figure. 1B). Thereafter, these animals experience a long anoestrus period (March to September) during which no ovulation occurs (Gómez-Brunet *et al.,* 2003; 2010c). In Payoya goats, the mean dates for the onset and end of ovarian activity are August 28 and January 29 respectively. In this breed, the anoestrus period occurs between February and August (Zarazaga *et al*., 2005). In Murciano-Grandina goats, Falagan *et al.* (1989) have described an anoestrus period from March to June, but provide no information regarding the onset of the breeding season. Most Spanish goats live at similar latitudes and are subjected to similar management production systems; thus it is very probable that the reproductive seasonality of other Spanish goat breeds. e.g., Florída, Verata, Blanca Celtibérica, are very similar to that of Malagueña and Payoya goats. Interestingly, while reproductive seasonality in Spanish sheep breeds differs widely from that seen in breeds from higher latitudes, the pattern of reproductive seasonality in Spanish goats is very similar to those reported in goats from other latitudes (Restall, 1992; Chemineau *et al.*, 1992a; Duarte *et al.*, 2008).

Like rams, bucks from temperate regions show marked seasonal variation in their reproductive activity (Ortavant *et al.*, 1985). For example, in Alpine and Saanen bucks maintained at 46ºN, the reduction in day length in autumn is accompanied by an increase in testosterone secretion, an increase in the libido, and an increase in quantitative and qualitative sperm production (Delgadillo *et al.*, 1991; Delgadillo and Chemineau, 1992). In males from subtropical Mexico (at 26ºN), testicular weight, testosterone secretion and semen quality and quantity vary seasonally, with higher values during the breeding season (May/June - December/January) (Delgadillo *et al.*, 1999). In local Greek bucks (at 40-42ºN), testosterone secretion is low from January to May, and high from June to November (Todini *et al.*, 2007). The males of Spanish breeds (e.g., Murciano-Granadina, Verata, Malagueña, Florída) from southern Spain (36-38º N) also show reproductive seasonality (Figure 3), but in contrast to those from higher latitudes, maximum testicular activity (testicular diameter and testosterone secretion), sexual behaviour (measured as reaction time) and semen production occur in spring/early summer (April-July), with minima in autumn/winter (Roca *et al.*, 1991, 1992; Perez and Mateos, 1994, 1996; Arrebola *et al.*, 2010). The ancestral practice of reproductive activity management in these Mediterranean areas is based on the use of the male effect in April-May-June. At this time, males have sufficient libido and produce enough sperm to mate successfully with the females. This contrast to the goat management at higher latitudes (for example, in France at 46º) and at subtropical latitudes of Mexico (26ºN) where the males need light treatment to encourage sexual activity to induce in the females synchronous ovulations outside of the breeding season oestrus (Delgadillo *et al.*, 2001; Chemineau *et al.*, 2008).

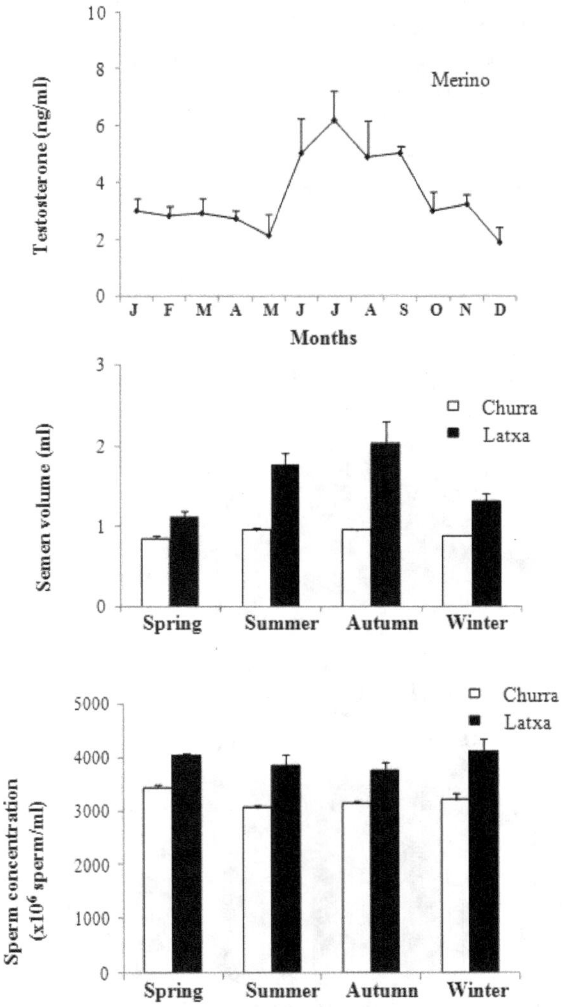

Figure 2. Seasonal changes (mean ± SEM) in testosterone secretion, semen volume and sperm concentration in different Spanish sheep breeds (Adapted from Beltran de Heredia and Gabiña, 2004; Santiago-Moreno et al., 2005; Mazariegos et al., 2010).

As in sheep, the photoperiod is the main factor controlling reproductive seasonality in goats (Malpaux *et al.*, 2001; Chemineau *et al.*, 2008). However, nutritional status and sociosexual factors can interact with the photoperiod and modulate reproductive activity in both sexes (Walkden-Brown *et al.*, 1999; Scaramuzzi and Martin, 2008). For example, in Spanish Payoya goats, the anoestrus period is reduced by one month in animals fed at maintenance level compared to those fed over twice the maintenance level. The shorter anoestrus of does on the better diet is due to both delayed entry into anoestrus and an earlier resumption of ovarian cyclicity (Zarazaga *et al.*, 2005). In this breed, the response of luteinizing hormone (LH) secretion to photoperiod variation is modified by the feeding level such that nutritional deprivation reduces secretion of pituitary LH, which in turn leads to anovulation and anoestrus (Zarazaga *et al.*, 2011a,b). In bucks, a higher feeding level does not modify the onset or the end of reproductive activity (as measured by changes in testosterone secretion), but sexual behaviour in late spring is improved (Zarazaga *et al.*, 2009b). In subtropical does a high nutrition level increases reproductive activity during anoestrus (Urrutia-Morales *et al.*, 2009).

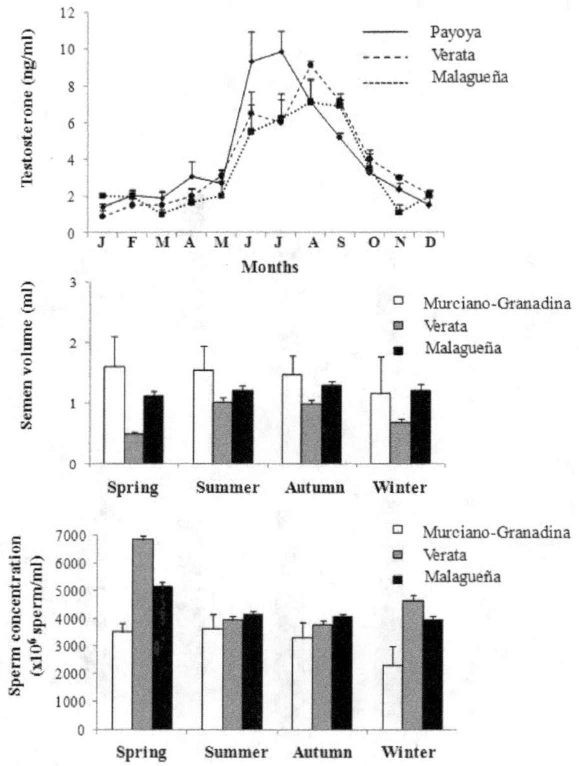

Figure 3. Seasonal changes (mean ± E.S.M) in testosterone secretion, semen volume and sperm concentration in different Spanish goats breeds (Adapted from Perez and Mateos, 1994; 1996; Arrebola et al., 2010; Gómez-Brunet et al., unpublished data).

PHOTOPERIODIC CONTROL OF THE CIRCANNUAL REPRODUCTIVE RHYTHM OF SHEEP AND GOATS: REFRACTORY STATES AND ENDOGENOUS RHYTHMS

Sheep and goats are considered short-day breeders since, under natural environmental conditions, reproductive activity begins during the shortening days of late summer and early autumn, and ends during the lengthening days of late winter and early spring (Hafez, 1952), Further, under artificial conditions, short days (SD) stimulate reproductive activity while long days (LD) inhibit it (Chemineau *et al.*, 1992b.). The photoperiod has been shown the main factor driving the annual reproductive cycle in these species. This was achieved by subjecting them to numerous and varied light–controlled experiments using inversion of the annual photoperiod cycle, the use of 2-3 month alternating SD/LD treatments, the acceleration of annual photoperiod changes, medium-term maintenance under a constant LD or SD photoperiod (which induce refractoriness), and long-term maintenance under constant LD or SD photoperiods (to verify the existence of an endogenous circannual rhythm of reproduction) (Malpaux *et al.*, 2006).

Regardless of the light regimen followed, the pineal gland regulates the change in the reproductive function via its endocrine activity, which is synchronised by the dark-light cycle. Photoperiod information perceived by the eyes is transmitted via a multi-synaptic pathway to the pineal gland, where the light is translated into a daily cycle of melatonin secretion characterised by high levels of secretion at night and low levels during the day (Arendt, 1998). The duration of the nocturnal secretion of melatonin, reflecting the length of the night, regulates the pulsatile secretion of gonadotropin releasing hormone (Gn-RH) from the hypothalamus. In sheep, long-duration melatonin presence (SD) stimulates the Gn-RH system and LH. Thus, changes in Gn-RH secretion induce corresponding changes in LH secretion, in turn determining the presence or absence of ovulation in the female and sexual activity in the male (Karsch *et al.*, 1984; Malpaux *et al.*, 2001; Chemineau *et al.*, 2010).

In general, mammals regulate their reproductive responses to seasonal change via two mechanisms: through a direct response to changes in the photoperiod, or through an endogenous circannual rhythm synchronised by photoperiod cues. Studies in sheep (Karsch *et al.,* 1989) have shown that each reproductive transition is generated endogenously – the reflection of a circannual reproductive rhythm synchronised by photoperiod. In addition, in many seasonal species it has been shown that the specific effects of SDs and LDs are not permanent, and that photorefractoriness occurs when animals are subjected to a constant photoperiod, with spontaneous reversion of the reproductive physiological condition to the previous photoperiodic state (Lincoln *et al.*, 2005).

Endogenous reproductive rhythm in sheep

In sheep from higher temperate latitudes, refractoriness to inhibitory LD and stimulatory SD photoperiods plays a central role in regulating seasonal reproductive transitions (Karsch *et al.*, 1986). In fact, artificially constant SDs from the winter solstice onwards do not prevent the breeding season from ending (Robinson and Karsch, 1984; Malpaux et al., 1987), nor do artificial constant LDs from the summer solstice onwards prevent the onset of the breeding season (Robinson *et al.*, 1985). In highly seasonal sheep breeds, the existence of an endogenous reproductive rhythm is supported by studies in which ewes were exposed to constant short (8h light/day) or long days (16h light/day) or an equinoctial (12h light/day) photoperiod over a long period; the animals alternated between periods of sexual rest and sexual activity (Karsch et al., 1989). Indeed, the annual reproductive cycle is the product of an endogenous circannual rhythm that is not directly driven by the photoperiod. However, it is synchronised by it through its control of the circadian rhythm of melatonin secretion (Woodfill *et al.*, 1994). The duration of elevated melatonin levels over the long days of spring-summer provides the signal for day length to synchronise the rhythm of reproductive neuroendocrine activity, allowing reproductive activity to commence in the autumn (Malpaux *et al.*, 1989; Barrell *et al.*, 2000). The ever shorter days between the summer solstice and the autumn equinox are the critical signal involved in timing the end of reproductive activity in mid-winter (Malpaux *et al.*, 1988) which contributes to ensuring the proper duration of the breeding season (Malpaux and Karsch, 1990; O'Callaghan *et al.*, 1991a).

As mentioned earlier, Spanish Mediterranean sheep breeds have a long breeding season. However, the extent to which these breeds use the photoperiod to regulate their annual reproductive cycle remains unclear. To address this, our group examined whether refractoriness to long and short days in Spanish Manchega sheep, which show reduced seasonality, determines the onset and the end of the breeding season respectively, as occur in highly seasonal breeds. Ovulatory activity was monitored via the measurement of plasma P4 (collected twice per week) in intact Manchega ewes exposed to artificial LDs (16 h light/day) from the summer to the winter solstice (June to December), to simulated SDs (Melatonin implants) from the winter to the summer solstice (December to June), or maintained under natural photoperiod (NP) conditions for 40ºN (from 10 h 17' to 16 h 3' light per day from the winter to summer solstice). The LD Manchega ewes responded similarly to sheep breeds from higher latitudes studied by Robinson *et al.* (1985). The onset of seasonal ovulatory activity occurred at the same time in ewes maintained under constant LDs as in those under NP conditions (16 and 5 July respectively; Figure 4) (Gómez-Brunet *et al.*, 2008). Interestingly, in the SD ewes, two animals, showed continuous cyclical ovulatory activity over the course of the experiment, while the six remaining ewes entered anoestrus two months later than their natural photoperiod counterparts (16 May and 15 March, respectively; Figure 5). (Gómez-Brunet *et al.*, 2001). Indeed, the SD Manchega ewes responded differently to highly seasonal SD ewes raised at high latitudes (>52ºN) (Robinson and Karsch, 1984; Malpaux *et al.*, 1987), but in a fashion similar to less seasonal ewes kept at 39-40ºN (Minton, 1990). Altogether, these results show that in Spanish Manchega sheep, refractoriness to LDs regulates the onset of the breeding season, as it does in highly seasonal sheep breeds from higher latitudes. However, also indicate that in less seasonal breeds, refractoriness to SDs might not be the most important physiological mechanism regulating the onset of anoestrus (unlike in highly seasonal breeds). Differences in seasonality among sheep breeds might therefore be related, at least in part, to the degree to which refractoriness to short days governs the transition into anoestrus.

Refractoriness is conceptually considered to be the first step in the expression of an endogenous circannual reproductive rhythm (Malpaux *et al.*, 2006). Thus, in a second study, we determined the ovulatory activity (via the measurement of plasma P4 concentrations collected twice per week) in two groups of intact Manchega ewes subjected to either a constant photoperiod of LDs (16 h light/day) or NP conditions for 40ºN for 16 months, between June (Year 1) and October (Year 2). As expected, under the NP conditions the ewes showed well–defined seasonality in their cyclical ovulatory activity (Figure 6A). During the first year all females showed ovarian cycles from July to March; thereafter, they experienced an anoestrus period of 112 days, before starting cyclicity again by July in Year 2. The LD ewes also showed seasonal variation in their ovulatory activity but the timing of these changes was profoundly modified by the photoperiod treatment (Figure 6B). In these animals, the onset of ovulatory activity in the first year of study was no different to that seen in their NP counterparts (July 2 and 3 respectively). However, the end of ovarian cyclical activity, and the subsequent behaviour of these animals in the following cycle, differed. Five ewes ended cyclical ovulatory activity between December 21 and February 18 in Year 1; then, after an anoestrous period of about 196 days, four of these animals started ovarian cycles by May 23 before stopping again by August 31 (Year 2), and then remaining anoestrous until the end of the experiment. The other three ewes showed continuous cyclical ovulatory activity during the first year of the study but stopped cyclicity by July of the second Year. Thereafter, one of the ewes experienced progesterone cycles at irregular intervals until September before becoming definitively anoestrous, while the other two remained anoestrous for 153 days from July until the end of the experiment (Figure 6B). These results show that Spanish Mediterranean breeds of sheep, with their reduced seasonality, appear to use an endogenous circannual rhythm to generate their reproductive transitions (Gómez-Brunet *et al.*, 2008), as do highly seasonal breeds from higher temperate latitudes. However, in the Spanish breeds, the circannual ovarian reproductive cycle is differently expressed, and they are much more sensitive to changes in the photoperiod than breeds of higher latitudes. The long days of spring may be critical for resetting the endogenous reproductive rhythm of Spanish Mediterranean sheep breeds, as they are in breeds from higher temperate regions (Malpaux *et al.*, 1989; Barrell *et al.*, 2000), but with a shorter lag time (about 3-4 months shorter) between the perception of the photoperiod signal and the onset of the breeding season.

Endogenous reproductive rhythm in goats

In goats from temperate, subtropical and Mediterranean latitudes, it is well established that artificial LDs inhibit and artificial SDs stimulate reproductive activity (Bisonnette 1941; Chemineau *et al.*, 1992b; Delgadillo *et al.*, 2004; Duarte *et al.*, 2010; Zarazaga *et al.*, 2011a). In Spanish Payoya goats, the mean intervals between the shift from long to short days, and from short to long days, to either stimulate or inhibit ovulatory activity (measured by determining plasma LH concentrations in ovariectomised does subcutaneously treated with a constantly-releasing oestradiol 17β implant [OVX+E]) are approximately 66 and 22 days respectively (Zarazaga *et al.*, 2011a). These intervals are shorter than those described for Saanen dairy goats from higher temperate latitudes (Chemineau *et al.*, 1986), probably due to breed differences.

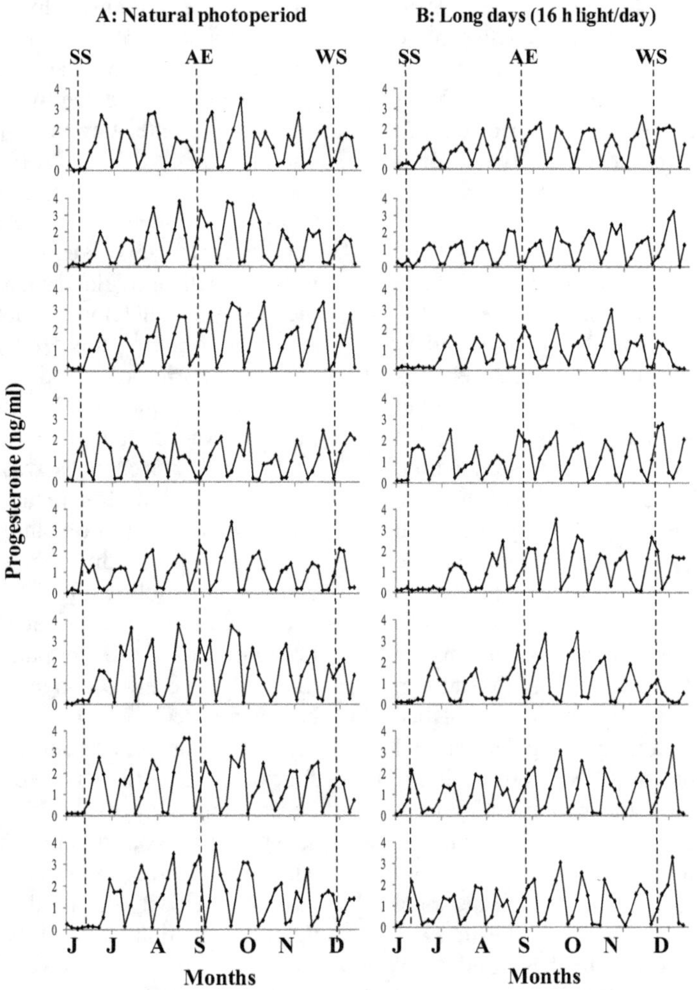

Figure 4. Ovulatory activity of Manchega ewes raised at latitude 40° 25'N maintained either under the natural photoperiod (A) or long days (16 h light/day) (B) between June 22 and December 31. WS = winter solstice; SS = summer solstice; AE: autumn equinox (Adapted from Gómez-Brunet et al., 2008).

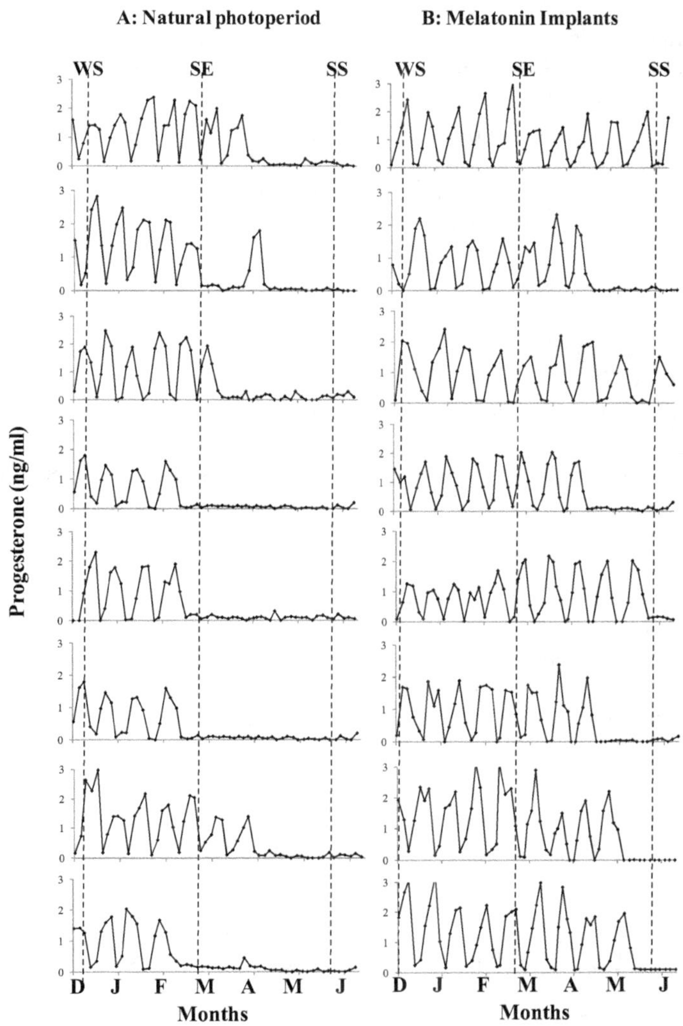

Figure 5. Ovulatory activity of Manchega ewes raised at latitude 40º 25'N and maintained under the natural photoperiod (A) or implanted with melatonin (to simulate a short day photoperiod) between December 23 and June 30. WS = winter solstice; SS = summer solstice; AE: autumn equinox (Adapted from Gómez-Brunet et al., 2001).

In contrast to ewes, which show important variations in their seasonality depending on breed and latitude of origin, most goat breeds are strongly seasonal irrespective of their origin. In the Mediterranean area, differences exist between sheep and goats in terms of the timing and duration of the seasonal reproductive cycle, with a more restricted breeding season seen in does (Mascarenhas *et al.*, 1995; Gomez-Brunet *et al.*, 2003; Zarazaga *et al.*, 2005) than sheep (Forcada *et al.*, 1992; Avdi *et al.*, 1993; Gómez-Brunet *et al.*, 2008). Although studies on the physiological mechanisms underlying the photoperiodic control of seasonal breeding activity in does have been limited (BonDurant *et al.*, 1981; Gebbie *et al.*, 1999), it is likely that, under natural environmental conditions, each reproductive transition is generated endogenously as the reflection of a circannual reproductive rhythm synchronised by photoperiod in a fashion similar to that seen in ewes. To identify the mechanisms involved in the photoperiodic control of the seasonal cycle of ovulatory activity in Spanish goat breeds, cyclicity was determined in Malagueña does exposed to the same artificial photoperiod conditions as described above for Manchega ewes. In an initial study to determine whether the onset of the breeding season results from refractoriness to the inhibitory effects of LDs, and to determine whether the end of the breeding season results from refractoriness to the stimulatory effects of the SDs, does at 40ºN were maintained under NP conditions, or exposed to artificial constant LDs (16 h light/day) from June to December, or received a series of subcutaneous melatonin implants (MI) from December to June to mimic an SD signal. The results showed that the onset of seasonal cyclical ovulatory activity in intact does maintained under constant LD at the summer solstice did not prevent the onset of the breeding season, which occurred at the same time as in the goats kept under the NP conditions (October 3 and September 23 respectively; Figure 7). Further, treatment with melatonin implants from the winter solstice onward did not prevent the end of the breeding season; these SD does became anoestrous at the same time as their counterparts under NP conditions (March 1 and February 23 respectively; Figure 8). Thus, the onset of cyclic ovulatory activity in Malagueña goats is not driven by decreasing day length from the summer solstice onward, and increasing day length after the winter solstice is not the signal required for the end of ovulatory cyclicity (onset of anoestrus). These data support the idea that, as in goat breeds originating from or adapted to either northerly (52ºN) or subtropical (26ºN) latitudes (Gebbie *et al.*, 1999; Delgadillo *et al.*, 2011), refractoriness to LDs and SDs regulates the transition between breeding and anoestrus in Spanish goats (Gómez-Brunet *et al.*, 2010c).

In a second study, variations in the ovulatory activity of Malagueña goats subjected to either a constant LD photoperiod or the NP conditions at 40ºN for two years, were examined to evaluate whether, as in sheep, an endogenous circannual rhythm is involved in regulating seasonal changes in reproductive activity. As expected, the NP Malagueña goats showed marked seasonality in their ovulatory activity (Figure 9A), which lasted from September to February as preoviusly recorded (Gómez- Brunet *et al.*, 2003; 2010c). The LD experimental goats also expressed alternating periods of ovulatory/anovulatoy activity over the course of the study, but these periods varied greatly with respect to those seen in the NP goats (Figure 9B; Gómez Brunet *et al.*, unpublished). Therefore, in Malagueña goats also an endogenous rhythm that, irrespective of the photoperiod, appears to be responsible for reproductive transitions under natural conditions, thus allowing them to initiate their reproductive rhythm in the early autumn and finish in winter.

The results of our studies on the role of photoperiod in regulating seasonal reproduction in Spanish Mediterranean sheep and goats, which show marked species-specific differences in their seasonal reproductive patterns (ewes begin their ovarian cycles three months earlier than goats), suggest that both possess an endogenous rhythm responsible for reproductive transitions that does not rely of the natural photoperiod. However, these rhythms are differently expressed in the two species..

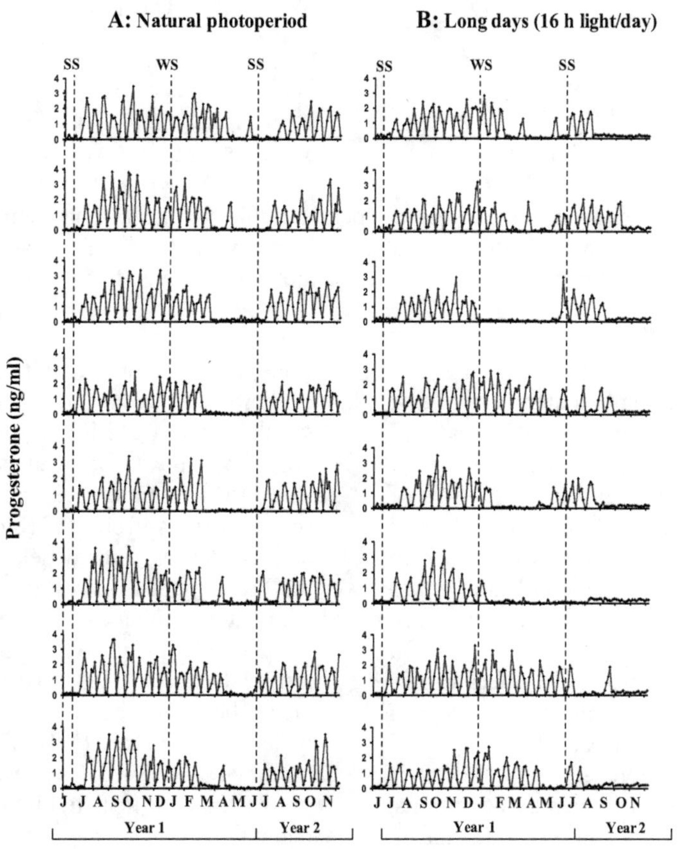

Figure 6. Ovulatory activity of Manchega ewes raised at latitude 40º 25'N and maintained under natural (A) or long days (16 h light/day) (B) for 16 months between June 22 (Year 1) and October 31 (Year 2). WS = winter solstice; SS = summer solstice. (Adapted from Gómez-Brunet et al., 2008).

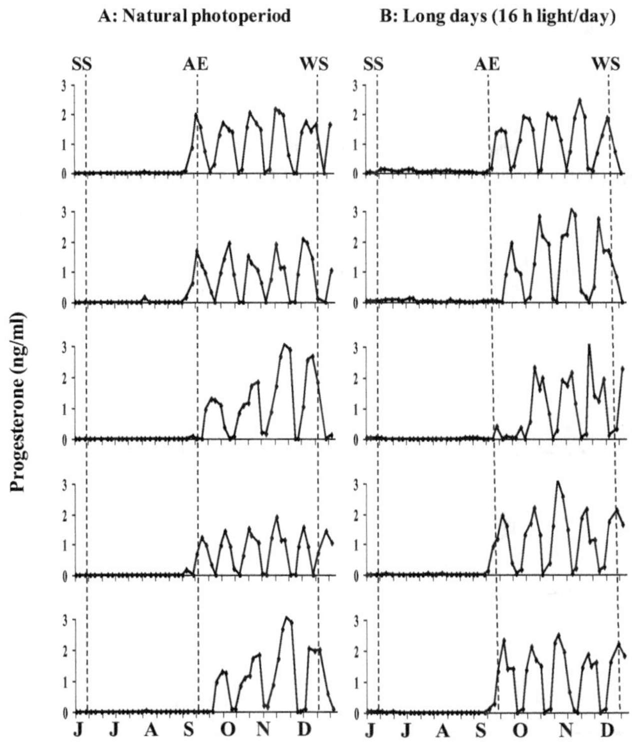

Figure 7. Ovulatory activity of Malagueña does raised at 40º25'N and maintained either under the natural photoperiod (A) or long days (16 h light/day) (B) between June 22 and December 31. WS = winter solstice; SS = summer solstice; AE: autumn equinox (Adapted from Gómez-Brunet et al., 2010c).

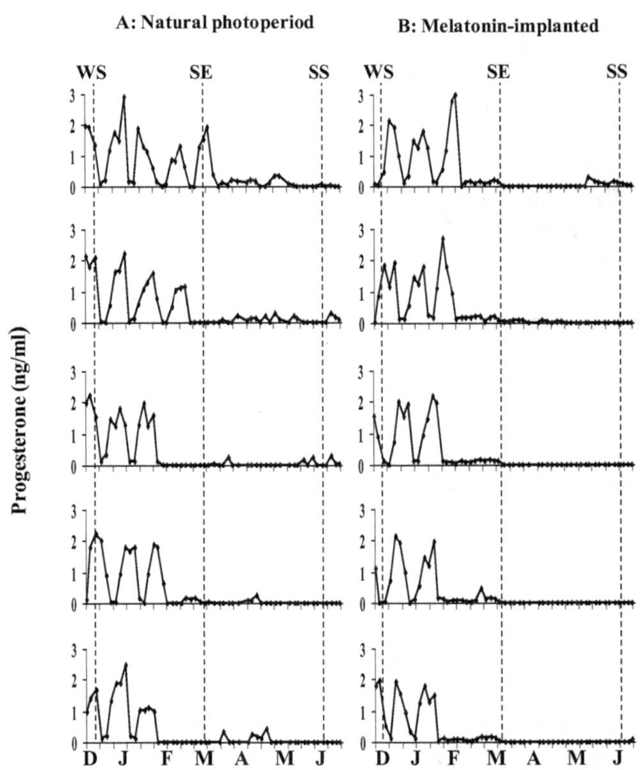

Figure 8. Ovulatory activity of Malagueña does raised at 40º 25'N and maintained under the natural photoperiod (A) or implanted with melatonin (to simulate a short day photoperiod) between December 23 and June 30. WS = winter solstice; SS = summer solstice; SE = spring equinox (adapted from Gómez-Brunet et al., 2010c).

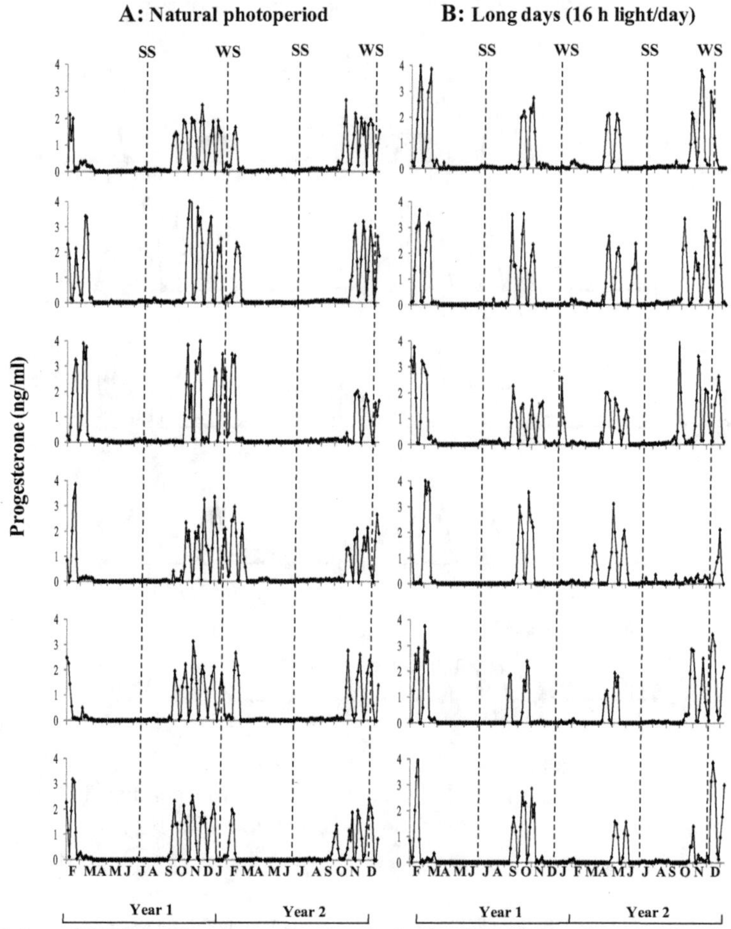

Figure 9. Ovulatory activity of Malagueña does raised at latitude 40°25'N and maintained under natural (A) or long days (16 h light/day) (B) for 23 months between February 12 (Year 1) and December 31 (Year 2). WS = winter solstice; SS = summer solstice. (Gómez-Brunet et al., unpublished data).

CONTROLLING OVULATORY ACTIVITY IN SHEEP AND GOATS OUTSIDE OF THE BREEDING SEASON

Over the last few decades, a number of ways of achieving reproductive activity outside of the normal breeding season have been developed for use with sheep and goats. Some are based on the administration of exogenous hormones (progesterone/progestagen, equine chorionic gonadotropin (eCG), and/or prostaglandin $F_{2\alpha}$ ($PGF_{2\alpha}$) or its analogues, while others involve environmental manipulation (e.g., day length control) or exposures to males (the male effect).

The most commonly used hormonal treatment involves the administration of a synthetic progestagen (FGA: fluorogestone acetate), eCG (equine chorionic gonadotropin), hCG (human chorionic gonadotropin) and cloprostenol (a $PGF_{2\alpha}$ analogue) (Corteel *et al.,* 1988: Gómez-Brunet *et al.*, 2007). This induces and synchronises ovulations rapidly, ensuring high fertility (60% lambing/kidding) with just one round of AI performed 46-48 h after treatment.

Since the seasonality of reproduction in sheep and goats is controlled by day length, out of season reproduction can also be achieved using strategies based on changing the length of the photoperiod. Treatments based on the alternation of LDs and SDs can be used in either closed or open barns by imposing an artificial light regimen (Chemineau *et al.*, 2008). Under field conditions, SD effects are easily provided by melatonin implants. This advances the onset of the breeding season and induces ovulatory activity in females throughout seasonal anoestrus (Chemineau *et al.,* 1988; 1992b), but it does not synchronise it well enough to be able to use AI.

Exposing seasonally anovulatory females to a male after a period of isolation has been shown to induce ovulation in both sheep and goats (Ungerfeld *et al.*, 2004; Chemineau *et al.,* 2006; Delgadillo *et al.*, 2009). In most females this male effect triggers a short ovarian cycle of 5–7 days in length, followed by a second ovulation occurring 7–9 days later associated with oestrus behaviour and the appearance of a corpus luteum (CL) of normal lifespan. The response to the male effect depends of many factors, but the depth of anoestrus and breed are reported the most important. In general, the male effect appears to be more effective (regardless of breed) at the end of the anoestrous period, and can be used to advance the breeding season by about a month. Moreover, while in highly seasonal sheep breeds the response varies widely, especially in mid-anoestrus, in breeds with reduced seasonality the male effect is even positive at the beginning and middle of anoestrus (Chanvallon *et al.*, 2009). In Spanish breeds, the male effect not only advances the breeding season and induces ovulation during anoestrus in mature animals, it causes the same in prepuberal ewes (López-Sebastián *et al.*, 1985b; Falagan, 1987; Folch, 1990). In goats the response varies within breeds during the seasonal anoestrus period, and among breeds from different latitudes (Walkden-Brown *et al.*, 1999). For instance, in highly seasonal breeds, the male effect can only advance the onset of the breeding season by a few weeks; it does not satisfactorily induce full sexual activity in the middle of the anoestrous period. However, in breeds with moderate seasonality, such as the Creole goats of Guadeloupe Island, the introduction of the male induces highly fertile ovarian activity in anovulatory females throughout the year (Walkden-Brown *et al.*, 1999). In Spanish goats, continuous exposure to bucks extends the seasonal ovulatory period. In fact, in Malagueña does isolated from males, the onset of seasonal ovulatory activity occurs about one month later than in Payoya goats kept continuously with males (September/October vs. August) (Gómez-Brunet *et al.*, 2003, 2010c; Zarazaga *et al.*, 2005). Moreover, most Spanish does exposed to males during anoestrous ovulate (87-92%) and eventually kid (85%) (Mateos Rex and Perez Llano, 1993; Folch *et al.*, 1993). Nutrition and sociosexual signals are also known to interact. In Spanish ewes, the oestrous response to the male effect during anoestrus, and the fertility obtained as a result of natural breeding, are reduced in those with poor nutritional status (74% and 32% respectively) compared to those with a medium (95% and 67% respectively) or high (95 and 77% respectively) status (Folch, 1990). In goats, the oestrus response to the male effect is reduced in does of low body weight compared to those of medium and heavy body weight (Veliz *et al.*, 2006). Further, the proportion of does showing signs of oestrus over the first 15 days after their introduction to sexually active bucks is greater if they receive nutritional supplementation for seven days before the introduction of the bucks (De Santiago-Miramontes *et al.*, 2008).

In both species, oestrus induced by a single injected dose (20-25 mg) of progesterone (P4) before allowing the male effect, is limited to a single peak of female sexual activity that occurs three days after exposure to the males, followed by ovulation and the formation of a normal CL (Chemineau, 1985; Pearce *et al.*, 1987). Notwithstanding, the degree of synchronisation achieved is again insufficient for successful AI.

The following lines describe three approaches used to control reproductive activity and optimise fertility in local Spanish sheep and goat breeds. The first is based in the use of melatonin implants combined with the male effect, the second uses the male effect in combination with progesterone (P4) and cloprostenol, and the third uses the male effect in combination with cloprostenol alone.

Melatonin implants and the male effect

Subcutaneous melatonin implants containing 18 mg of melatonin (hormone that communicates photoperiodic information from the pineal gland to the neuroendocrine system) induce high plasma concentrations of melatonin over the full 24 h of the day without suppressing the endogenous secretion of the pineal hormone during the night. It therefore provides an SD-like response by lengthening the duration of the melatonin signal (O'Callaghan *et al.*, 1991b). Several countries of the European Union (UK, France, Greece, Portugal and Spain) authorise the commercial use of melatonin implants, which are widely used in combination with the male effect to advance the onset of the breeding season in anoestrous ewes and goats, and/or to improve reproductive performance after natural mating.

In sheep breeds from the temperate regions of northern Europe (>45°N), melatonin implants have been traditionally inserted around the time of the summer solstice in order to advance the breeding season (Haresign *et al.*, 1990). In Mediterranean sheep breeds the onset of the breeding season starts earlier than in breeds living at higher latitudes, thus melatonin implants are usually inserted at about the time of the spring equinox (Chemineau *et al.*, 1988). The use of melatonin implants in tropical or subtropical latitudes is limited, as most ewes in these areas have a year-long breeding pattern.

In Spain, melatonin treatment is largely used with ewes under intensive, semi-extensive and extensive production systems. The treatment is usually given in conjunction with the male effect. The females are isolated from rams at least two months prior to implantation, and then the males introduced about 35 days after implantation. Both sexes remain together for a mating period of 40-45 days. In Spanish Rasa Aragonesa sheep, subcutaneous melatonin implants containing 18 mg of melatonin (Melovine®, CEVA Salud Animal, S.A., Barcelona, Spain), given in early March (spring), influence the ovarian response to the male effect, leading to modifications in the mating pattern and consequently the lambing curve (Abecia *et al.*, 2006). The proportion of ewes mated within the first 20 days after introduction to the rams, and the proportion of ewes lambed within the first 17 days of the lambing period (145–161 days after ram introduction), is greater in melatonin-treated than non-treated ewes (Figure. 10A). Others authors have also reported the efficacy of melatonin- implantation in Manchega and Rasa Aragonesa ewes at around the winter solstice (December/January, when most females are cycling) in delaying the onset of anoestrous and advancing the onset of reproductive activity in the absence of males (Gómez-Brunet *et al.*, 2001; Forcada *et al.*, 2002; Abecia *et al.*, 2010). From a practical point of view, melatonin-implantation around the winter solstice is very interesting since it can open up possibilities for developing breeding strategies in Mediterranean ewes showing reduced seasonality.

Numerous studies performed under field conditions with commercial have shown the efficacy of melatonin implants to improve the reproductive variables of Spanish sheep breeds (López-Sebastian and Inskeep, 1991; Folch *et al.*, 1991; Gómez-Brunet *et al.*, 1995; Martin *et al.*, 2002; Palacios *et al.*, 2006; Palacin *et al.*, 2006; Abecia *et al.,* 2007). For example, the fertility and litter size of treated ewes are improved; in general the number of lambs produced is increased by 15-30%, although this varies depending on the breed, management system and time of melatonin implantation (Table 1). Males treated with melatonin implants during the non-breeding season experience an increase in their scrotal circumference and sexual behaviour, and produce more – and more fertile - sperm (Bravo and Roy *et al.*, 2003; Garcia- Pastor *et al.*, 2004). Treating Rasa Aragonesa and Manchega rams with melatonin implants increases net lamb production under field conditions (Palacín *et al.*, 2008), possibly by increasing sperm progressive motility from days 45-90 after implantation (Casao *et al.*, 2010).

Studies on the effects of melatonin implants in Spanish Mediterranean goats have been limited. Although its effectiveness is still not fully established, some studies have shown its efficacy in advancing the breeding season and improving reproductive performance. In Payoya goats, melatonin implants inserted in spring (March) induce cyclicity during the anoestrus period (April/May) in a breeding system in which the bucks remain permanently with the does. However, this effect is not permanent and goats return to ovarian inactivity, before ovarian activity restarts in September (Zarazaga *et al.*, 2009a). In Florida goats (Figure 10B), melatonin implantation of the does in March also influences the response to the introduction of the buck, advancing the lambing pattern and improving fertility and prolificacy (Santos *et al.*, 2002).

The male effect, progesterone and $PGF_{2\alpha}$

In cycling animals, PGF_{2a} or its analogues (e.g., cloprostenol) causes rapid luteolysis of the CL and achieves well-synchronised behavioural oestrus (Acritopoulou and Haresign., 1980). Thus, the synchronization required for breeding females by natural mounting or AI when using the male effect alone or when combined with a progesterone injection, may be achieved by the administration of cloprostenol during the normal luteal phase induced by the male effect.

A study conducted during the non-breeding season (March-April) with Rasa Aragonesa ewes managed in extensive production flocks (at 42ºN, NE Spain), showed that the male effect combined with an i.m-injection of 25 mg progesterone (P4) at ram introduction, followed by 100 µg of cloprostenol 14 or 16 later, induced oestrus and ovulation, and was associated with lambing rates of about 70% after natural mating (López-Sebastian and Inskeep, 1988).

Table1. Fertility (%), prolificacy and fecundity of different Spanish sheep breeds treated-or not-with melatonin implants

Breed	Aptitude	Melatonin-Treated	N animals		Fertility (%)		Prolificity		Fecundity		Reference
			C	M	C	M	C	M	C	M	
Merino	Meat	January	167	174	56.3	72.4	1.03	1.16	0.57	0.83	Martín et al., 2002
Merino	Meat	March	150	175	51.3	64.5	1.07	1.11	0.55	0.72	Gómez-Brunet et al., 1995
Manchega	Mixta	April	42	42	59.5	69	1.04	1.34	0.62	0.93	Palacin et al., 2006
Churra	Milk	October	17	21	23.5	47.5	1.00	1.00	0.23	0.47	Palacios et al., 2006
Churra	Milk	April	55	48	68	71	1.07	1.20	0.73	0.85	López & Inskeep, 1991
Rasa Aragonesa	Meat	February	486	254	54	63	1.33	1.36	0.71	0.86	Folch et al., 1991
Rasa Aragonesa	Meat	March	100	100	38	53	1.06	1.10	0.40	0.58	López & Inskeep, 1991

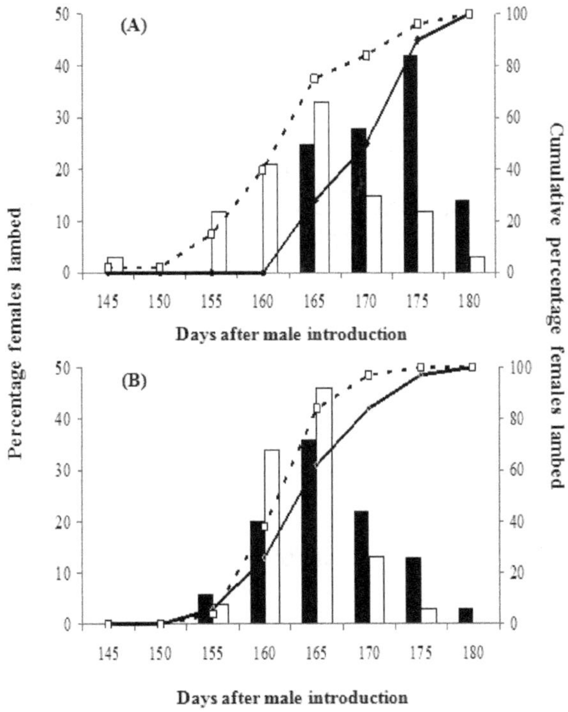

Figure 10. Proportion (%) and cumulative percentage of Melatonin-treated (□ and ---□---) and untreated (■ and—■—) Rasa Aragonesa ewes (A) and Florida goats (B) that lambed after introduction to males. Melatonin implants were administered in March. (Adapted from Santos et al.., 2002 and Abecia et al., 2006).

Monitoring follicular dynamics in goats by ultrasonography has shown that early lysis of the CL (on day 5–6 of the oestrous cycle) leads to ovulation occurring sooner, as well as a more tightly synchronised period of ovulations, than when lysis occurs at a more advanced luteal stage (on days 11–16) (Gonzalez-Bulnes *et al.*, 2005). Our group assessed the effectiveness of the IMA.PRO2® method at inducing and synchronizing ovulation in Murciano-Granadina (a high production dairy breed) does during the non-breeding season (April- June). The females were subjected to the male effect plus a single i.m-dose of 25 mg of P4 in olive oil at the time of male introduction, and to early lysis of the induced CL by a single i.m-dose of 75 µg cloprostenol given 9 days after exposure to the bucks (López Sebastián *et al.*, 2007). Oestrus behaviour was observed in 87.5% of the does a 37.0 ± 1.4 h ho after the administration of the cloprostenol, with the preovulatory LH surge occurring 40.5 ± 1.6 h after its injection. In addition, the pregnancy rate achieved under field conditions in inseminated does treated via the IMA.PRO2® method, without the need for the detection of oestrus, was higher than in females treated using the conventional method (intravaginal progestagens for 11 days plus 350 IU of eCG and 75 µg of cloprostenol on day 9 after sponge insertion) (65.2 and 49.4% respectively).

In Spain, as in other Mediterranean countries, goat farming for milk and cheese production has become an important industry involving the use of high yielding breeds subjected to selection programmes, with the use of AI an essential part of the production strategy. The degree of oestrus and ovulatory synchronisation obtained with the IMA.PRO2® method, the high fertility rates it achieves, and its allowing systematic timed AI without the need for the detection of oestrus, has led to its becoming extensively employed in intensive and semiextensive dairy flocks in Spain.

Following the recent forbidding of the i.m-injection of P4 in the EU, our group tested the value of P4 intravaginal gelatine capsules as a replacement for P4 injection at buck introduction as part of the IMA.PRO2®method (Gómez- Brunet et al., 2011). During the non-breeding season (April), Murciano-Granadina goats were treated with intravaginal P4 capsules (120 mg in palm oil) or with 25 mg i.m-P4 in olive oil at the time of introduction to the bucks. Oestrus behaviour and the LH surge occurred at 53.5 ± 3.0 h and 58 ±3.9 h respectively after buck introduction in does treated with the intravaginal P4 capsules, and at 63.3 ± 1.7 h and 69.0 ± 2.5 h respectively in those treated with i.m P4 (Figure 11). The pregnancy rate in inseminated does at 50 h after the i.m-injection of 75 µg of cloprostenol, given 9 days after buck introduction, were 60.6% in the IMA-PRO2®-treated goats and 58.5% in those treated with the intravaginal P4 capsules, confirming the effectiveness of the P4 capsules in replacing the i.m. P4 injection in the IMA.PRO2®method.

The male effect and $PGF_{2\alpha}$

The size of the response to the male effect is known to depend on the strength of seasonality in both sexes. Depending on the breed (and its latitude of origin) and on the depth of the sexual shutdown that occurs in the anoestrous period, great variation is seen in the response to the male effect during this time (Walkden-Brown *et al.*, 1999). In some breeds, photoperiod treatment of the males, females or even of both sexes may be necessary to optimise the response to the male effect (Flores *et al.*, 2000; Pellicer-Rubio *et al.*, 2007).

In the general context of a more restrictive use of hormone administration, our group is assessing the potential of prostaglandin-AI protocols based on the use of the male effect combined with cloprostenol administration alone at different times of the buck-induced luteal phase (López-Sebastian et al., unpublished data). We have characterised the response of Murciano-Granadina goats to the male effect (using vasectomised males) in terms of the detection of oestrus and variations in daily plasma progesterone from day 0 (male introduction day) to day 21. The results show that 80% of goats responded to the male effect, and of these 73.6% responded with a short P4 cycle followed by a normal P4 cycle, while 26.3% responded with a normal P4 cycle. Regardless of the cycle type induced, fertile ovulation occurred at 7.4 ± 0.3 days.

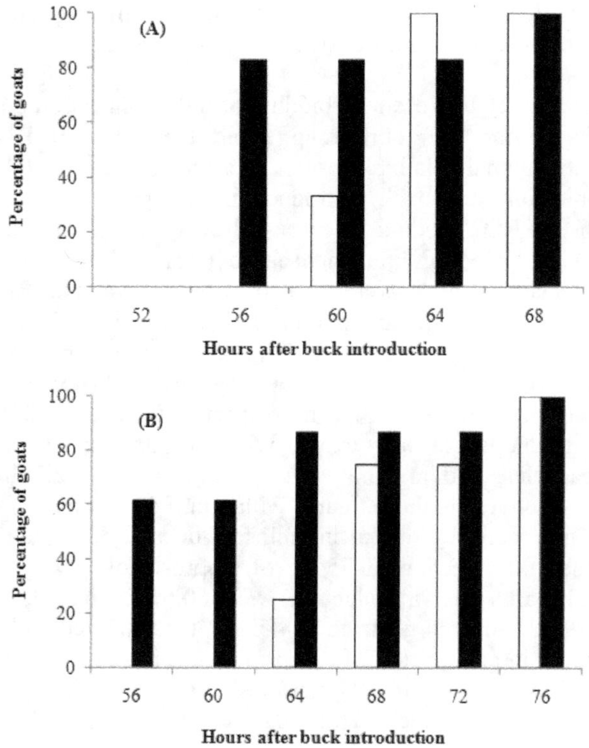

Figure 11. Cumulative percentage of Murciano-Granadina does showing oestrous behaviour (A) and the preovulatory LH surge (B) at different time points after treatment with a single i.m-dose of 25 mg of P4 in olive oil (white bars), or with intravaginal P4 capsules (120 mg in palm oil) (black bars) at the time buck introduction (Adapted from Gómez-Brunet et al., 2011).

CONCLUSIONS

This review indicates that local sheep and goat breeds from mid-high latitudes show reproductive seasonality, mainly regulated by variation in the photoperiod. Spanish sheep breeds are less seasonal than breeds from higher latitudes, but goat breeds show seasonality very similar to that of breeds from higher or even lower latitudes. The manipulation of the photoperiod has provided evidence that, at Spanish Mediterranean latitudes, these two species appear to use an endogenous circannual rhythm to control their reproductive transitions irrespective of the natural photoperiod, although it is differently expressed in each. The good response of Spanish breeds to the male effect in inducing ovulation outside of the normal breeding season has encouraged research into methods based on its use to control reproductive activity. In addition to their simplicity, these methods provide an interesting alternative to classic methods of reproductive control in the general context of reducing the use of hormones in animal production systems.

ACKNOWLEDGEMENTS

This work was supported by CICYT grant PB96-0435, and INIA grant RTA-2008-004.

REFERENCES

Abecia, J.A., Palacin, I., Forcada, F., Valares, J.A. 2006. The effect of melatonin treatment on the ovarian response of ewes to the ram effect. Domestic Animal Endocrinology. 31: 52-62.

Abecia, J.A., Valares, J.A., Forcada, F., Palacín, I., Martin, S., Martino, A. 2007. The effect of melatonin on the reproductive performance of three sheep breeds in Spain. Small Ruminant Research. 69: 10–16.

Abecia, J.A., Valares, J.A., Palacin, I., Forcada, F. 2010. The effect of melatonin replacement and month of treatment after the winter solstice on reproductive activity of two Mediterranean breeds of sheep. Biological Rhythms Research.42: 5-14.

Acritopoulou, S., Haresign, W. 1980. Response of ewes to a single injection of analogue of PGF2α given at different stages of the estrous cycle. Journal of Reproduction and Fertility. 58: 219–23.

Arendt, J. 1998. Melatonin and pineal gland: influence on mammalian seasonal and circadian physiology. Review of Reproduction. 3: 3-22.

Arrebola, F.A., Pérez-Martin, C.C., Santiago-Moreno, J. 2010. Limitation of seasonality in reproductive parameters of Mediterranean bucks, using photoperiod treatment. Small Ruminant Research.89: 31-35.

Arroyo, L.J., Gallegos-Sanchez, J., Vila-Godoy, A., Berruecos, J.M., Perera, G., Valencia, I. 2007. Reproductive activity of Pelibuey and Suffolk ewes at 19° north latitude. Animal Reproduction Science. 102: 24-30.

Avdi, M., Driancourt, M.A., Chemineau, P. 1993. Variations saisonnières du comportement d´estrus et de l´activité of ovulation chez les brebis Chios et Serres en Gréce. Reproduction Nutrition and Development. 33: 15-24.

Avdi, M., Banos, G., Stefos, K., Chemineau P. 2004. Seasonal variation in testicular volume and sexual behaviour of Chios and Serra rams. Theriogenology. 62: 275-282.

Barrell, G.A., Thrun, L.A., Browwn, E., Viguié, C., Karsch, F.J. 2000. Importance of photoperiodic signal to entrainment of the circannual reproductive rhythm of the ewe. Biology of Reproduction. 63: 769-774.

Beltran de Heredia, I., Gabiña D. 2004. Producción y calidad de semen de los moruecos adultos de raza Latxa. Pequeños Rumiantes. 5(1):8-12.

Bissonnette, H. 1941. Experimental modification of breeding cycle in goats. Physiological Zoology. 14: 329-338.

BonDurant, R.H., Darie, B.J., Munro, C.J., Stabenfeldt, G.H., Wang, P. 1981. Photoperiodic induction of fertile oestrus and changes in LH and progesterone concentrations in yearling dairy goats (*Capra hircus*). Journal of Reproduction and Fertility. 63:1-9.

Bravo, J.A., Roy, T.J. 2003. Influencia de los implantes de melatonina sobre las características espermáticas y actividad sexual del morueco en estación no sexual. SEOC. Producción ovina y caprina.28. 152-154.

Bronson, F.H. 1989. Seasonal Strategies: Ultimate Factors. *In*: Bronson FH (Ed). Mammalian Reproductive Biology, London, pp, 28-59.

Casao, A., Vegas, S., Palacin, I., Perez- Pe, R., Laviña, A., Quintın, F.J., Sevilla, E., Abecia, J.A., Cebrian-Perez, J.A., Forcada, F., Muiño-Blanco, T. 2010. Effects of melatonin implants during non-breeding season on sperm motility and reproductive parameters in Rasa Aragonesa rams. Reproduction in Domestic Animals. 45 (3): 425- 432.

Castel, J.M., Ruiz, F.A., Mena, Y., Sanchez-Rodriguez, M. 2010. Present situation and future perspectives for goats production systems in Spain. Small Ruminant Research. 89: 207-2010.

Chanvallon, A., Sagot, L., Potier, E., Scaramuzzi, R.J., Fabre-Nys, C. 2009. Interaction among seasonality of reproduction, genotype and the ram effect. Archiva Zootechnica 12 (3): 73-81.

Chemineau, P. 1985. Effects of a progestagen on buck-induced short ovarian cycles in the Creole meat goat. Animal Reproduction Science. 9: 87-94.

Chemineau, P. 1986. Seasonal behaviour and gonadal activity during year in the tropical Creole meat goat. I. Female oestrus behaviour and ovarian activity. Reproduction Nutrition and Development. 26: 441-452.

Chemineau, P., Normant, F., Ravault, J.P., Thimonier, J. 1986. Induction and persistence of pituitary and ovarian activity in the out-of-season lactating dairy goat after a treatment combining a skeleton photoperiod, melatonin and the male effect. Journal of Reproduction and Fertility. 78: 497–504.

Chemineau, P., Pelletier, J., Guerin, Y., Colas, G., Ravault, J.P., Touré, G., Almeida, G., Thimonier, J., Ortavant, R. 1988. Photoperiodic and melatonin treatments for the control of seasonal reproduction in sheep and goats. Reproduction Nutrition and Development. 28: 409-422.

Chemineau, P., Daveau, A., Maurice, F., Delgadillo, J.A. 1992a. Seasonality of oestrus and ovulation is not modified by subjecting female Alpine goats to a tropical photoperiod. Small Ruminant Research 8, 299-312.

Chemineau, P., Malpaux, B., Delgadillo, J.A, Guerin, Y., Ravault, J.P, Thimonier, J., Pelletier, J. 1992b: Control of sheep and goat reproduction: use of light and melatonin. Animal Reproduction Science. 30: 157-184.

Chemineau, P., Pellicer-Rubio, M.T., Lassoued, N., Khaldi, G., Monniaux, D., 2006. Male-induced short oestrus and ovarian cycles in sheep and goats: a working hypothesis.

Reproduction Nutrition and Development. 46: 417-429.

Chemineau, P., Guillaume, D., Migaud, M., Thiéry, J.C., Pellicer-Rubio, M.T., Malpaux, B. 2008. Seasonality of reproduction in mammals: intimate regulatory mechanism and practical implications. Reproduction in Domestic Animals. 43 (Supplement 2): 40-47.

Chemineau, P., Bodin, L., Migaud, M., Thiéry, J.C., Malpaux, B. 2010. Neuroendocrine and genetic control of seasonal reproduction in sheep and goats. Reproduction in Domestic Animals. 45 (Supplement 3): 42-49

Corteel, J.M., Leboeuf, B., Baril, G. 1988. Artificial breeding of adult goats and kids induced with hormones to ovulate outside the breeding season. Small Ruminant Research. 1: 19–35.

De Santiago-Miramontes, R., Muñoz-Gutierrez, M., Malpaux, B., Scaramuzzi, R.J., Delgadillo, J.A. 2008. Ovulation rate in female anoestrus goats exposed to the male effect is increased by nutritional supplementation under grazing conditions. Animal Reproduction Science. 105: 409-416.

Delgadillo J.A., Leboeuf B., Chemineau P. 1991. Decrease in seasonality of sexual behaviour and semen production in bucks by exposure to short photoperiod cycles. Theriogenology. 36(5):755-770.

Delgadillo, J.A., Chemineau, P. 1992. Abolition of seasonal release of luteinizing hormone and testosterone in Alpine male goats (*Capra hircus*) by short photoperiod cycles. Journal of Reproduction and Fertility. 94: 45-55.

Delgadillo, J.A., Canedo, G.A., Chemineau, P., Guillaume, D., Malpaux, B. 1999. Evidence for an annual reproductive rhythm independent of food availability in male Creole goat in subtropical northern Mexico. Theriogenology. 52: 727-737.

Delgadillo, J.A., Carrillo, E., Morán, J., Duarte, G., Chemineau, P., Malpaux, B. 2001. Induction of sexual activity of male Creole goats in subtropical northern Mexico using long days and melatonin. Journal of Animal Science.79: 2245-52.

Delgadillo, J.A., Fitz-Rodriguez, G., Duarte, G., Véliz, F.G, Carrillo, E., Flores, J.A, Vielma, J., Hernandez, H., Malpaux, B., 2004. Management of photoperiod to control caprine reproduction in the subtropics. Reproduction Fertility and Development. 16: 471-478.

Delgadillo, J.A., Gelez, H., Ungerfeld, R., Hawken, P.A, R., Martin, G.B. 2009. The ''male effect'' in sheep and goats: visiting the dogmas. Behavioral Brain Research 200: 304–314.

Delgadillo, J.A., De la Torre-Villegas, S., Arellano-Solis, V., Duarte, G., Malpaux, B. 2011. Refractoriness to short and long days determines the end and onset of the breeding season in subtropical goats. Theriogenology. 76 (6): 1146-1151.

Duarte, G., Flores, J.A., Malpaux, B., Delgadillo, J.A. 2008. Reproductive seasonality in female goats adapted to a subtropical environment persists independently of food availability. Domestic Animal Endocrinology, 35: 262-370.

Duarte, G., Nava-Hernandez, M.P., Malpaux, B., Delgadillo, J.A. 2010. Ovulatory activity of female goat adapted to the subtropics is responsive to photoperiod. Animal Reproduction Science. 120: 65-70.

Falagan, A.1987. Influencia del efecto macho sobre la aparición de la pubertad en corderas Segureñas en función de la época de nacimiento. ITEA. 72: 331-333.

Falagan, A., Gonzalez, C., Lopez-Sebastian, A. 1989. Periodos de anoestro de la cabra Murciano-Granadina en la región de Murcia. ITEA .9:298–300.

Flores, J.A., Véliz, F.G., Pérez-Villanueva, J.A., Martinez De La Escalera, G., Chemineau, P., Poindron, P., Malpaux, B., Delgadillo, J.A. 2000. Male reproductive condition is the limiting factor of efficiency in the male effect during seasonal anoestrus in female goats. Biology of Reproduction. 62(5): 1409–1414.

Folch, J. 1990. Utilización práctica del efecto macho para la provocación de celos y ovulaciones en ganado ovino. ITEA. 86A: 145-163.

Folch, J., Alabart, J.L., Garabayo, A., Echegoyen, E. 1991. Utilización de la melatonina para provocar la cubrición de las corderas en primavera. ITEA. 11:142-144.

Folch, J., Gonzalez F., Alabart, J.L. 1993. Efecto macho en cabras Blanca Celtibérica. ITEA. 12: 454-456.

Folch, J., Alabart, 2000. Características reproductivas de la oveja Rasa Aragonesa. Ovis.68: 27-36.

Forcada, F., Abecia, J.A., Sierra, I. 1992. Seasonal changes in oestrous activity and ovulation rate in Rasa Aragonesa ewes maintained at two different body conditions levels. Small Ruminant Research. 8: 313-324.

Forcada, F., Abecia, J.A., Zúñiga, O., Lozano, J.M. 2002. Variation in the ability of melatonin implants inserted at two different times after the winter solstice to restore reproductive activity in reduced seasonality ewes. Australian Journal of Agriculture Research. 53: 167–173.

Forcada, F., Abecia, J.A. 2006. The effect of nutrition on the seasonality of reproduction in ewes. Reproduction Nutrition and Development. 46: 355-365.

García Pastor, L., Gonzalez, J.M., Figueras, I., Callejas, M., Cebrián, L.M., Espada, M. 2004. Implantes de melatonina en moruecos. Efecto sobre algunos índices reproductivos en un rebaño de raza Aragonesa. SEOC. Producción ovina y caprina.30: 154-156.

Gebbie, F.E., Forsyth, I.A., Arendt, J.1999. Effects of maintaining solstice light and temperature on reproductive activity, coat growth, plasma prolactin and melatonin in goats. Journal of Reproduction and Fertility.116: 25-33.

Gómez-Brunet, A., López Sebastián, A. 1991. Effect of season on plasma concentrations of prolactin and cortisol in pregnant, non-pregnant and lactating ewes. Animal Reproduction Science. 26: 251-268.

Gómez-Brunet A., López-Sebastian A., Muñiz-Herrando, E., Cabellos, B. 1992. Función luteal y secreción de LH durante el anoestro post-parto en ovejas Manchegas: Influencia de la época del parto y tipo de destete. Investigación Agraria. Producción y Sanidad Animales.7 (3): 169-183.

Gómez-Brunet, A., Lopez-Sebastián, A., Picazo, R.A., Cabellos, B., Goddard, S. 1995. Reproductive response and LH secretion in ewes treated with melatonin implants and induced to ovulate with the ram effect. Animal Reproduction Science. 39: 23–34.

Gómez-Brunet, A., Santiago-Moreno, J., Chemineau, P., Malpaux, B., del Campo, A., López-Sebastian, A. 2001. Effect of melatonin implantation at the winter solstice on the end of the breeding season in Mouflon (*Ovis gmelini musimon*) and Manchega sheep (*Ovis aries*). Rencontres and Recherches in Ruminants. 8: A375.

Gómez-Brunet, A., Santiago-Moreno, J., Micheo, J.M., Sanchez, A., Gonzalez-Bulnes, A., López-Sebastian, A. 2003. Variación anual de la actividad ovulatoria en la cabra de raza Malagueña. Producción ovina y caprina. 28: 178-180.

Gómez-Brunet, A., Santiago-Moreno, J., Montoro, V., Garde, J., Pons, P., Gonzalez-Bulnes, A., López-Sebastían A. 2007. Reproductive performance and progesterone secretion in estrus-induced Manchega ewes treated with hCG at the time of AI. Small Ruminant Research.71: 117-122.

Gómez-Brunet, A., Santiago-Moreno, J., del Campo, A., Malpaux, B., Chemineau, P., Tortonese, D.J., Gonzalez-Bulnes, A., López-Sebastian, A. 2008. Endogenous circannual cycles of ovarian activity and changes in prolactin and melatonin secretion in wild and domestic female sheep maintained under a long-day photoperiod. Biology of Reproduction. 78: 552-562.

Gómez-Brunet, A., Santiago-Moreno, J., Chemineau, P., Malpaux, B., López-Sebastian A. 2010a. Melatonin secretion during postnatal development in wild and domestic female lambs. Animal Reproduction Science. 119: 24-30.

Gómez-Brunet, A., Santiago-Moreno, J., López-Sebastían A. 2010b. Estacionalidad reproductiva en los ungulados silvestres. In: Ungulados silvestres de España: biología y tecnologías reproductivas para su conservación y aprovechamiento cinegético. Monografias INIA. Serie Medioambiental. 2: 63-84.

Gómez-Brunet, A., Santiago-Moreno, J., Toledano, Diaz, A., López-Sebastían A. 2010c. Evidence that refractoriness to long and short daylengths regulates seasonal reproductive transitions in Mediterranean goats.

Reproduction in Domestic Animals. 45: 338-343.

Gómez-Brunet, A., Toledano, A., Coloma' M.A., Velázquez' R., Castaño, C., Carrizosa, J.A., Urrutia, B., Santiago J., López, A. 2011. Induction and synchronization of ovulation by combining the male effect, intravaginal progesterone capsules and cloprostenol, in goats during the non-breeding season. In: Proceedings of EAAP Congress, Stavanger, Norway. A6.

Gonzalez L.J., Murphy, B.D., Foote, W.C., Ortega E. 1992. Circannual estrous variation and ovulation rate in Pelibuey ewes. Small Ruminant Research. 8: 225-232.

Gonzalez-Bulnes, A., Diaz-Delfa, C., García-García, R.M., Urrutia, B., Carrizosa, J.A., López-Sebastián, A. 2005. Origin and fate of preovulatory follicles after induced luteolysis at different stages of the luteal phase of the oestrous cycle in goats. Animal Reproduction Science. 86: 237–45.

Gonzalez-López, J. 1993. La reproducción en la oveja Merina. Caracterización de su fisiología. Ovis. 28. Ed. Luzán S.A. pp 71.

Gonzalez-Lopez, J. 1995. Caracterización del control reproductivo en la oveja merina. Ovis 41. Ed. Luzán S.A. pp 71

Greyling, J.P.C. 2000. Reproduction traits in the Boer goat doe. Small Ruminant Research. 36: 171-177.

Hafez, E.S.E. 1952. Studies on the breeding season and reproduction of the ewe. Journal of Agricultural Science. 42: 189–265.

Haresign, W., Peters, A.R., Staples, L.D. 1990. The effect of melatonin implants on breeding activity and litter size in commercial sheep flocks in the UK. Animal. Production. 50: 111–121.

Karsch, F.J., Bittman, E.L., Foster, D.L., Goodman, R.L., Legan, S.J., Robinson, J.E. 1984. Neuroendocrine basis of seasonal reproduction. Recent Progress in Hormone. Research. 40: 185-232.

Karsch, F.J., Bittman, E.L., Robinson, J.E., Yellon, S.M., Wayne, N.L., Olster, D.H., Kaynard, A.H. 1986. Melatonin and photorefractoriness: loss of response to the melatonin signal leads to seasonal reproductive transitions in the ewe. Biology of Reproduction. 34: 265-274.

Karsch, F.J., Robinson, J.E., Woodfill, C.J.I., Brown, M.B. 1989. Circannual cycles of luteinizing hormone and prolactin secretion in ewes during prolonged exposure to a fixed photoperiod: Evidence for an endogenous reproductive rhythm. Biology of Reproduction. 41: 1034-1046.

Lincoln, G.A., Johnston, J.D., Anderson, H., Wagner, G., Hazlerigg, D.G. 2005. Photorefractoriness in mammals: Dissociating a seasonal timer from the circadian-based photoperiod response. Endocrinology. 146: 3782-3790.

López-Sebastián, A., Alonso de Miguel, M., Gómez-Brunet, A.1985a Características del comienzo de la pubertad en corderas Manchegas mediante la estimulación por machos en estación desfavorable. Investigación Agraria. Serie Ganadera.22: 167- 181.

López-Sebastián, A., Gómez-Brunet, A., Pinilla, A., Cabellos, B.1985b. Influencia de la época de nacimiento y el efecto de los machos sobre la aparición de la pubertad en corderas Manchegas. ITEA 58: 45-49.

López-Sebastián, A., Inskeep, E.K. 1988. Effects of progesterone pretreatment and duration of ram exposure on synchronization of estrus, conception and pregnancy by prostaglandin during seasonal anestrous. Animal Reproduction Science.17:185–95.

López-Sebastián A., Inskeep, E.K.1991. Response of ewes of Mediterranean sheep breeds to subcutaneous implants of melatonin. Livestock Production Science. 27: 177–184.

López-Sebastián, A., Palacio, M., Gómez-Brunet A. 1991. Techniques for the improvement of fertility and prolificacy of anoestrus ewes in Spain. FAO/IAEA. Isotope aided studies on sheep and goat production.1: 157-164.

López-Sebastián, A., Gonzalez-Bulnes, A., Carrizosa J.A., Urrutia, B., Díaz-Delfa, C., Santiago-Moreno, J., Gómez-Brunet, A. 2007. New estrus synchronization and artificial insemination protocol for goats based on male exposure, progesterone and cloprostenol during the non-breeding season. Theriogenology. 68:1081-1087.

Malpaux, B. 2006. Seasonal regulation of reproduction in mammals. In Knobil and Neill´s Physiology of Reproduction. Third Edition. Ed. D Neill pp 2231-2281. Elsevier. Amsterdam.

Malpaux, B., Robinson, J.E., Brown, M.B., Karsch, F.J. 1987. Reproductive refractoriness of the ewe to inductive photoperiod is not caused by inappropriate secretion of melatonin. Biology of Reproduction. 36:1333-1241.

Malpaux, B., Wayne, N.L, Karsch, F.J. 1988.Termination of the breeding season in the Suffolk ewe: involvement of an endogenous rhythm of reproduction. Biology of Reproduction. 39: 254-263.

Malpaux, B., Robinson, J.E., Wayne, N.L, Karsch, F.J. 1989. Regulation of the onset of the breeding season of the ewe: importance of long days and of an endogenous reproductive rhythm. Journal of Endocrinology. 122: 269-278.

Malpaux, B., Karsch, F.J. 1990. A role for short days in sustaining seasonal reproductive activity in the ewe. Journal of Reproduction and Fertility. 90: 355-362.

Malpaux, B., Migaud, M., Tricoire, H., Chemineau, P.,2001. Biology of mammalian photoperiodism and the critical role of the pineal gland and melatonin. Journal of Biological Rhythms. 16: 336-347.

Martin, A., Deletang, F., Martin, S., Martino, A. 2002. Utilización de implantes de melatonina en corderas de raza Merino en cubrición de Febrero. SEOC. Producción ovina y caprina 27: 1073-1076.

Mascarenhas, R., Simoes, A., Robalo, J. 1995. Cyclic reproductive activity and efficiency of reproduction in Serrana goats. Animal Reproduction Science. 38:223-229.

Mateos Rex, E., Pérez LLano, B. 1993. El efecto macho en el manejo reproductivo del ganado caprino. Mundo Ganadero. 9: 82-86.

Mazariegos, V. Vázquez, J.M., Garrido, C., Salavador, S., De la Fuente, F. 2010. Influencia de la raza, época del año y ritmo de recogida sobre las características del semen fresco en ganado ovino. SEOC. Producción ovina y caprina, 35. 150-154.

Minton, J.E. 1990. Role of photorefractoriness in onset of anoestrus in Rambouillet x Dorset ewes. Journal of Reproduction and Fertility. 89 (1): 261-268.

O´Callaghan, D., Karsch, F.J, Boland, M.P., Roche, J.E. 1991a. Role of short days in timing the onset and duration of reproductive activity in ewes under atificial photoperiods. Biology of Reproduction. 44: 23-28.

O'Callaghan, D., Karsch, F.J., Boland, M.P., Roche, J.F. 1991b. What photoperiodic signal is provided by a continuous-release melatonin implant? Biology of Reproduction. 45: 927–933.

Ortavant, R., Pelletier, J., Ravault, J.P., Thimonier, J., Volland-Nail, P. 1985. Photoperiod: main proximal and distal factor of the circannual cycle of reproduction in farm mammals. Oxford Review of Reproductive Biology. 7: 306-345.

Palacin, I., Abecia, J.A., Forcada, F., Martin, S., Martino, A. 2006. Eficacia del tratamiento con melatonina durante el anoestro en ovejas rasa Aragonesa en tres diferentes sistemas de explotación. SEOC. Producción ovina y caprina. 31:348-351.

Palacin, I., Abecia, J.A., Forcada, F., Casao, A., Cebrian-Perez, J.A., Muiño-Blanco, T., Palacios, C., Pontes, J.M. 2008: Effect of exogenous melatonin treatment on out of season ram fertility. Italian Journal of Animal Science. 7: 199-206.

Palacios, C., Palacin, I., Abecia, J.A., Forcada, F., Martin, S., Martino, A. 2006. Utilización práctica de los implantes de melatonina para la cubrición de corderas de 4 razas ovinas. SEOC. Producción ovina y caprina. 31: 350-352.

Pearce, D.T., Oldham, C.M., Haresign, W., Gray, S.J.1987. Effects of duration and timing of progesterone priming on the incidence of corpora lutea with a normal life-span in Merino ewes induced to ovulate by the introduction of rams, Animal Reproduction Science.13: 81-89.

Pellicer-Rubio, M.T., Leboeuf, B., Bernelas, D., Forgerit, Y., Pougnard, J.L., Bonne, J.L., Senty, E., Chemineau, P. 2007. Highly synchronous and fertile reproductive activity induced by the male effect during deep anoestrus in lactating goats subjected to treatment with artificially long days followed by a natural photoperiod. Animal Reproduction Science. 98: 241–258.

Pérez Llano B., Mateos Rex, R. 1994. Desarrollo testicular y producción de testosterona en machos cabrios de las razas Verata y Malagueña. Investigación Agraria: Producción y Sanidad Animal. 9 (1): 13-20.

Perez, B., Mateos, E. 1996. Effect of photoperiod on semen production and quality in bucks of Verata and Malagueña breed. Small Ruminant Research. 22: 163-168.

Perez, J.P., Gil, J.M., Sierra, I. 2007. Technical efficience of meat sheep production systems in Spain. Small Ruminant Research. 69: 237-241.

Pope, W.F., Mc Clure, K.E., Hogue, D.E., Day, M Y. 1989. Effect of season and lactation on post-partum fertility of Polipay, Dorset, St Croix and Tharghe ewes. Journal of Animal Science. 67: 1167-1174.

Poulton, A.L, Robinson, T.J. 1987. The response of rams and ewes of three breeds to artificial photoperiod. Journal of Reproduction and Fertility. 79: 609-626.

Restall, B.J. 1992. Seasonal variation in reproductive activity in Australian goats. Animal Reproduction Science. 27: 305-318.

Rivera, G.M., Alanis, G.A., Chaves, M.A., Ferrero, S.B., Morello, H.H. 2003. Seasonality of estrus and ovulation in Creole goats of Argentina. Small Ruminant Research. 48: 109-117.

Robinson, J.E., Karsch, F.J. 1984. Refractoriness to inductive day lengths terminates the breeding season of the Suffolk ewe. Biology of Reproduction 31: 656-663.

Robinson, J.E., Wayne, N.L., Karsch, F.J. 1985. Refractoriness to inhibitory day lengths initiates the breeding season of the Suffolk ewe. Biology of Reproduction. 33: 324-334.

Roca, J., Martinez, E., Vazquez, J.M., Ruiz, S., Coy, P. 1991. Influence of season on testicle size and libido in male coats from the Mediterranean area. Animal Production. 52: 317-321.

Roca, J., Martinez, E., Vazquez, J.M., Coy.P. 1992. Characteristic and sexual variations in the semen of Murciano-Granadina goats in the Mediterranean area. Animal Reproduction Science. 29: 255-26.

Rosa, H.J.D., Bryant, M.J. 2003. Seasonality of reproduction in sheep. Small Ruminant Research. 48: 155-171.

Santa Maria, A., Cox, I., Muñoz, E., Rodriguez, R., Caldera, L. 1990. Estudio del ciclo sexual, estacionalidad reproductiva y control del estro en la cabra Criolla en Chile. Final Research Coordination Meeting FAO, Bogotá, Colombia 19-23, September 1988, pp 363-385. International Agency Vienna, Austria.

Santiago-Moreno, J., Gómez-Brunet, A., Gonzalez-Bulnes, A., Villar, D., López-Sebastián, A. 2000a. Attainment of puberty in the European mouflon and the domestic Manchega ewe (*Ovis aries*). Reproduction Domestic. Animal. 35: 49-52.

Santiago-Moreno, J., Gonzalez-Bulnes, A., Gómez-Brunet A., López-Sebastían A. 2000b. Effect of season on the resumption of postpartum cyclic ovarian activity in Mouflon (*Ovis gmelini musimon*) and Manchega sheep (*Ovis aries*). Small Ruminant Research. 36: 69-74.

Santiago-Moreno, J., López-Sebastián, A., González-Bulnes, A., Gómez-Brunet, A., Chemineau, P. 2000c. Seasonal changes in ovulatory activity, plasma prolactin, and melatonin concentrations in Mouflon (*Ovis gmelini musimon*) and Manchega (*Ovis aries*) ewes. Reproduction Nutrition and Development. 40: 421–430.

Santiago-Moreno J., Gómez-Brunet A., Gonzalez-Bulnes, A., Toledano-Díaz, A., Malpaux, B., López-Sebastían, A. 2005. Differences in reproductive pattern between wild and domestic rams are not associated with inter-specific annual variations in plasma prolactin concentrations. Domestic Animal Endocrinology. 28: 416-429.

Santiago-Moreno, J., Gómez-Brunet, A., Toledano-Díaz, A., Picazo, R., Gonzalez-Bulnes, A., López-Sebastián, A. 2006. Seasonal endocrine changes and breeding activity in Mediterranean wild ruminants. Reproduction in Domestic Animals. 41 Supplement 2: 72-81.

Santos, R., Martín, S., Deletang, F., Martino, A. 2002. Utilización de los implantes de melatonina en Cabra Florida en régimen intensivo. SEOC. Producción ovina y caprina.27: 1103-1107

Scaramuzzi, R.J., Martin, G.B. 2008. The importance of interaction among nutrition, seasonality and socio-sexual factors in the development of hormone-free methods for controlling fertility. Reproduction in Domestic Animals. 43 (Supplement 3): 129-136.

Setchell, B.P. 1992. Domestication and reproduction. Animal Reproduction Science. 28:195-202

Shelton, M. 1978. Reproduction and breeding of goats. Journal of Dairy Science. 61: 994-1010.

Spanish Ministry for Agriculture, Food and Fisheries, 2009. Online Statistical Agribusiness Yearbook, 2008. Ministerio de Agricultura, Pesca y Alimentación. Madrid 2009.

Thimonier, J., Mauléon, P. 1969. Variations saisonnières du comportement d´oestrus et des activités ovariennes et hypophysaires chez les ovins. Annales de Biologie Animale Biochimie, Biophysique.9: 233-250.

Thimonier, J., Terqui, M., Chemineau P. 1986. Conduite de la reproduction des petits ruminants dans les différents parties du monde. In : Proceedings of an International Symposium on the Use of Nuclear Techniques in Studies of Animal Production and Health in Different Environments, International Atomic Energy Agency, Vienna, pp 135-147.

Todini, L., Malfatti, A., Terzano, G.M., Borghese, A., Pizzillo, M., Debenedetti, A. 2007. Seasonality of plasma testosterone in males of four Mediterranean goats and in three different climatic conditions. Theriogenology. 67: 627-630.

Ungerfedl, R., Fosberg, M., Rubianes, E. 2004. Overview of the response of anoestrus ewes to the ram effect. Reproduction Fertility and Development. 16: 479-490.

Urrutia-Morales, J., Meza-Herrera, C.A., , Escobar Medina, F.J., Gamez-Vazquez, H.G., Ramirez-Andrade, B.M., Diaz Gómez, M.O., Gonzalez-Bulnes, A 2009. Relative roles of photoperiodic and nutritional cues in modulatinarian activity in goats. Reproductive Biology. 9(3): 283-294.

Valls Ortiz, M. 1983. Frequent lambing of sheep flocks in Spain: Productivity and management consequences. Livestook Production Science.10: 49-58.

Veliz, F.G., Poindron, P., Malpaux, B., Delgadillo, J. A. 2006. Positive correlation between the body weight of anestrus goats and their response to the male effect with sexually active bucks. Reproduction Nutrition and Development. 46: 657:661.

Walkden-Brown, S.W., Martin, G.B., Restall, B.J.1999. Role of male- female interaction in regulating reproduction in sheep and goats. Journal of Reproduction and Fertility, Supplement 54: 243– 257.

Wheeler, A.G., Land, R.B. 1977. Seasonal variation in oestrus and ovarian activity of Finnish-Landrace Tasmanian-Merino and Scottish-Blackface ewes. Animal Production. 24: 363-376.

Woodfill, C.J.I., Wayne, N.L, Moenter, S.M., Karsch, F.J. 1994. Photoperiodic synchronization of a circannual reproductive rhythm in sheep: Identification of season-specific time cues. Biology of Reproduction. 50: 965-976.

Yenikoye, A.1984. Annual variation in estrual behavior, rate and possibilities for ovulation in Peulh ewes from Niger. Reproduction Nutrition and Development. 24: 11-19.

Zarazaga, L.A., Guzmán, C., Domínguez, C., Pérez, M.C., Prieto, R. 2005. Effect of plane of nutrition on seasonality of reproduction in Spanish Payoya goats. Animal Reproduction Science. 87: 253-257.

Zarazaga, L, A., Gatica, M.C., Celi, I., et al., 2009a. Effect of melatonin implants on sexual activity in Mediterranean goat females without separation from males. Theriogenology. 72:910–918.

Zarazaga, L.A., Guzmán, J.L., Domínguez, C., Pérez, M.C., Prieto, R. 2009b. Effects of season and feeding level on reproductive activity and semen quality in Payoya buck goats. Theriogenology. 71: 1316-1325.

Zarazaga, L., Celi, I., Guzmán, J.L., Malpaux, B. 2011a. The response of luteinizing hormone secretion to photoperiod is modified by the level of nutrition in female Mediterranean goats. Animal Reproduction Science. 126: 83-80.

Zarazaga, L.A., Celi, I., Guzmán, J.L., Malpaux, B. 2011b. The role of nutrition in the regulation of LH secretion by the opiodergic, dopaminergic and serotonergic systems in female Mediterranean goats. Biology of Reproduction. 84: 447-456.

Zeder M.A., Hesse, B. 2000. The initial domestication of goats (*Capra hircus*) in the Zagros Mountain 10000 years ago. Science. 287: 2254-2257.

5

CONCENTRATION OF SOME ELEMENTS IN BLOOD SERUM OF NONLACTATING GOATS IN A SUBTROPICAL REGION OF SOUTHWEST OF MÉXICO STATE

A. Olmedo-Juárez[a], R. Rojo[a*], A.Z.M. Salem[b], J.F. Vázquez-Armijo[a], S. Rebollar-Rebollar[a], B. Albarran[a] and J. Lugo[c]

[a] *Centro Universitario UAEM-Temascaltepec, Universidad Autónoma del Estado de México, km 67.5 Carr. Toluca-Tejupilco, Estado de México, 51300, México.*
[b] *Facultad de Medicina Veterinaria y Zootecnia, Universidad Autónoma del estado de México Instituto Literario No. 100, Toluca , 50000 Edo. de México, México"*
[c] *Facultad de Ciencias Laboratorio de Edafología y Ambiente, Universidad Autónoma del Estado de México, Instituto Literario No. 100, Toluca , 50000 Edo. de México, México*
*E-mail: dr_rojo70@*yahoo.com.mx
**Corresponding Author*

SUMMARY

The objective of this study was to determine the mineral levels in blood serum of nonlactating adult goats in the south of the State of México during the dry and rainy seasons. Eighty four multiparous (>2 calving and non-lactating) does were sampled (LW 39 ± 8 kg). Blood samples were collected by jugular vein punction, followed by separation of blood serum by adding 10% trichloroacetic acid for later laboratory analysis. Data were analyzed as a completely randomized factorial experiment. There was as effect ($P<0.05$) of season on serum concentrations of P, and there was an interaction ($P<0.01$) of season*region for Ca, Na, Cu and Zn. Mg was unaffected by treatments. Results suggest possible deficiencies in the diets of P, Ca, Cu and Zn.

Key words: season; blood serum in goats; mineral status.

INTRODUCTION

Production of goats frequently occurs in regions with low income levels and the milk and meat products are often destined for self consumption (Devendra, 1991; Hernández, 2000 Gallegos et al., 2005). This activity is often carried out by traditional families under traditional extensive systems and use of modern technologies is very low (Devendra, 1988; Hernández, 2005; Cotler et al., 2006). The State of México accounts for 2% of Mexico's national goat inventory (Medrano, 2000; Rebollar et al., 2007), and the south-western region of the state is the most important contributor to Mexican goat meat production. Several studies have shown that the mineral profiles of the local forages very lead to deficiencies in mineral consumption according to established animal requirements. In particular, mineral levels are reported to be from marginal to deficient relative to requirements for production (NRC, 2007).

Several studies have been completed on the mineral status of cattle and sheep (Dominguez and Huerta 2007; Morales et al., 2007), but there is very little information about goats making it difficult to assess their mineral status, or to determine a mineral supplement (Meschy, 2000). Thus the objective was to determine the mineral status of adult goats based upon their blood serum levels in four localities of the southwest region of the state of Mexico during the dry and rainy seasons.

MATERIALS AND METHODS

Area of samples collection

The study was completed in the Tejupilco, Amatepec, Luvianos and Tlatlaya localities of south-western Mexico State. This region is located between 18° 45' 30" north latitude, and 100° 36' 45" west longitude, at 1330 mosl. Sandy soils is predominant (Table 1) and the climate is classified as tropical sub-humid with rain during summer (García, 1986). Average temperature and relative humidity during the experimental period were 25 ± 5; 33 ± 4 °C and 55 ± 3; 45 ± 4 % for the rainy and dry seasons respectively (Meteorological experimental station of Centro Universitario UAEM-Temascaltepec). In this region of 3534 km^2, the landscape has a very steep slope and the vegetation is relatively homogeneous. During the rainy season, forage biomass is suitable to provide sufficient feed for animals, but during the dry season forage is scarce and a small quantities of grain are also fed to animals.

Animals

Eighty four clinically healthy non-lactating goats from four goat flocks (Tejupilco, Amatepec, Luvianos, Tlatlaya, 21 goats per locality) were used in a grazing study in the in dry and rainy season in the southwest of México state. The average live weights were 39 ± 8 kg. The genotype of the animals was variables degrees of cross breeding. The flocks were kept extensively and fed native forages resources. During the rainy season the goats consumed a variability of forages such as: grasses; *Cynodon plectostachyus, Andropogon gayanus*, browse shrubs; Q*uercus laeta, Pithecellobium dulce, Lysiloma acapulsencis, Leucaena leucocephala, Heliocarpus velutinus, Guazuma ulmifolia,* which provide sufficient feed intake for the animals to maintain live weight but, during dry season, these forages are scarce and animals are fed corn stover and some tree foliages such as *Quercus laeta, Pithecellobium dulce* and *Lysiloma acapulsenci.* Producers normally feed the animals small quantities of grains such as corn or sorghum.

Sample collection and preparation

Blood samples were collected from the 84 non-lactating goats described above during the dry (April-May) and rainy (August-September) seasons of 2008. Same goats were sampled in each season. Blood samples were collected by jugular vein puncture with a syringe and needle, and then drawn by vacuum into evacuated tubes. Blood samples were then centrifuged at 3000 rpm for 15 min and plasma was placed into polyethylene tubes and frozen at -20°C. Samples were then prepared for mineral assay according to Fick et al. (1979). Concentrations of Ca, Mg, Na K, Cu and Zn were determined by atomic absorption (Perkin Elmer 210) and P concentration was determined by colorimetry using a spectrophotometer with visible UV.

Statistical analyses

A completely randomized design with a factorial arrangement of treatments (Steel and Torrie, 1980) was used. Factors were season and locality according to the model: $Y_{ijk} = \mu + S_i + M_j + S*M_{ij} + E_{ijk}$, where: Y_{ijk} = response variable (P, Na, K, Ca, Mg, Cu, Zn), μ= General mean, S_i= dry and rainy season effect, L_j= Effect of locality (Tejupilco, Luvianos, Amatepec, Tlatlaya), $S*L_{ij}$= interaction of season and locality and E_{ijk}= Error term. Data obtained were analyzed using Statistical Analysis System (SAS, 2006) software and differences among means were ranked using Tukey's test (Steel and Torrie, 1980).

RESULTS

Phosphorous (P), calcium (Ca), magnesium (Mg) and potassium (K) concentration in blood serum

Amatepec locality had the highest P values (4.79 mg/dl), while the rest had similar values ranging from 3.05 to 4.33 mg/dl (Table 2). The Ca concentration was higher during the rainy season (11.20 mg/dl), than during the dry season (9.54 mg/dl). For the rest of the localities, values ranged from 7.81 to 9.54 mg/dl. There were no differences among season or localities of Mg, and values ranged from 2.44 to 2.85 mg/dl. Luvianos locality had the highest Na ($P<0.01$) during the dry season (411.76 mg/dl) and Tejupilco region had the lowest value during the dry season (322.0 mg/dl). The K concentration differed ($P<0.01$) during dry season with Tlatlaya having the highest K value (29.50 mg/dl) and Tejupilco the lowest (17.31 mg/dl).

Table1. Some physicochemical characteristics in the soil of a subtropical region in the south-western of Mexico State

pH	OM	N %	CIC cmol/kg	AD	Texture	EC (µHOS)
5.37	2.94	0,612	45.722	1.0817	sandy	48

OM: Organic matter, N: Nitrogen, ICC: Cationic interchange capacity, AD: Apparent density, EC: Electric conductivity. (Ramírez, 2009).

Table 2. Macromineral concentration (mg/dl) in blood serum of non lactating goats.

Season	Regions	P	K	Ca	Mg	Na
Dry	Tejupilco	3.06 [b]	17.31 [d]	9.85 [b]	2,82	322.00 [c]
	Luvianos	3.29 [b]	23.96 [b]	7.8 [d]	2.51	411.76 [a]
	Amatepec	3.19 [b]	23.49 [b]	9.54 [b]	2.85	378.33 [ab]
	Tlatlaya	3.05 [b]	29.50 [a]	8.77 [bcd]	2.75	385.23 [ab]
Rainy	Tejupilco	3.46 [ab]	25.84 [ab]	8.79 [bcd]	2.47	377.90 [ab]
	Luvianos	3.92 [ab]	18.50 [cd]	8.36 [bcd]	2.71	360.57 [bc]
	Amatepec	4.79 [a]	21.94 [bc]	11.20 [a]	2.68	401.00 [ab]
	Tlatlaya	4.33 [ab]	23.95 [b]	9.33 [bc]	2.44	385.38 [ab]
Season		0.0001	0.1281	0.0261	0.1102	0.356
Regions		0.1272	<.0001	<.0001	0.5776	0.0005
Season*Regions		0.1828	<.0001	<.0001	0.169	<.0001
SEM		0.307	0.933	0.269	0.134	10.51

Means in same column with different superscripts differ (P<0.05).
SEM= standard error of the mean.

Cupper (Cu) and Zinc (Zn) concentration in blood serum

Tlatlaya had the highest serum Cu concentration during dry season compared with the rainy season (0.192 and 0.080 mg/dl, respectively). The rest of localities had similar Cu concentration rangeing from 0.082 to 0.111 mg/dl. The Zn concentration had the same trend, but their was a season*locality (P<0.0004), with Luvianos having higher concentration during the dry season (0.117 mg/dl) than during the rainy season (0.056 mg/dl) whereas, for the rest of localities, values ranged from 0.059 to 0.096 mg/dl (Table 3).

DISCUSSION

Phosphorous (P), calcium (Ca), magnesium (Mg) and potassium (K) concentration in blood serum

Circulating blood levels of P in small ruminants have been well documented. Levels of P in adult cattle and sheep was range between 4 to 8 mg/dl (NRC, 2007), and McDowell (1997) reported P concentrations in grazing ruminants of 3 to 8 mg/dl, which are considered inadequate. Dominguez and Huerta (2007) reported similar levels, which ranged from 10.7 to 12.1 mg/dl, in animals in an extensive grazing system, which were considered adequate. In our study, there were lower P levels than reported in the literature and it is suggested that the goats could experience productive and reproductive deficiencies caused by the lack of P which could compromise many hormonal, metabolic and structural functions, as well as reduced appetite, which could compromise efficient utilization of the diet (Ternouth, 1990).

Table 3. Microminerals concentration (mg/dl) in blood serum of non lactating goats.

Season	Regions	Cu	Zn
Dry	Tejupilco	0.090 [bc]	0.096 [c]
	Luvianos	0.111 [bc]	0.117 [bc]
	Amatepec	0.125 [b]	0.159 [a]
	Tlatlaya	0.192 [a]	0.150 [ab]
Rainy	Tejupilco	0.106 [bc]	0.098 [c]
	Luvianos	0.091 [bc]	0.056 [d]
	Amatepec	0.082 [bc]	0.128 [abc]
	Tlatlaya	0.080 [c]	0.162 [a]
Season		<.0001	0.0034
Regions		0.0006	<.0001
Season*Regions		<.0001	0.0004
SEM		0.010	0.307

Means in same column with different superscripts are differ (P<0.05).
SEM= standard error of the mean.

Calcium is a mineral which is involved in many biological functions in the body. While the structural part of the skeleton contains 98 to 99% of Ca within the body (Underwood, 1981), Ca also functions in soft tissues. Several research studies report blood Ca values of 8 to 12 mg/dl (NRC, 2007; Khan et al., 2007), which contrasts with our findings. Recently, Khan et al. (2007) reported Ca levels between 6.36 and 9.71 mg/dl in adult ewes grazing extensively in semiarid regions of Pakistan. These levels are similar to those we found in Luvianos. Levels reported by Dominguez and Huerta (2007), from adult ewes in the Toluca valley, were similar to those at Amatepec during the dry and rainy seasons.

NRC (2007) for small ruminants reported desirable blood Mg values 1.8 to 3.5 mg/dl, which are lower than those of the non-lactating goats in our four localities. However, Dominguez and Huerta (2007) reported Mg levels of ewes at Toluca valley ranging from 2.1 to 2.9 mg/dl, which are similar to our results.

Blood Na concentrations in small ruminants have been widely reported under conditions where animals were fed exclusively on forages, and it is often used as an indicator of animal mineral status (McDowell, 1997). The NRC (2007) for small ruminants reported normal blood Na levels of 322 to 350 mg/dl. Our goats had higher values suggesting that reproductive or productive problems related to deficiencies of Na were unlikely. This was likely because the goats received common salt on a daily basis as a normal practice (Kawas et al., 1997).

Considering the values reported by the NRC (2007) regarding K (15 to 20 mg/dl), goats from our region would not likely be deficient in this mineral. Dominguez and Huerta (2007), reported levels of 26.1 mg/dl in adult ewes, which are similar to those in our study. However K is a mineral not frequently studied as deficiencies are not common in ruminants (Underwood, 1981), and few studies have been completed using blood serum K as an indicator of K status.

Cupper (Cu) and Zinc (Zn) concentration in blood serum

Among the trace minerals, Cu is one which can be impacted by antagonists which is why blood concentrations are not necessarily a good indicator of animal status (Underwood and Suttle, 1999). In grazing cattle, Cu levels from 0.08 to 0.150 mg/dl were reported (McDowell, 1997; NRC, 2007) but, in goats, Galbraith (1997) reported levels of 0.090 to 0.138 mg/dl. Khan et al. (2007), working with goats grazing native pastures in a semi arid region of Pakistan found that blood Cu levels in summer fluctuated from 0.071 to 0.078 mg/dl, whereas in winter values varied from 0.114 to 0.146 mg/dl. These findings, overall, suggest that the blood Cu values in this study range from deficient to marginal. Copper deficiency, after P, was found to be an important mineral limitation for grazing cattle in tropical regions (McDowell, 1997). Moreover, secondary deficiencies of Cu could be due to interference with other minerals provided in excess, such as Fe, S or Mb in the diet. Immature forages had high levels of S that interferes with Cu absorption (McDowell, 1997).

Normal values of Zn reported by NRC (2007) range between 0.08 and 0.120 mg/dl, whereas Ahmed et al. (2001) determined serum levels of 0.120 to 0.256 mg/dl in Nubian goats. In our study, the goats from Tejupilco, Amatepec and Tlatlaya were above this range, while does from Luvianos could be Zn deficient as a result of levels below the normal range, in particular during the rainy season. Indeed, low Zn levels were associated with high Ca, mainly at Luvianos during the rainy season.

CONCLUSION

Levels of blood P and Ca from the four regions of south western Mexico State during the dry and rainy season had low values, indicating a probable deficiency of these elements. Potassium, and in particular Na, were above levels reported in the literature. With respect of Cu and Zn, non-lactating goats could have reproductive problems due to their deficient to marginal levels in blood serum of these minerals.

REFERENCES

Ahmed, M.M.M., Hamed, T.F.M., Barri, M.E.S. 2001. Variation of zinc and copper concentrations in the plasma of Nubian goats according to physiological state. Small Ruminant Research. 39:189-193.

Cotler, H., Fregoso, A., Damián J.L. 2006. Caracterización de los sistemas de producción en la Cuenca Lerma-Chapala a escala regional. Instituto Nacional de Ecología. Dirección de Manejo Integral de Cuencas Hídricas. Dirección General de Investigación de Ordenamiento Ecológico y Conservación de Ecosistemas. p 28.

Devendra, C. 1988. Goat production in small farm systems. In: Workshop on Small Ruminant Research and Development. Noviembre 2-4, Cairo, Egipto.

Devendra, C. 1991. Goat meat production potential in the developing countries. En: VII Reunión Nacional sobre Caprinocultura. Octubre 23-25, Monterrey, Nuevo León.

Domínguez, V.I., Huerta, M.B. 2007. Mineral concentration and interrelationship in soil, forage and blood serum of sheep during two seasons in the Toluca Valley, México. Agrociencia. 42:173-183.

Fick, K.R., McDowell, L.R., Miles, P.H., Wilkinson, N.S.J.D., Funk, J.D., Conrad. J.H. 1979. Methods of Mineral Analysis for Plant and Animal Tissues, second ed. Department of Animal Science, University of Florida, Gainesville. pp. 1-80.

Galbraith, H., Chigwada, W., Scaife, J.R., Humprhries, W.R. 1997. The effect of dietary molybdenum supplementation on tissue copper concentrations, mohair fiber and carcass characteristics of growing Angora goats. Animal Feed Science and Technology. 67:83-90.

Gallegos, S.J., Germán, C.G.A., Camacho, R.J.C. 2005. Manual del participante. La cabra. Fondo de tierras e instalación del joven emprendedor rural. Colegio de Postgraduados. Secretaría de la Reforma Agraria. México, D. F. p 59.

García, M.E. 1986. Apuntes de climatología, 5ª Edición. Enriqueta García de Miranda, México, D. F. p. 155.

Hernández, Z.J.S. 2000. La caprinocultura en el marco de la ganadería poblana (México): contribución de la especie caprina y sistemas de producción. Archivos de Zootecnia. 49:341-352.

Hernández, Z.J.S., Herrera, M.G., Rodero E.S., Vargas S.L., Villareal, O.E., Reséndiz, R.M., Carreón, L.L., Sierra, A.C.V. 2005. Tendencia en el crecimiento de cabritos criollos en sistemas extensivos. Archivos de Zootecnia. 54: 429-436.

Kawas, J., Armienta, G., Olivares E. 1997. Seasonal changes of mineral concentrations of tropicals grasses in Mexico. In: XVIII International Grassland Congress. Winnipeg. Manitoba & Saskatoon, Saakatchewan. Canada. p. 123.

Khan, Z.I., Hussain, A., Ashraf, M., Ashraf, M.Y., McDowell, L.R. Huchzermeyer, B. 2007. Copper nutrition of goats grazing native and improved pasture with the seasonal variation in semiarid region of Pakistan. Small Ruminant Research. 67:138-148.

Khan, Z.I., Hussain, A., Ashraf, M., Ashraf, M.Y., McDowell, L.R. 2007. Macromineral status of grazing sheep in semiarid region of Pakistan. Small Ruminant Research. 68:279-284.

McDowell, L.R., Velásquez, P.J., Valle, G. 1997. Minerales para rumiantes en pastoreo en regiones tropicales. Universidad de Florida Gainesville. USA. p 81.

Medrano, J. A. 2000. Recursos animales locales del centro de México. Archivos de Zootecnia. 49:385-390.

Meschy, F. 2000. Recent progress in the assessment of mineral requirements of goats. Livestock Production Science. 64:9-14.

Morales, A. E, Domínguez, V. I., Gonzales, R. M., Jaramillo, E. G., Castelán, O. O., Pescador, S. N., Huerta, B. M. 2007. Mineral diagnosis in forage and blood serum of dairy cattle in two seasons in the Toluca Valley, México. Técnica Pecuaria México. 45 (3), 329-344.

NRC 2007. Nutrient Requirements of Small Ruminants. NRC- NAP- Washington. pp. 347.

Ramírez, O. A. 2009. Diagnóstico del estado mineral en suelo, planta, animal de la caprinocultura en Amatepec, México. Tesis de licenciatura. Centro Universitario UAEM, Temascaltepec, México. 68 p.

Rebollar, R.S., Hernández, M.J., García, J.A., García, M.R., Torres, G., Bórquez, J.L., Mejía, P. 2007. Canales y márgenes de comercialización de caprinos en Tejupilco y Amatepec, estado de México. Agrociencia. 41:363-370.

SAS Institute. 2006. SAS User´s Guide: Statistics. Ver 9.0. SAS Institute. Cary, N.C. USA. pp. 956.

Steel, R. G. D. y Torrie J. H. 1980. Bioestadística: Principios y procedimientos. McGraw-Hill. México. pp. 181-184.

Ternouth, J.H., Sevilla, C.C. 1990. Australian Journal Agricultural Research, 41, 175. Animal Science 34 Supplement.1:130-133.

Underwood, J. 1981. Microminerales. Los minerales en la nutrición del ganado. 2ᵈᵃ Ed. Editorial Acribia, Zaragoza, España. pp.209.

Underwood, J. Suttle, F. 1999. The detection and correction of mineral imbalances. The Mineral Nutrition of Livestock. 3Th Ed. CABI Publishing. Wallingford, UK. pp. 600.

6

DETERMINANTS OF THE ADOPTION OF SMALL RUMINANT RELATED TECHNOLOGIES IN THE HIGHLANDS OF ETHIOPIA

Getahun Legesse[1,2*], Marianna Siegmund-Schultze[1], Girma Abebe[2] and Anne Valle Zárate[1]

[1]Hohenheim University, Department of Animal Production in the Tropics and Subtropics, D-70593 Stuttgart, Germany
[2]Hawassa University, Department of Animal and Range Sciences, Box – 05, Awassa, Ethiopia
[*]Current address: University of Manitoba, Faculty of Agricultural and Food Sciences, Department of Animal Science, 12 Dafoe Road, Winnipeg, MB, Canada - R3T 2N2; E-mail: Getahun_Legesse@umanitoba.ca, getahunleg@yahoo.com.
**Corresponding author*

SUMMARY

This paper takes up the case of two market-sheds in the southern Ethiopian highlands (namely Adilo and Kofele) to examine the factors affecting the adoption of small ruminant related technologies in mixed-farming systems. A survey was conducted using semi-structured questionnaires with 155 randomly selected small ruminant keepers between May and June 2006. Farmers in each site initiated new practices like small ruminant fattening and managing a household 'veterinary kit'. Logistic regression analysis revealed that size of land and livestock holdings significantly affected the adoption of small ruminant technologies in both study sites. Farmer variables such as gender, literacy, age and family size appeared to influence adoption only in one location. In the densely populated area, Adilo, the adoption of more intensive feeding technology of commercial concentrates decreased with increasing farm size only up to a point. Younger farmers, female farmers and literate household heads were more likely to adopt the utilization of commercial concentrates. In relatively resource rich Kofele, treating small ruminants via the household veterinary kit increased with number of livestock, however with farm size only up to the point at which it reached a maximum. The present study showed that location or production system remarkably affects the options of interventions and determines their adoption.

Key words: small ruminants; technologies; adoption; Ethiopia; mixed-farming, logistic regression.

INTRODUCTION

In Ethiopian mixed crop livestock systems, sheep and goats are normally kept in small flocks. As a result of the increasing human population in most highland areas, the practice of tethering and housing these small ruminants is on the rise. Consequently, the feeding and fodder production issue is becoming more and more imperative (Peacock 2005). In addition to feed shortage, the losses caused by diseases are also substantial. There is an extensive evidence of helminth parasites becoming a major constraint to productivity of small ruminants in the country that resulted in the loss of millions of dollars (Biffa *et al.* 2006; Abebe *et al.* 2010). These problems require appropriate interventions in order to increase the contribution of the small ruminant sector to the country in general and to the resource-poor farmers in particular.

Without due consideration of socioeconomic and biophysical factors that may affect the adoption of small ruminant technologies, it will be almost impossible to design proper development programs and enhance their contribution to the rural households' economy. Farmers do not normally adopt new practices for two seemingly obvious reasons. They are either unwilling or unable (Nowak 1992). Evaluation of factors influencing the adoption of possible technologies could play a decisive role in developing feasible and sustainable programs. According to Nederlof and Dangbégnon (2007), technologies that are forwarded to resource-poor farmers should meet a variety of needs of these people and be acceptable from a socio-cultural perspective besides technical and economic considerations.

One of the major aims of participatory research is to identify possible intervention areas. Observing the pattern of the farmers' attitude towards adopting certain agricultural technologies would give a basis for the selection of viable interventions. Batz *et al.* (2003) noted that understanding the factors that have determined adoption in the past offers relevant information about the characteristics which will facilitate the quicker and wider adoption of forthcoming technologies. This study attempted to examine factors affecting the adoption of small ruminant related technologies in two mixed-farming systems of southern Ethiopia primarily based on the cases of commercial concentrates and household veterinary kits.

MATERIALS AND METHODS

This study was part of a broader study on the productive and economic performance of small ruminants in mixed farming systems of southern Ethiopia. The study was carried out in two small ruminant market-sheds, namely Adilo and Kofele. Adilo represents the most densely populated areas in the country (more than 500 people km^{-2}) while Kofele represents Ethiopian highlands populated at medium level (about 200 people km^{-2}).

The annual rainfall distribution at both study sites is generally bimodal. During the four years from 2000 to 2003, annual rainfall ranged between 1055 and 1194 mm at the two meteorological sites around Adilo (Shone and Durame) (NMSA 2004). Adilo is an area characterized by land scarcity and food deficiency. The altitude of most of the villages around Adilo is between 1600 and 2000 m above sea level. Rainfall in the mentioned period was about 1200 mm in Kofele (NMSA 2004). Kofele is a typical highland area with a relatively cool climate, food sufficiency and fairly high agricultural potential in terms of land availability and soil fertility. The altitude of the district ranges from 2000 to 3050 m above sea level, and the mean monthly minimum and maximum temperatures for the years 2000 to 2003 were 7.5 and 19.6°C, respectively (NMSA 2004).

Both market-sheds are among the major suppliers of small ruminants, mainly sheep, to the neighboring big cities (*e.g.* Shashemene, Hawassa) and Addis Ababa, particularly during holidays (Legesse *et al.* 2008; Legesse *et al.* 2010). The characteristics of transactions related to small ruminant production, the factors affecting the sale price of small ruminants and the financial profitability of the traditional small ruminant enterprises in the two sites were recently described (Legesse *et al.* 2010). Based on genetic analysis, Gizaw *et al.* (2007) reported that sheep in Kofele and Adilo are from the same fat-tailed Arsi-Bale breed. Nearly all goats in the study area are also from local breeds. Besides small ruminants, farmers also keep cattle, chicken and equines and grow crops like enset (*Ensete ventricosum*), maize (*Zea mays*) and potato (*Solanum tuberosum*). The majority of the people living in Kofele and Adilo are Muslims and evangelical Christians, respectively. Ethnically, nearly all respondents in Kofele were Oromos while the majority of the Adilo respondents belonged to the Kembata and Hadiya ethnic groups.

A single-visit survey on the adoption of small ruminant related technologies was conducted using semi-structured questionnaires in each study site with all randomly selected small ruminant keepers participating in a flock and household monitoring which involved a total of 155 households (*i.e.* 90 in Adilo and 65 in Kofele) (Legesse 2008). Criteria for sampling households for the monitoring were having at least three small ruminants (sheep and/or goats), one or more cattle and willingness to participate. The survey was administered between May and June

2006. Since previous studies indicated that the two market-sheds have differences in the ranking of small ruminant problems (Legesse *et al.* 2008), the questionnaire included site-specific questions.

Prior to the adoption survey, two farmer workshops were organized in March 2006 in the study sites with the small ruminant keepers participating in a flock and household monitoring. Eighty-two farmers from Adilo (out of the total 90) and fifty-five farmers from Kofele (out of the total 65) who were willing and able to attend participated in the workshops. Other than the participating farmers, the team included three researchers, an assistant and two enumerators in each respective site. A checklist was prepared before the workshops. Core observations from a previously conducted diagnostic survey, group discussions and the then on-going flock monitoring as well as potential improvement measures were presented to the farmers. The topics were discussed under the guidance of the moderator (one of the researchers), in order to get the views and perspectives of the small ruminant keepers on a certain issue. Another member of the research team made notes in a discrete way. The enumerators served as translators. Each workshop lasted for about three hours.

The number one problem in Adilo small ruminant market-shed was found to be feed shortage while small ruminant diseases ranked first in Kofele (Legesse *et al.* 2008). In spite of this, fattening small ruminants (particularly sheep) targeting holiday markets has been a common practice in Adilo. In addition to home produced feedstuffs, a large number of small ruminant keepers are using commercial concentrates (mainly wheat bran and oilcake) to fatten their animals. Hence, the utilization of commercial concentrates in the household was taken as a dependent variable to be tested for its adoption in the area. Despite the provision of basic veterinary services principally by government-owned livestock clinics in both sites for small ruminants, Kofele farmers commonly administer injections and/or other medicaments by themselves. Therefore, such practice was taken as a dependent variable to be tested for its adoption in the area. Logistic regression was employed to determine predictors for the adoption of the selected practices. The explanatory variables which were used in the model and their definitions are given in Table 1. The choice of explanatory variables has been made based on findings of past studies (Feder *et al.* 1985; Gebremedhin *et al.* 2003; Feleke and Zegeye 2006). The explanatory variables which are assumed to affect the dependent variables will be briefly described as follows.

Gender of household head: male headed and female headed households could make different decisions whether or not to adopt different types of agricultural technologies. However, male-headed households are assumed to have higher access to information sources than female-headed households and may be generally more likely to adopt a technology.

Literacy: could increase the farmer's ability to acquire, analyze and use information relevant to the adoption of agricultural technology.

Radio: radio ownership is a proxy for access to information, which is expected to have a positive influence on adoption of technologies.

Social participation: this is the involvement of the household head in local administration, community leadership and/or his/her link with non-governmental organizations (NGO). Farmers with extra social activities are hypothesized to have more information networks and are likely to adopt agricultural technologies more easily than those with no social network.

Farmer's age: this variable could have a positive or negative effect on a farmer's decision to adopt agricultural technology. Older age is usually associated with more farming experience and could positively influence the adoption of some technologies. Contrary to this, younger farmers are more likely to take risk and adopt agricultural technologies compared with their older counterparts. Therefore, the effect of age could be positive or negative.

Family size: the effect of family size could also be positive or negative. It influences adoption positively through supply of labor especially during peak seasons of labor demand or negatively through competition for resources, which may be more severe with very high family size.

Farm size: defined as the total farmland owned and/or rented by the household is an indicator of resource endowment; it is expected to be associated with the decision to adopt agricultural technologies positively.

Livestock ownership: Livestock ownership is considered as a proxy for wealth status of a household. Livestock are assets that could be used in the production process or exchanged for cash. Thus, it is expected to positively influence the adoption of agricultural technologies.

Table 1. The definitions of categorical and continuous variables considered in the study.

Variable	Definition
Categorical (binary) variables	
GENDER	The gender of the head of the household (1 if the head is male)
LITERACY	Literacy dummy (1 if the farmer can at least read and write)
RADIO	The presence of a radio in the household (1 if there is any functional radio)
SOCIAL_PAR	Social participation (1 if the household head participates in any social activity)
Continuous variables	
AGE	The age of the head of the household (in years)
AGE_SQ	Square of the age of the head of the household
FSIZE	The size of the family (in caput)
FSIZE_SQ	Square of the size of the family
LAND	The size of the farm (in ha)
LAND_SQ	Square of the size of the farm
TLU	Tropical Livestock Unit owned by the household (in units[a])
TLU_SQ	Square of TLU owned by the household

[a]Tropical Livestock Unit (0.7 TLU=1 head of cattle; 0.5 TLU=1 head of horse, donkey or mule; 0.1 TLU=1 head of sheep or goat).

According to the logistic model, the probability of a small ruminant keeper adopting a specified technology in the respective area is given by:

$$\eta_i = \log[P_i/(1-P_i)] = \beta_o + \beta_1 X_{1i} + \beta_2 X_{2i} + \ldots + \beta_n X_{ni}$$

Where:

η is a function of the mean of the dependent variable known as a link function;

i denotes the ith observation;

P_i is the probability of the ith farmer adopting a specified technology;

X_{1i}, X_{2i}, …, X_{ni} are the identified variables contributing to the decision of adopting the technology;

β_0 is the intercept;

β_1, β_2, …, β_n are regression parameters (the coefficients associated with each explanatory variable).

The analysis was performed using PROC GENMOD procedure of the Statistical Analysis System version 9.1.3. (SAS 2004). The squared terms of the continuous variables were introduced in the model to capture whether the pattern of the influence of the variables changed with increasing values (Gujarati 2003); the non-significant squared parameters were excluded from the final model.

RESULTS AND DISCUSSION

Farmer feedback workshops

Farmers in both sites confirmed the general observations from the survey and the results from the monitoring phase. Adilo small ruminant keepers highlighted that the major bottleneck of small ruminant production in the study area was feed shortage. During the workshop, farmers stated that there was a need in the area for cheaper feedstuffs (mainly wheat bran and oilcake). Farmers further indicated that fodder trees like *Sesbania sesban* could be important feed sources for small ruminants especially during the dry season in which they mostly face feed scarcity. A woman member of the participants underlined the implication of improved feeding by saying: "a well-managed ewe can give twins and triplets in short intervals". The farmers exhibited that they did not only know the factors constraining their small ruminant production but also some alternative strategies. Targeting on such strategies may possibly increase the adoptability of technologies.

Sheep fattening is not a common practice around Kofele. On the reasons why they do not fatten sheep, Kofele farmers explained that sheep keeping was generally a secondary activity since they had reasonably large farm size and alternative farm activities. But, with increasing pressure on grazing land and diminishing cultivated land, farmers started to give more attention to sheep. Some farmers already started fattening sheep in recent times. Others started integrating sheep with maize, one of the major staple food crops in the area, by means of systematic haltering techniques using ropes (but without tying the animal to a particular spot). Such animals could graze in the maize field without damaging the crop. According to Rogers (2003), relative advantage, which can be defined as the degree to which a technology is perceived as better than the idea it replaces, is one of the factors that would determine the adoption likelihood. The advantages should also be worth the additional costs. The attitude of the Kofele farmers towards sheep fattening through the utilization of commercial concentrates clearly confirmed that notion.

In a previous study, parasites and diseases were ranked as the first critical constraint for small ruminant production in Kofele (Legesse *et al.* 2008). Farmers confirmed the result. According to the workshop participants, the major setback in this regard was the pervasiveness of liver fluke (*Fasciola hepatica*) as a result of marshy grazing areas. They use anthelmintic drugs but the effectiveness of the drugs was reported to be limited. That could be due to the use of expired and forged drugs bought from illegal sources in the absence of alternatives or development of resistance as a result of using some medicaments inappropriately and repeatedly. The farmers indicated that veterinary technicians were assigned for each locality by the time of the workshop. But these technicians lacked necessary equipment and medicaments. And they indicated that it might be good to offer training on basic animal health care for the farmers themselves. Avoiding grazing their animals in the marshy areas was not considered as an alternative since those areas were the only grazing land available for most; other forms of land were generally reserved for crop production.

The control of the intermediate snail population is indicated as a good opportunity for the reduction of transmission of fasciola in the fasciolosis endemic areas (Asrat 2004). A study conducted by Eguale and Tilahun (2002) further noted that Endod (*Phytolacca dodecandra*, the African soapberry plant) is a potential plant for the control of fasciola transmitting snails, particularly *Lymnaea truncatula* and *Lymnaea natalensis*. Endod might, therefore, provide the less costly means of snail control though its production on commercial scale has not been achieved yet (Asrat 2004). With this background knowledge, the farmers in Kofele were asked whether treating the marshy grazing area for example with Endod could be a potential solution. The farmers responded that the marshy area in the district is too vast for such an approach. Sheep strains resistant to flukes were also not known by the farmers. So the only partial solution the farmers reported to apply was chemical deworming (*i.e.* providing anthelmintic drugs to the animals).

Determinants of the adoption of feed technologies in Adilo area

Table 2 presents the means and standard deviations of the variables used in the logistic regression. For a binary indicator variable, the mean represents the percentages of farmers of each group with the attribute.

Table 2. Descriptive statistics of the independent variables identified as affecting the adoption of technologies in the two study sites.

Variable[a]	**Adilo** (n=90)		**Kofele** (n=65)	
	Mean (Range)[b]	std[c]	Mean (Range)	std[c]
GENDER	0.83	0.37	0.82	0.39
LITERACY	0.63	0.48	0.17	0.38
RADIO	0.47	0.50	0.50	0.50
SOCIAL_PAR	0.61	0.49	0.59	0.50
AGE	47.7 (28-77)	12.0	50.5 (25-75)	15.4
FSIZE	8.4 (3-19)	2.8	8.4 (2-20)	3.7
LAND	0.84 (0.125-2.9)	0.45	3.0 (0.15-8.5)	1.7
TLU	3.3 (0.4-7.2)	1.5	9.5 (2.5-28.7)	5.2

[a]The definitions of variables (and the units) are provided in Table 1.
[b]For continuous variables the range of the values is given.
[c]std=standard deviation.

For example, the LITERACY variable in Adilo indicates that 63% of the respondents could at least read and write while this percentage was only 17% for the Kofele farmers. For a continuous variable, *e.g.* AGE, the variable represents the mean age of the respondents (*i.e.* 48 in Adilo and 50 in Kofele). The average land holding of Kofele respondents was more than threefold of that of Adilo farmers. The livestock units calculated for Kofele small ruminant keepers were also three times as big as the average units in Adilo.

In Adilo small ruminant market-shed, the feed shortage being the most important problem, the utilization of commercial concentrates reported by 41% of the farmers, was taken to trace the potential factors that could influence the adoption likelihood of technologies or practices. Both linear and quadratic coefficients were significant for farmer's age in Adilo (*i.e.* the linear term negative and quadratic term positive, Table 3), implying that the utilization of commercial concentrate decreased with increasing age only up to a point. This minimum utilization was reached at about 55 years of age. Observations and the farmer feedback seminar also showed that the younger farmers were the ones commonly practicing the fattening, which mostly involved commercial concentrates.

Similarly, both the linear and quadratic coefficients in Adilo were significant for farm size, implying the probability of the adoption of commercial concentrates decreased with increasing farm size only up to the point at which it reaches a minimum. This minimum use of commercial concentrates was reached at 1.75 ha. Farmers with small farm size did not seem to have much alternative other than tethering their animals and feeding commercial feedstuffs. On the other hand, both the linear and quadratic coefficients were significant for livestock holding in terms of TLU (Tropical Livestock Unit; see Table 1 for the definition), implying that adoption increased with livestock holding up to reaching a maximum and then declining as TLU increases further. This maximum was reached at a TLU of about 4. Owning a higher number of livestock (especially cattle) may enable farmers to afford buying commercial concentrates. However, those who own higher numbers of livestock may have other alternatives like fattening cattle, which was also a common practice in the area. Nearly 90% of the farmers in Adilo reported that they had exercised cattle fattening in the past, slightly lower than those reported sheep fattening (92%).

Female farmers were more likely to adopt the utilization of commercial concentrates. Fattening small ruminants, especially the one involving commercial concentrates, requires follow up; women farmers who spend most of their time in their vicinity seemed to increase their benefit by the increased utilization of concentrates rather than taking the animals out for search of better grazing land. Household heads who could read and write were more likely to adopt commercial concentrates, which may be an indication of their access to information or ability to record gain or loss that might in turn affect their decision to utilize certain inputs. Households with bigger families were more inclined to utilize commercial concentrates, probably attempting to improve the income status of the household to satisfy the needs of their bigger families through the practice. The small ruminant fattening that involves commercial concentrates is not exclusively related to wealth in the area although it showed an association with livestock holding. Further investigation of the fattening practice in various wealth categories may help to better understand the situation.

Table 3. Logistic estimates of the adoption of commercial concentrate utilization among small ruminant keepers in Adilo.

Variable[a]	Estimate	Standard error	Significance
INTERCEPT	14.8668	5.1775	0.0041
GENDER	-2.0719	0.9110	0.0162
LITERACY	1.6833	0.7647	0.0194
RADIO	0.3803	0.5474	0.4858
SOCIAL_PAR	0.8795	0.6600	0.1735
AGE	-0.6666	0.2044	0.0002
AGE_SQ	0.0063	0.0020	0.0003
FSIZE	0.2165	0.1042	0.0344
LAND	-6.2657	2.0220	0.0025
LAND_SQ	1.8899	0.7701	0.0138
TLU	1.6452	0.9233	0.0610
TLU_SQ	-0.1920	0.1165	0.0839

[a]The definitions of variables are provided in Table 1.

Two-third of the respondents started to use commercial concentrates for sheep fattening after the year 2000. The earliest adopter started back in 1983. All farmers noted the ever-increasing price of commercial concentrates as a crucial problem associated to its utilization for small ruminant production or traditional fattening. None of them mentioned the availability as a problem indicating its steady availability in the local market. All small ruminant keepers except one believed that forming a cooperative in order to bring commercial concentrates from the primary producers to the locality could be a solution to tackle the affordability of feedstuffs.

The respondents in Adilo were also asked about the utilization of fodder trees for small ruminant feeding as an alternative feed option. Almost all (97%) farmers in Adilo stated that they use native fodder trees (*e.g. Erythrina spp.*) as a feed for small ruminants. More than 41% of them reported using exotic fodder trees (*e.g. Sesbania sesban*). Farmers were also asked about their criteria to choose and grow a certain fodder tree species. More than 92% of the farmers indicated that they consider the yield and palatability of the species. The resistance or tolerance of the fodder tree to diseases and/or moisture stress was cited by more than 85% of the farmers while about 81% of them reported the consideration of managerial and input requirement for the specific fodder tree. Growing backyard horticultural crops being the commonest practice in Adilo, two-third of the farmers said that the matter of shading would be among their criteria of choice. Only 42% of the farmers in Adilo indicated that they would consider the recommendation of extension agents in their decision process. Less than 7% of the respondents reported they did not have any particular criteria. Mekoya (2008) noted that the successful introduction of fodder trees requires the consideration of farmers' multiple criteria in addition to the awareness of growing and feeding fodder trees and resource availability.

One of the factors that affected the utilization of commercial concentrates among farmers in Adilo was its rising price (Legesse *et al.* 2008). The affordability of a new practice or technology should, therefore, be the first factor to be considered. Adugna (2003) reported that the crossbred dairy goats that were originally promoted to improve the welfare of poor farm households found their way into the hands of better-off farmers that could assume the associated risks. Improving the availability of cash in the household and designing a mechanism to have the concentrates at cheaper price may be a prerequisite for their broader utilization in the area.

Gender has been reported to affect the adoption of certain interventions. In Marsabit Mountain of northern Kenya, more women than men ranked labor-saving as an important attribute of a new technology (Ngutu and Recke 2006). Some of the decisions women make to adopt technologies may be also related to their access to resources, benefits and decision making power. Traditionally, Ethiopian families are headed by men if both partners are alive. The adoption of commercial concentrates was higher by women heads in the current study, most of whom were widows. This indicates the potential of rural women to be involved in promising activities if they are given the decision making power.

The issue of poor quantity and quality feedstuffs particularly during the dry season is not only the primary concern in Adilo but also in most mixed-farming systems of the country. Studies indicated the presence of endemic fodder trees that has reasonably high forage potential and can effectively serve as a cheap source of protein supplement (Larbi *et al.* 1993; Mekoya *et al.* 2008). Daily weight gains of sheep and goats almost doubled with *Erythrina abyssinica* leaf supplementation (Larbi *et al.* 1993). But, in addition to quantitative information on nutritive value and animal performance from indigenous and exotic fodder trees, the preference criteria of farmers should be considered. Though there have been several research and development efforts on exotic multi-purpose trees in Ethiopian highlands in the last three decades, their adoption is reported to be limited (Mengistu 1997; Mekoya *et al.* 2008).

Determinants of the adoption of household veterinary kits in Kofele area

The small ruminant keepers in Kofele previously reported that diseases are the major threat for production in their area (Legesse *et al.* 2008), causing substantial morbidity and mortality of their small ruminants. To tackle the problem, farmers are commonly buying anthelmintic drugs and other medical kits (*e.g.* syringes, needles) and administer them to their animals. Farmers were asked whether they had been treating their animals by themselves or not. The contribution of the farm and farmer variables to adoption of this practice was estimated (Table 4).

Table 4. Logistic estimates of the adoption of treating small ruminants through own application of veterinary kits among small ruminant keepers in Kofele.

Variable[a]	Estimate	Standard error	Significance
Intercept	-3.4398	1.8659	0.0653
GENDER	1.4176	1.0514	0.1661
LITERACY	0.8379	1.1088	0.4428
RADIO	-2.4659	1.2326	0.0257
SOCIAL_PAR	0.5732	1.1420	0.6138
AGE	0.0390	0.0295	0.1760
FSIZE	-0.1706	0.1312	0.1869
LAND	1.4704	0.7180	0.0288
LAND_SQ	-0.2608	0.0973	0.0032
TLU	0.2844	0.1361	0.0192

[a]The definitions of variables are provided in Table 1.

Both, the linear and quadratic coefficients were significant for farm size (*i.e.* the linear term positive and quadratic term negative), implying treating small ruminants in the household increased with farm size only up to the point at which it reaches a maximum. This maximum of treating small ruminants was reached at about 4 ha. Since most of the reported diseases are somehow related to the grazing of animals in bottomlands (*e.g.* marshy areas), those farmers with large land holding might have a better chance to graze their animals in safer areas. Those farmers who own higher number of livestock were significantly inclined to undertake the mentioned practice in the household. The adoption of the practice was negatively related to the presence of a functional radio in the household. A possible explanation of this result is that farmers who owned a radio might have got better information through the radio programs about the risk of utilizing unauthorized or expired drugs, since most of the drugs farmers buy from open markets are with unknown expiry dates and dosages. Radio owners might show preference to go to veterinary clinics. The practice of providing treatment to small ruminants was not significantly related to gender, literacy, age and family size.

About 81% of the Kofele respondents indicated that they made use of veterinary clinic services for their livestock; and 62% of the farmers reported they specifically used the service for their small ruminants. Those who did not make use of veterinary clinic services were asked for their reasons. Half of them mentioned that veterinary clinic services and drugs were too expensive; problems related to the accessibility of the clinics were stated by nearly half of (47%) the farmers. A slightly higher proportion of external parasite control was reported to be provided in the household than by the clinic service while slightly higher number of farmers indicated to have internal parasite control in the clinic (Figure 1). Farmers usually go to the veterinary clinic for general medication (*i.e.* non-parasitic diseases). Eighty-nine percent of the respondents indicated they use ethno-veterinary medicines (*e.g.* using natural plants like tobacco) for their sheep.

About three-fourth of the small ruminant keepers reported providing drugs and injections for their small ruminants without any consultation with veterinarians. When they were asked to indicate the reasons for doing so, 59% of them said that veterinary clinic services were unaffordable. Twenty-nine percent of these households also perceived that they knew the diseases and the treatment it needed very well. The long distance of the clinic from their village was also mentioned by nearly 10% of them as a reason for avoiding those services. More than ninety percent of all respondents however reported they did not receive any formal training about those treatments and general livestock disease control.

The farmers were asked how many times they were visited by a veterinarian or a development agent in the past year. Three fourth of the respondents reported they were not at all visited by these professionals while one fourth of them said they were visited once per year. All farmers except one claimed that the agricultural development agent in their community did not have reliable knowledge on small ruminant disease control. No farmer ever knew of any intervention from NGOs in the livestock health sector in the area. Nearly two-third of the farmers thought that training selected farmers in the community on various aspects of livestock health as community health workers could be one solution for disease related problems.

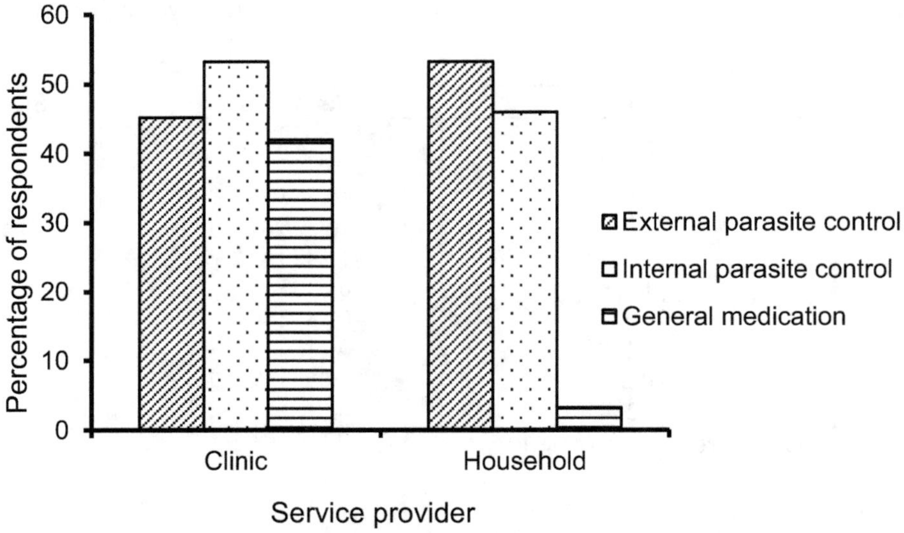

Figure 1. Types of reported medication for small ruminants provided in the clinic or in the household of farmers in Kofele.

According to Curran and MacLehose (2002), community animal health services have a considerable potential for enhancing human health and wealth in addition to their positive impact on livestock health and productivity. In the pastoral areas of the Afar region of Ethiopia, community health workers were reported to achieve better results in the Pan African Rinderpest Campaign than the government teams in terms of vaccination coverage rate and effectiveness (Peeling and Holden 2004). After assessing the impact of a community-based animal health worker project in pastoralist districts of Ethiopia, Admassu *et al.* (2005) reported a significant reduction in the impact of diseases. In small ruminants, there was a significant reduction in the impact of mange, helminthosis, contagious caprine-pleuropneumonia, contagious ecthyma and non-specific diarrhoea. Through the development of strong institutions to support and regulate community initiatives, they can be even more fully exploited (Peeling and Holden 2004). The geographic, societal and cultural set-up of the smallholder farmers in the highlands of Ethiopia is certainly different from those in pastoral systems. The case in Kofele, however, indicates the potential of community-based animal health services. Strengthening the farmers' initiatives through appropriate training of farmers may improve the health and productivity of their livestock. Community-based animal health workers can specially offer preventive and curative services for relatively easier problems such as internal and external parasitism (Catley *et al.* 2004). Previous studies indicated that community-based animal health workers were considered to be highly accessible and more affordable, available and reliable than other service providers (Admassu *et al.* 2005). According to Bamikole and Ikhatua (2009), community animal health workers at village level may also facilitate free flow of information about beneficial small ruminant management practices among farmers.

CONCLUSIONS

The present study has shown that location or production system remarkably affects the options of interventions and the adoption of a specific technology. The study sites are apart a hundred kilometers and they both practice rainfed crop-livestock farming and are also known for their small ruminant markets. Nevertheless, the existing practices that may be an entry point for further intervention and strengthening are apparently different. The relative importance of the factors determining small ruminant keepers' decisions to adopt a specific technology was different in the two sites. In addition to biophysical factors, differences in ethnicity, religion and socioeconomic conditions might have a direct or indirect influence on the decision-making process in the adoption of a new practice. Therefore, making blanket recommendations of technologies for small ruminant keepers in Ethiopian highlands will make the recommended practices unsuitable for local or household conditions and ultimately results in the non-adoption of the technologies forwarded. When opting for more tailor-made solutions, the results of the logistic regressions can be used to identify and address those farmers that carry more of the system- or topic-specific characteristics of adopters.

Acknowledgements

The willingness of the farmers in the study area to participate in the investigation and the support received from DAAD (German Academic Exchange Service) and NORAD (Norwegian Agency for Development Cooperation) project is gratefully acknowledged.

REFERENCES

Abebe, R., Gebreyohannes, M., Mekuria, S., Abunna, F., Regassa, A. 2010. Gastrointestinal nematode infections in small ruminants under the traditional husbandry system during the dry season in southern Ethiopia. Tropical Animal Health and Production. 42: 1111–1117.

Admassu, B., Nega, S., Haile, T., Abera, B., Hussein, A. Catley, A. 2005. Impact assessment of a community-based animal health project in Dollo Ado and Dollo Bay districts, southern Ethiopia. Tropical Animal Health and Production. 37: 33-48.

Adugna, T. 2003. Factors influencing adoption of crossbred dairy goats: the case of three districts in eastern Ethiopia. Proceedings of the 11th annual conference of the Ethiopian Animal Production Society. Addis Ababa, Ethiopia.

Asrat, M. 2004. Infection prevalence of ovine fasciolosis in irrigation schemes along the Upper Awash River Basin and effects of strategic anthelmintic treatment in selected upstream areas. MSc dissertation, Addis Ababa University, Addis Ababa, Ethiopia.

Bamikole, M.A., Ikhatua, U.J. 2009. Compilation and adoption of ethno-veterinary medicine, traditional and other management practices by small ruminant farmers in Edo State Nigeria. Tropical Animal Health and Production. 41: 1549–1561.

Batz, F.J., Janssen, W., Peters, K.J. 2003. Predicting technology adoption to improve research priority-setting. Agricultural Economics. 28: 151-164.

Biffa, D., Jobre, Y., Chakka, H. 2006. Ovine helminthosis. A major health constraint to productivity of sheep in Ethiopia. Animal Health Research Reviews. 7(1): 1-12.

Catley, A., Leyland, T., Mariner, J.C., Akabwai, D.M.O., Admassu, B., Asfaw, W., Bekele, G., Hassan, H.S. 2004. Para-veterinary professionals and the development of quality, self-sustaining community-based services. Revue Scientifique et Technique. 23(1): 225-252.

Curran, M.M., MacLehose, H.G. 2002. Community animal health services for improving household wealth and health status of low-income farmers. Tropical Animal Health and Production. 34(6): 449-470.

Eguale, T., Tilahun, G. 2002. Molluscicidal effects of endod (Phytolacca dodecandra) on fasciola transmitting snails. SINET: Ethiopian Journal of Science. 25(2): 275-284.

Feder, G, Just, R.E., Zilberman, D. 1985. Adoption of agricultural innovation in developing countries: A survey. Economic Development and Cultural Change. 33(2): 255-98.

Feleke, S., Zegeye, T. 2006. Adoption of improved maize varieties in Southern Ethiopia: Factors and strategy options. Food Policy. 31: 442-457.

Gebremedhin, B., Ahmed, M.M., Ehui, S.K. 2003. Determinants of adoption of improved forage technologies in crop–livestock mixed systems: evidence from the highlands of Ethiopia. Tropical Grasslands. 37: 262-273.

Gizaw, S, Van Arendonk, J.A.M., Komen, H., Windig, J.J., Hanotte, O. 2007. Population structure, genetic variation and morphological diversity in indigenous sheep of Ethiopia. Animal Genetics. 38: 621-628.

Gujarati, D.M. 2003. Basic Econometrics. Tata McGraw-Hill, New Delhi, India.

Larbi, A., Thomas, D., Hanson, J. 1993. Forage potential of Erythrina abyssinica: intake, digestibility and growth rates for stall-fed sheep and goats in southern Ethiopia. Agroforestry Systems. 21: 263-270.

Legesse, G. 2008. Productive and economic performance of small ruminants in two production systems of the highlands of Ethiopia. PhD dissertation, Hohenheim University, Stuttgart, Germany. Cuvillier. Goettingen, Germany.

Legesse, G., Abebe, G., Siegmund-Schultze. M., Valle Zárate, A. 2008. Small ruminant production in two mixed-farming systems of southern Ethiopia: Status and prospects for improvement. Experimental Agriculture. 44(3): 399-412.

Legesse, G., Siegmund-Schultze, M., Abebe, G., Valle Zárate, A. 2010. Economic performance of small ruminants in mixed-farming systems of Southern Ethiopia. Tropical Animal Health and Production. 42: 1531–1539.

Mekoya, A., Oosting, S.J., Fernandez-Rivera, S., Van der Zijpp, A.J. 2008. Multipurpose fodder trees in the Ethiopian highlands: Farmers' preference and relationship of indigenous knowledge of feed value with laboratory indicators. Agricultural Systems. 96: 184-194.

Mekoya, A.K. 2008. Multipurpose fodder trees in Ethiopia: Farmers' perception, constraints to adoption and effects of long-term supplementation on sheep performance. PhD dissertation, Wageningen University, Wageningen, The Netherlands.

Mengistu, A. 1997. Conservation based forage development for Ethiopia. Self Help Development International / Institute for Sustainable Development. Addis Ababa, Ethiopia.

Nederlof, E.S., Dangbégnon, C. 2007. Lessons for farmer-oriented research: Experiences from a West African soil fertility management project. Agriculture and Human Values. 24: 369-387.

Ngutu, M.N., Recke, H. 2006. Exploring farmers' innovativeness: experiences with the adaptation of water-saving technologies for small-scale vegetable production around Marsabit Mountain in Northern Kenya. Experimental Agriculture. 42: 459-474.

NMSA (National Meteorological Services Agency). 2004. Unpublished meteorological data. NMSA, Addis Ababa, Ethiopia.

Nowak, P. 1992. Why farmers adopt production technology. Journal of Soil and Water Conservation. 47(1): 14-16.

Peacock, C. 2005. Goats-A pathway out of poverty. Small Ruminant Research. 60: 179-186.

Peeling, D., Holden, S. 2004. The effectiveness of community-based animal health workers, for the poor, for communities and for public safety. Revue Scientifique et Technique. 23(1): 253-276.

Rogers, E. M. 2003. Diffusion of innovations. 5th edition. Free Press, New York, USA.

SAS. 2004. SAS (Statistical Analysis System). Version 9.1.2. SAS Institute Inc., Cary, NC, USA.

7

A FORESIGHT REFLECTION ON SUSTAINABLE METHODS FOR CONTROLLING MAMMALIAN FARM ANIMAL REPRODUCTION

Philippe Chemineau

INRA, DEPE, 147 rue de l'Université, 75007 Paris France
E-mail: Philippe.Chemineau@paris.inra.fr
** Corresponding author*

SUMMARY

Controlling farm animal reproduction was/is one of the essential tools for domestication of species. It is still of high interest for genetic improvement, adjustment of production to feed availability, to market and reduction of unproductive periods. Detection of oestrous behaviour in cyclic females, synchronization of conceptions and increase of the potency of diffusion of sires are three common objectives among species. The various situations of reproductive systems, which are reviewed here, are very different among the various livestock systems in the world, because of intrinsic properties of species, but also because of the various degrees of intensification of the livestock systems themselves. A clear tendency appears to continue increasing productivity by improving reproductive efficiency, developing new and sustainable techniques without hormones, and continuing to develop AI and reproductive biotechnologies. Future areas of investment in research could be, first, the physiological and ethological bases of the socio-sexual inter-relationships between animals, second the genetic control of reproductive traits, third increasing the efficiency of classical and new reproductive biotechnologies and fourth engineering new and innovative reproductive techniques to be used in farm conditions. These reproductive techniques should be developed respecting the three pillars of sustainability: environment, economy and society. Thus, they should be included within the livestock systems in which they are supposed to be applied and which should be assessed for sustainability.

Key-words: Mammals; Reproduction; Sustainability.

INTRODUCTION

Controlling farm animal reproduction, in association with the link between animals and humans, is probably one of the key elements that have allowed the domestication of animals. After identification of a specific individual which bears interesting traits, control of reproduction allows to maintain them over the progeny, and to associate them with other useful traits. Very early during the history of domestication a change in the proportion males/females and in the age of animals was observed and this shows that the control of reproduction was used as tool for improvement (Zeder, 2001). Since that time, and especially over the last 50 years, the scientific progress in the area of estimating the genetic value of farm animals and in the development of artificial insemination with deep-frozen semen and other reproductive techniques, especially in the dairy cattle, have together allowed a dramatic increase in the rate of the genetic progress. However, one may question whether we have reached a limit in the application of these techniques of reproductive control and it appears interesting to focus on their sustainability in farm conditions. These are the objectives of the present paper.

Thus, after a rapid overview of the interest of controlling animal reproduction in farms, we will review rapidly the major reproductive systems existing in the different mammalian farm animal species and what could be the tendencies, weak signals and potential ruptures existing in these systems. We will end by examining the research areas in which to invest in the future and the necessity of developing an integrative approach in order to insert the reproductive technologies within livestock systems to enable them to be efficient and sustainable.

WHAT IS THE INTEREST IN CONTROLLING ANIMAL REPRODUCTION?

Controlling animal reproduction presents three major interests

Genetic progress

The first interest is clearly to increase the speed of genetic progress. Very early during the process of domestication of farm animal species, the interest of controlling animal reproduction appeared. Once the choice is made of selecting a male or a female with specific and interesting traits, the farmer wishes to spread these traits as widely as possible into their progeny. Thus, the first and still applied way of controlling reproduction is to choose the females to which a specific chosen male should be mated, in order to be sure that its progeny is identified and bred as an adult producing animal.

Since more than 50 years now, the extensive use of artificial insemination (AI) with deep-frozen semen in the bovine species (the « male side ») has led to a dramatic increase of the speed of selection, because of mainly three crucial advantages (van Arendonk, 2011). The first one is the possibility, given by progeny testing in farms, to evaluate the genetic value of young bulls, then to accurately select the best ones before their death. The second one is to largely spread the semen and favourable alleles of these best bulls over space and time to improve the genetic value of the entire flock, even though it has not participated in the progeny-testing scheme. The third advantage comes from the possibility, given by putting in place the progeny of the same bulls in different farms, of connecting herds and, so, being able to compare their genetic value without confounding it with the effects of the environment.

This is true for nearly all farm animal species, even if the amount of sperm produced by males and the minimum number of sperm cells required for AI, which determine the number of females which can be artificially inseminated per sire, varies very much between species (Chemineau et al., 2001). The difference between species also appears in the relative ability to deep-freeze semen, then to use it in farms at a reasonable cost compared to the income of the farmer. Thus, in intensive dairy cattle systems, a large number of females for a single sire are artificially inseminated with deep-frozen semen, while, in sheep, a much fewer number of ewes per ram are inseminated with liquid semen.

The same rationale can be applied for the « female side » using embryo transfer or, more recently, in vitro fertilisation with ovum-pick-up. But, obviously, it is much more difficult and costly to increase substantially the female progeny than the male progeny.

Choice of the periods of parturitions

The choice of the periods of parturition provides a way to cope with feed availability, livestock system or consumer's demand. The concentration of parturitions over a few weeks or a few days limits the time spent on them, reducing the costs and allowing improved care which limits early mortality. Thereafter, the presence of homogeneous groups of animals allows them to be fed more adequately to meet their needs thus providing less variable growth rate curves and more homogeneous groups for slaughter and the market. The social impact of controlling reproduction for farming family is also important in that it allows available time for other activities, or leisure. Thus, controlling reproduction is a way for the farmer to better find an adequate equilibrium between productivity, adaptation to the market and family life.

Reduction of unproductive periods

In a majority of farm animal species, advancing female puberty increases their lifetime productivity (swine, beef cattle, sheep), and also allows synchronizing their breeding period with that of adults (swine, sheep and goats). Reducing the duration of annual sexual rest (sheep and goats) may allow increasing productivity over the year, which may increase their productivity.

Common features among species for the objectives of controlling reproduction

In spite of large differences between species, there are common features among them concerning the objectives for controlling reproduction.

The first one is the detection of oestrous behaviour, and/or, eventually, associated ovulation(s), in cyclic females. This can be done simply through observation of animals by the farmer, or by automatic detection using the help of robotics.

The second one is the synchronisation of fertile oestrus and ovulations at a precise timing in the year. The « wish » of synchronization could be very precise (ovulations spread over less than 24 hours) in order to perform AI at fixed time(s) without oestrus detection, or more weak (over days) if synchronisation of parturitions is only wanted. Synchronization is associated, especially in sheep, to a moderate increase of ovulation rate to provide two lambs per lambing, but not more.

The third one is the increase of the potency of diffusion of favourable alleles of specific male and/or female sires. This objective can be reached in males by an improvement of sperm production, of semen technology, but also by better synchronization of ovulations and of AI conditions in females, which allow a reduction in the number of sperm inseminated to produce a progeny. In females, this objective, as stated earlier, is more complicated to reach because they produce fewer oocytes per ova shedding and because these oocytes are of more difficult access than sperm in males. Superovulation, ovum pick-up associated with in vitro fertilisation, are the main steps to improve for a better diffusion of female alleles.

WHAT IS THE ACTUAL SITUATION OF BREEDING SYSTEMS IN LIVESTOCK?

Reasons for the high variability among species regarding reproductive control.

The situation among farm animal species is extremely variable for four main reasons. The first one comes from the biological differences between species regarding their characteristics of reproduction. As it is an important trait for speciation, firstly in the wild, but also in our favourite farm animals, every species has developed differing oestrous cycle characteristics in terms of duration, presence all the year round and ovulation rate.

The second one comes from the more or less ability of their gametes (sperm and oocytes) to support the in vitro « survival » required mainly to provoke an increase in the diffusion rate or an ability to be artificially inseminated easily. For example, the techniques used for deep-freeze bull semen do not work as well in other species. In sheep, goats and boars, it is necessary to separate the seminal plasma before freezing (goats (Nunes et al., 1982)). In boars, it is necessary to increase sperm concentration of the ejaculate (Martinat-Botte et al., 2009). In sheep, the only way to get a high fertility rate when using deep-frozen semen is to inseminate directly into the uterine horns (Vallet et al., 1992).

The third one is due to the fact that some species of major economic importance (i.e. dairy cattle), have received a lot of attention in terms of research since 50 years, while some other ones remain explored only by a small set of scientists.

The fourth one is that, within one single species, there is a huge variability among livestock breeding systems over the world (De Roest, 2011). Regarding the number of farms or the number of animals, which use techniques for controlling reproduction, it is very obvious that the majority do not use sophisticated ones or do not use even a single one, but rather natural mating.

In spite of this difficult context, we have attempted to draw a very crude description of the major reproductive systems and fertility problems or questions, within species and within the type of production (i.e. meat or milk, essentially).

Cattle

In intensive dairy cattle, the most common system is represented by Holstein cows in which reproduction is done by AI with frozen semen after detection of natural oestrus. In this system, the intense selection for milk production is associated with increasing fertility problems (Dobson et al., 2008). Fertility, expressed in terms of non-return rates after first AI has decreased since more than 15 years and is now below 40% in most cases. These difficulties can be either increased by local conditions (heat stress, use of exogenous hormones to increase milk production), or decreased by adequate selection and/or more friendly management. In the more intensive units, the replacement rate is not even provided, which clearly shows that this system is not sustainable in the medium term. The first problem existing regarding reproduction in this system concerns oestrous detection, which needs to be automatized either at milking (by measuring hormones in milk (Asmussen, 2010)) or by watching behavioural activity. The second problem concerns fertility itself, as, even after good heat detection, conception rates are low and unpredictable. These two areas are the subjects of active research programs in various countries at the moment (Garnsworthy et al., 2008; Royal et al., 2008; Wathes et al., 2008).

In beef cattle, the situation is much more dependant on local feeding conditions (i.e. forage availability) and, as a consequence, the reproductive system is highly dependant on forage availability and on the management of body reserves of cows over the annual cycle (Blanc and Agabriel, 2008). The very large majority of conceptions are achieved by natural mating with a very low control of reproduction, hormonal treatments are used in a limited number of systems of developed countries. In most systems, the priority is to synchronise oestrus and ovulations at a period chosen so that calving will occur when forage availability is at maximum. The economic objective of obtaining one calf per year and per cow is probably not reached in many systems, especially those of the tropics or subtropics in which the rainy season is not well-defined from one year to another.

The situation of dual-purpose cattle is interesting because the farmer needs to find an equilibrium between the two above systems and because he can use the interactions between food availability, milk production and calf presence to drive reproductive activity of the cows (Alejandrino et al., 1999; Gonzalez-Stagnaro, 2008). Except in very harsh conditions when food availability limits the reconstitution of body reserves, the control of reproduction seems generally easier than in the two above systems. In this system, the farmer is generally close to the dairy females and better able to easily detect oestrus behaviour and then to organize mating or use AI.

Reproductive biotechnologies (embryo freezing and transfer, in vitro fertilization after ovum pick-up or oocyte maturation) could be used in both dairy and beef cattle, to be included in selection schemes to improve the speed of genetic progress. They are also useful to exchange genotypes with a high sanitary security from one country to another (exchange of living sires is generally prohibited nowadays)(Thibier, 2006).

Sheep

The majority of sheep breeds are reared for meat in extensive systems in which rams could be in permanent contact with ewes all over the year in less controlled ones. In the systems of temperate and subtropical areas, reproductive activity is driven by photoperiodic changes which impose the presence of an anoestrus season and lead animals to conceive only in autumn-winter, and lamb in winter-spring (Hafez, 1952). In tropical areas the driver is immediate food availability which induces more disperse and less predictable periods of lambing which can, then, be discordant with forage availability (Chemineau et al., 2007). In intensive meat systems, hormonal synchronisation using a sequence progestagens-gonadotropins, allows a precise timing in ovulation time after progestagen withdrawal, then the use of a single AI with liquid semen at a fixed time on a large number of females. In these systems, fertility after AI is generally higher than 55 lambing per cent, with a litter size between 1.5 and 2 .0. This allows a good productivity rate when females lamb thrice over two years (Benoit et al., 2009). In less intensive systems, or in organic farms, and especially in the less seasonal breeds, such as the breeds originating from the Mediterranean Basin, the « male-effect » (re-introduction of active rams among anoestrous ewes to induce synchronised and fertile ovulations) is used (Girard, 1813; Thimonier et al., 2000). But, to our knowledge, probably because of the less precise synchrony of ovulations, the male-effect is rarely associated with AI.

Milk sheep are more or less in the same reproductive situation as intensive meat sheep systems, but the objective of farmers is to get one lambing per year, not more. These breeds are generally reared in temperate or Mediterranean areas, thus submitted to photoperiodic variations, which induces seasonality of ovulatory and oestrous activities as in meat sheep. When AI is used, which is the case in the more intensive breeding schemes for milk production, hormonal synchronisation of oestrous behaviour and ovulations is largely used for getting a very precise timing of ovulations, then the insemination of liquid semen (Lagriffoul et al., 2010).

Reproductive biotechnologies (embryo freezing and transfer, in vitro fertilization after ovum pick-up or oocyte maturation) (Cognie et al., 2004; Lagriffoul et al., 2010) could also be used in sheep for the same purposes as cattle, but, to our knowledge, they are not extensively used at a commercial level, except for exchanges of genotypes and management of nucleus flocks.

Goats

The majority of world goat flocks are reared for meat and, as in sheep; goat-bucks are generally present all over the year with females. Photoperiodic driving of reproductive seasonality (i.e. conceptions from

September to January in the Northern hemisphere) is also very strong in temperate and subtropical breeds which experience a deep anoestrous season in spring and summer. As in sheep, an opportunistic strategy of reproduction based on response to feed availability is developed in tropical breeds (Chemineau et al., 2007). Dairy goats or dual-purpose breeds follow the same reproductive pattern as meat goats and, as in sheep, in intensive systems, could be managed using hormonal treatments with a single AI with deep-frozen semen at a fixed time after progestin withdrawal (Gonzalez-Stagnaro, 1984). This deep-frozen semen could come from buck sires maintained under accelerated light-rhythms, which allow all the year round a permanently very high semen production of exceptional quality for deep-freezing (Delgadillo et al., 1991; 1992; 1993). Under these conditions, fertility rates of intensive dairy goats are generally about 60 kidding per cent (Leboeuf et al., 2008). Liquid semen could also be used if photoperiodic treatments are applied to the sires in order to prevent the seasonal decrease in sperm fertilizing ability.

The buck-effect appears to be very efficient for inducing fertile oestrus and ovulations (Chemineau, 1983; Chemineau et al., 2006), especially when males are light-treated so that they are sexually active when females are in anoestrus (and untreated males in sexual rest)(Delgadillo and Velez, 2010; Delgadillo, 2011). Reproductive biotechnologies (embryo freezing and transfer, in vitro fertilization after ovum pick-up or oocyte maturation) (Ali Al Ahmad et al., 2008) could also be used in goats for the same purposes as cattle and sheep, but, to our knowledge, they are not extensively used at commercial level, except for exchanges of genotypes and very exceptional situations (« cleaning » a flock from a disease, for example).

Pigs

The rhythm of intensive pig production is governed by the reproductive rate of sows, which ovulate at a fixed time after weaning of the piglets (Soede et al., 2011). This allows the farmer to get a natural synchronisation of a group of females and the opportunity of using semen for AI. Liquid semen is used on a large scale because of the ability of boar sperm to survive up to five days without significant decrease in fertility rate (Fantinati et al., 2009). These conditions have led to the development of reproductive systems with an exceptional efficiency at a relatively low cost, thus allowing rapid and efficient breeding schemes (Tribout et al., 2010). Hormonal treatments are rarely used but could be applied to young sows to synchronise the first fertile oestrus, then allowing a good first synchronisation of the herd (van Leeuwen et al., 2011). AI with frozen semen and/or embryo transfer are still in their infancy and not used more than experimentally nowadays.

« Extensive » (familial) pig production is done in small units, where animals are close to the farmer which allows an efficient oestrus detection after weaning, then natural mating.

WHAT ARE THE TENDENCIES, WEAK SIGNALS AND POSSIBLE BREAKDOWNS?

Livestock systems are expected to follow a clear tendency of more and more intensification to follow the expected strong demand of animal products linked to the increase of a world population with higher incomes, especially in emerging countries (De Roest, 2011). However, it is also very clear that the majority of our intensive systems, developed over the last 50 years have reached (even passed) certain limits in terms of foot-print on the global environment (for example nitrate and phosphate in the water, green house gases emitted by enteric fermentation of ruminants and/or by the manure of pigs) or the welfare of animals (group housing of animals or wider space allowance). We need to modify these intensive systems in order to orient their production towards more sustainability and more friendliness to animals and environment.

Even though, for health reasons (overweight and cardiovascular diseases), it would be necessary to reduce our animal product consumption at a world-wide level (especially in industrialised countries), this will not probably be sufficient to cover the needs in animal products (Karl-Heinz, 2009; Paillard and Treyer, 2010; Paillard S., 2010; Ronzon T., 2011). Thus, the livestock industry is facing the challenge of increasing its production and, at the same time, reducing its impact on the environment.

Reproduction is concerned in three ways in this challenge: (a) to improve reproductive efficiency per producing female, which allows a better productivity thus reducing the impact per kg of animal product, (b) to develop new, innovative and efficient breeding technologies without hormones for controlling reproduction (Martin et al., 2004) (c) to develop AI and reproductive biotechnologies to be used in full association with genomic selection.

i. Increasing reproductive efficiency should be pursued as in the past, but not as a single trait to improve and not for increasing the average without taking into account the variability. For dairy species (cattle and goats), the purpose is still to get maximum fertility (i.e. the ratio between the number of females giving birth over the number of females inseminated). This is especially true, as we mentioned above, for intensive dairy cattle where fertility after AI is close to 40 calving per cent. Even if selection for fertility could be done while continuing selection on milk production (Barbat et al., 2010), we should focus on a better management of the whole lifetime of the female career by trying to find an optimization between milk production and fertility (Coyral-Castel et al., 2011). In beef cattle, a better management of body reserves and of their reconstitution at crucial times during the reproductive cycle may be an objective in order to get one calf per year and per cow (Blanc et al., 2006). For polytoquous species raised for meat production, such as sheep and pigs, prolificacy will still be a major component of farmer income, thus still being an objective to improve when possible (i.e. when prolificacy is low). However, in the most intensive systems, a precise control of the variability of litter size would be a reasonable objective to impair birth of small newborn, which have low chances for survival. In all livestock systems, increasing the reproductive efficiency is a very strong contributor to a decrease of environmental foot-print per kg of animal product provided to the consumer. However, this will need to be calculated at the system of production level, not only at the producing-female level, as it was generally done in the past.

ii. We need to develop new, innovative and efficient breeding technologies without hormones for controlling reproduction in the future. Since about 50 years, with the discovery of the very potent effects of steroids (progestagens) and gonadotropins, the industry has been able to develop highly efficient ways for controlling time and rates of conception. But, on the one hand, more and more discussions appear to question the use of such hormones in animal production, at least in Europe, for sanitary and ethical reasons. On the other hand, the producers themselves are tempted to develop new and hormone-free systems of reproduction of their animals. Intensive research in the area of socio-sexual relationships between animals has been done, especially in small ruminant species (sheep and goats) which allow, in a majority of livestock systems, the development of new and innovative systems based on the « male-effect » (Girard, 1813; Underwood, 1944; Shelton, 1960; Chemineau, 1983; Thimonier et al., 2000) to control out-of-season reproductive activity. Advances were also done in cattle to manipulate the post-partum anoestrus duration using temporary calf removal (Stumpf et al., 1992; Soto Belloso et al., 1997; Fagundes et al., 2006; Escrivao et al., 2009), associated or not, with the presence of bulls. This seems an interesting way to follow (Bonavera et al., 1990; Monje et al., 1992; Gazal et al., 1999; Molina et al., 2002; Berardinelli and Joshi, 2005b; a; Berardinelli et al., 2005; 2007; Berardinelli and Tauck, 2007; Miller and Ungerfeld, 2008). In cycling animals, the challenge is to find efficient and low-cost ways of detecting oestrous behaviour. This is clearly a very active area in terms of applied research, at least in cattle and sheep.

iii. Finally, it is still very important to develop AI and reproductive biotechnologies to be used in full association with genomic selection (Merton, 2011). Artificial insemination is still a highly efficient way to perform genetic improvement, as explained above, and it is still necessary to improve its efficiency and enlarge its domains of application in the future. This means that liquid semen could be used in association with deep-frozen semen. Also we need to continue to decrease the minimum number of spermatozoa to be inseminated in a variety of species. But with the new techniques of genomic selection, it will probably be important to continue the investment in basic and applied reproductive biotechnologies for decreasing inter-generation interval, thus getting as soon as possible (embryo) tissue for genomic evaluation, then collecting semen and oocytes from very young animals, once their value known (Merton, 2011). The continuous development of these reproductive biotechnologies should be done in close association with the programs developed simultaneously in genomic selection.

IN WHICH AREAS DO WE NEED TO INVEST IN THE FUTURE?

Research programs in the area of reproductive biology, with the aim of controlling farm animal reproduction, need to be developed into four main directions.

i. The first direction could be a better knowledge of the physiological and ethological bases of the socio-sexual inter-relationships between animals. Recent results obtained in dairy goats using sexually active males to induce out-of-season oestrus and ovulations (Delgadillo et al., 2009) showed the potency of these socio-sexual relationships. They also demonstrated that we have probably underestimated them as a means to develop new and friendly techniques for controlling reproduction. The identification of the relative responsibility of pheromonal communication *versus* behavioural activity in the response to the « male effect » could be of major interest. This could be used to better develop techniques coming from the cue(s) which is (are) mainly responsible in the response to the presence of males (Delgadillo, 2011). Putative pheromones have been identified in goats as potentially being the oxidative products of degradation of the fatty acids synthetized and secreted by sebaceous glands of buck neck (Iwata et al., 2003; Okamura and Mori, 2005). However, to our knowledge, if these compounds have been demonstrated to be involved in male attractiveness, the formal demonstration of their efficiency to stimulate LH secretion and to induce ovulations remains to be done. In beef cattle systems of production, the same observation could be done using temporary removal of calves (Berardinelli and Joshi, 2005a; Kawashima et al., 2008) and bull presence (Zalesky et al., 1984; Larson et al., 1994; Landaeta-Hernandez et al., 2004; Landaeta-Hernandez et al., 2006) to shorten the post-partum anoestrus. This is an area already explored (Custer et al., 1990; Burns and Spitzer, 1992; Fernandez et al., 1996; Fike et al., 1996; Bolanos et al., 1998; Rekwot et al., 2000a; Rekwot et al., 2000b; Ruas et al., 2001; Rekwot et al., 2004; Calegare et al., 2009; Fiol et al., 2010; Tauck et al., 2010), but which need continuous investigation. Attempts to identify putative pheromones were also done in cattle as a basis for the observed effects of bull presence on resumption of post-partum anoestrus (Tauck et al., 2006; Tauck and Berardinelli, 2007; Tauck et al., 2010). In swine, the efficiency of piglet weaning for induction of fertile oestrus could be an interesting model to better understand what are the underlying mechanisms involved in this effect of weaning. In all of these species, it would be of special interest to better know what are the sensory cues involved in the perception of the partner and to better understand the importance of sexual and/or social activity of the partner in eliciting a reproductive response. Finally, the identification of the physiological mechanisms used by responding animals to trigger their reproductive activity would also be interesting.

In all farm mammalian species, a better knowledge of the mechanisms used by males (or females) to detect the oestrous females could also be of high interest (Delgadillo et al., 2009). Pheromones associated to oestrus have been identified in wild species (elephant;(Lazar et al., 2002; Goodwin et al., 2006)), but, to our knowledge, not in the major farm species. This identification should be a major area to invest as it could be used for developing bio probes able to replace detector males in large bovine farms, especially in dairy cattle (Le Danvic C., 2011).

ii. The second direction could be an important investment in the genetic control of reproductive traits. Since the sequencing of the genomes of our main farm domestic species and with the dramatic development of genomic selection it is now possible to take into account the less heritable traits within programs of genetic improvement. This is the case for fertility traits which are among the lowest heritable ones and which are, moreover, difficult to phenotype. Fertility (i.e. non-return rate) of intensive dairy cows is already taken into account in selection schemes (Berglund, 2008; Barbat et al., 2010). The other different traits described above (seasonality, post-partum duration, oestrous behaviour, superovulation, etc.) could be further explored in terms of their genetic control (AlShorepy and Notter, 1997; Hanocq et al., 1999), then identify the genes and alleles responsible for favourable phenotypes and used in selection schemes.

iii. The third direction is a continuous effort in developing the efficiency of « classical » reproductive biotechnologies (AI, ET, IVF, etc.), simultaneously with an investment in new ones (cloning and homologous recombination)(Merton, 2011). As explained above, the use of genomic selection will probably provoke an renewal of interest for biotechnologies which may allow access to very young animals, even to the embryo itself, on which only a small set of cells could be used to evaluate marker polymorphisms (Le Bourhis et al., 2011). Once known the young animals will be interesting to use for producing progeny as soon as possible.

A renewal of interest is also expected for elite females for which estimation of genomic value will be much easier, and which could be an interesting way to get high value progeny from them. New biotechnologies of reproduction, which are defined here as those that aim to modify genome sequence and/or genome expression, are also an area in which the most advanced laboratories may invest. At least in Europe, because of the reluctance of the society towards these techniques, there is little possibility for using them in field conditions in the near future, but rather to use them as tools for exploring the role of genes suspected to control a specific trait and the importance of epigenetics in building the adult phenotypes.

iv. The fourth direction is clearly directed towards engineering of new and innovative reproductive techniques to be used in farm conditions. The oldest techniques such as the sequence progestagens- PMSG and/or the use of prostaglandins (Macmillan, 2010; Islam, 2011) which are now used on millions of females each year, have required at least ten years and strong efforts to pass from the discovery in the laboratory to a correct use in field conditions. This will be the case for the new and sustainable techniques described above which would require specific engineering for a correct application to farm conditions. This engineering is clearly part of the research and should be done by the laboratories themselves, which simultaneously invest on more basic research.

Engineering for oestrous detection is still an area of much interest in dairy cattle where monitoring the twice-daily endocrine status of milking females was proposed as an automatic technique for detecting the end of a luteal phase and the onset of the following follicular phase associated with oestrus behaviour (Asmussen, 2010). As stated above, the identification of putative pheromones associated to oestrous behaviour (Le Danvic C., 2011) could be the beginning of an engineering program using either olfactory binding proteins or/and pheromone receptor protein able to detect automatically the female status as trained dogs are able to do it (Fischer-Tenhagen et al., 2011). Quantification of animal behaviour could also be attempted automatically either by detecting it visually with video cameras, then estimating the status of females, or by equipping detector females or males with electronic apparatus able to register movements, then estimate their predictive value for oestrous behaviour (Lee et al., 2008). Recently, an electronic apron was developed in sheep to equip detector rams. This counts the number of mounts for a specific female in a given period of time, then calculates a probability for it to be in oestrus (Bocquier et al., 2006).

The use of new, non-hormonal molecules could be of high interest in controlling farm animal reproduction. The molecules could originate from plants containing native steroids (especially progesterone) or unknown compounds, from plants which are known in the local pharmacopeia to have reproductive effects (in animals or humans). But these molecules can also originate from research discoveries and could be peptides and/or antibodies able to stimulate the endogenous hormones (Hervé et al., 2004).

Engineering in the areas of genetics and biotechnologies is also extremely important in order to develop efficient techniques, which could be applied either in field or applied laboratory conditions. Two points are of special importance: the efficiency of the developed techniques (for example the number of living calves per 100 oocytes collected and fertilized in vitro) and their costs (for example the price of sexing semen or embryos).

NECESSITY OF A SYSTEMIC APPROACH OF REPRODUCTIVE SYSTEMS WITHIN INTEGRATED FARMING SYSTEMS

In addition (or in combination) of being technically efficient, reproductive techniques should be developed respecting the three pillars of sustainability: environment, economy and society. Thus, they should be included within the livestock systems in which they are supposed to be applied, taking into account these three elements. Thus the livestock system itself rather than the reproductive technique, should be assessed for sustainability (Gamborg and Sandoe, 2005; Neeteson-van Nieuwenhoven et al., 2006).

Respecting the environment is the first objective for sustainable reproductive systems. In that sense, it could be considered that reducing the use of exogenous hormones could be one objective to reach, in order to reduce the potential effects on the ecosystems, especially on biodiversity. Reduction of exogenous hormones could come from either a substitution of xenobiotic molecules by natural ones (for example synthetic progestagens replaced by progesterone), or the development of a new technique of induction (for example « male effect » to replace PMSG/eCG). However, as explained above, sustainability should not be only assessed at the level of the technique itself, but rather at the livestock system level, to prevent the adoption of a new technique, which may have positive effects per se, but adverse effects for the whole system (Baxter et al., 2011).

The second pillar is economic and it is obvious that a new technique for controlling farm animal reproduction should have an economic interest for the farmer in order to be adopted on a large scale (Moyo et al., 2009). This economic interest is not easy to estimate as it may include an advantage in terms of labour or in terms of image of the livestock system itself, which may provide an advantage to market the products (for example organic farms which prohibit the use of exogenous hormones for oestrus synchronization). However, as for the environment, the economic sustainability should be appreciated at the level of the livestock system or even larger to appreciate its efficiency (for example the potential exclusion of organic farms from artificial insemination, then genetic progress) (Brocard and Portier, 2008; Pellicer-Rubio et al., 2008; Benoit et al., 2009; Nauta, 2009).

Finally, sustainability should also be social, in order to validate the fact that the considered technique is socially acceptable (for example not going against animal welfare) and could be accepted by the farmers themselves and by the social stakeholders who develop some interest in the area (Costa-Neto, 2000; Cristofori et al., 2005; Mwacharo and Drucker, 2005; Kosgey et al., 2006; Getachew et al., 2010; Murage and Ilatsia, 2011). Finally, in the social acceptance, we could also raise the importance of new paradigms, partly coming from scientific observations of nature and farmers' practices, which may play a role in the development of new techniques for controlling reproduction (Mesa and Machado, 2009).

CONCLUSION

As for many other domains of agriculture, the control of farm mammalian reproduction is facing a new and somewhat contradictory challenge: it should continue to improve productivity (i.e. fecundity of females) in order to reduce the indirect environmental foot-print of livestock by producing more with the same livestock, but, at the same time to reduce the direct impact of the livestock systems on the environment and on welfare of animals. The exploration of new and more friendly ways of controlling animal reproduction is not only a necessity, but seems also feasible, especially by investing in a better knowledge of the ethological and physiological bases of the socio-sexual relationships between animals. The identification of sexual pheromones, associated with a better knowledge of the sensory cues used by females to perceive sexual male activity may lead to engineering of new and sustainable bio techniques able to assess the behavioural or/and endocrine status of the females, and to induce their activity during anoestrus. Replacement of exogenous hormones seems also feasible if an investment is done into the identification of new sources of natural hormones, for example in plants.

However, we should remind ourselves here that sustainability should be assessed at the level of the whole livestock system and not just at the level of the reproductive technique itself. Such an assessment is not trivial and needs a good overview of the considered system to achieve high efficiency.

ACKNOWLEDGEMENTS

I want to thank Prof. Cledwyn Thomas for his help in preparing the English version of this manuscript and Dr Jose Alberto Delgadillo-Sanchez for his help in preparing the Spanish version of the abstract.

REFERENCES

Alejandrino, A.L., Asaad, C.O., Malabayabas, B., De Vera, A.C., Herrera, M.S., Deocaris, C.C., Ignacio, L.M., Palo, L.P., 1999. Constraints on dairy cattle productivity at the smallholder level in the Philippines. Preventive Veterinary Medicine. 38, 167-178.

Ali Al Ahmad, M.Z., Chebloune, Y., Bouzar, B.A., Baril, G., Bouvier, F., Chatagnon, G., Leboeuf, B., Pepin, M., Guibert, J.M., Russo, P., Manfredi, E., Martin, J., Fieni, F., 2008. Lack of risk of transmission of caprine arthritis-encephalitis virus (CAEV) after an appropriate embryo transfer procedure. Theriogenology. 69, 408-415.

AlShorepy, S.A., Notter, D.R., 1997. Response to selection for fertility in a fall-lambing sheep flock. Journal of Animal Science. 75, 2033-2040.

Asmussen, T., 2010. Herd Navigator or "how to benefit from frequent measurements". ICAR Technical Series. 291-293.

Barbat, A., Le Mezec, P., Ducrocq, V., Mattalia, S., Fritz, S., Boichard, D., Ponsart, C., Humblot, P., 2010. Female Fertility in French Dairy Breeds: Current Situation and Strategies for Improvement. Journal of Reproduction and Development. 56, S15-S21.

Baxter, E.M., Lawrence, A.B., Edwards, S.A., 2011. Alternative farrowing systems: design criteria for farrowing systems based on the biological needs of sows and piglets. Animal. 5, 580-600.

Benoit, M., Tournadre, H., Dulph, J.P., Laignel, G., Prache, S., Cabaret, J., 2009. Is intensification of reproduction rhythm sustainable in an organic sheep production system? A 4-year interdisciplinary study. Animal. 3, 753-763.

Berardinelli, J.G., Joshi, P.S., 2005a. Initiation of postpartum luteal function in primiparous restricted-suckled beef cows exposed to a bull or excretory products of bulls or cows. Journal of Animal Science. 83, 2495-2500.

Berardinelli, J.G., Joshi, P.S., 2005b. Introduction of bulls at different days postpartum on resumption of ovarian cycling activity in primiparous beef cows. Journal of Animal Science. 83, 2106-2110.

Berardinelli, J.G., Joshi, P.S., Tauck, S.A., 2005. Postpartum resumption of ovarian cycling activity in first-calf suckled beef cows exposed to familiar or unfamiliar bulls. Animal Reproduction Science. 90, 201-209.

Berardinelli, J.G., Joshi, P.S., Tauck, S.A., 2007. Conception rates to artificial insemination in primiparous, suckled cows exposed to the biostimulatory effect of bulls before and during a gonadotropin-releasing hormone-based estrus synchronization protocol. Journal of Animal Science. 85, 848-852.

Berardinelli, J.G., Tauck, S.A., 2007. Intensity of the biostimulatory effect of bulls on resumption of ovulatory activity in primiparous, suckled, beef cows. Animal Reproduction Science. 99, 24-33.

Berglund, B., 2008. Genetic improvement of dairy cow reproductive performance. Reproduction in Domestic Animals 43, 89-95.

Blanc, F., Agabriel, J., 2008. Modelling the reproductive efficiency in a beef cow herd: effect of calving date, bull exposure and body condition at calving on the calving-conception interval and calving distribution. Jorunal of Agricultural Science. 146, 143-161.

Blanc, F., Bocquier, F., Agabriel, J., D'Hour, P., Chilliard, Y., 2006. Adaptive abilities of the females and sustainability of ruminant livestock systems. A review. Animal Research. 55, 489-510.

Bocquier, F., Gaubert, J.L., Blanc, F., Viudes, G., Maton, C., Debus, N., Teyssier, J., 2006. Utilisation de l'identification électronique pour la détection automatisée du comportement sexuel chez les ovins : perspectives pour la détection des chaleurs chez la brebis, 13. Rencontres Recherches Ruminants. 2006/12/06-07, INRA-Institut National de la Recherche Agronomique (FRA), Institut de l'Elevage (FRA), Paris (FRA), pp. 155-162.

Bolanos, J.M., Forsberg, M., Kindahl, H., Rodriguez-Martinez, H., 1998. Biostimulatory effects of estrous cows and bulls on resumption of ovarian activity in postpartum anestrous Zebu (Bos indicus) cows in the humid tropics. Theriogenology. 49, 629-636.

Bonavera, J.J., Schiersmann, G.C.S., Alberio, R.H., Mestre, J., 1990. A note on the effects of 72-hour calf removal and or bull exposure upon postpartum reproductive-performance of angus cows. Animal Production. 50, 202-206.

Brocard, V., Portier, B., 2008. Impacts of compact calvings and once-a-day milking in grassland based systems. Biodiversity and animal feed: future challenges for grassland production. Proceedings of the 22nd General Meeting of the European Grassland Federation, Uppsala, Sweden, 9-12 June 2008, 789-791.

Burns, P.D., Spitzer, J.C., 1992. Influence of biostimulation on reproduction in postpartum beef-cows. Journal of Animal Science. 70, 358-362.

Calegare, L., Alencar, M.M., Packer, I.U., Ferrell, C.L., Lanna, D.P.D., 2009. Cow/calf preweaning efficiency of Nellore and Bos taurus x Bos indicus crosses. Journal of Animal Science. 87, 740-747.

Chemineau, P., 1983. Effect on estrus and ovulation of exposing creole goats to the male at 3 times of the year. Journal of Reproduction and Fertility. 67, 65-72.

Chemineau, P., Cognie, Y., Thimonier, J., 2001. La maîtrise de la reproduction des mammifères domestiques. , In: Thibault, M.C.L.C. (Ed.), La reproduction chez les mammifères et l'homme, Ellipses, Paris, pp. 792-801.

Chemineau, P., Malpaux, B., Brillard, J.P., Fostier, A., 2007. Seasonality of reproduction and production in farm fishes, birds and mammals. Animal. 1, 419-432.

Chemineau, P., Pellicer-Rubio, M.T., Lassoued, N., Khaldi, G., Monniaux, D., 2006. Male-induced short oestrous and ovarian cycles in sheep and goats: a working hypothesis. Reproduction Nutrition Development. 46, 417-429.

Cognie, Y., Poulin, N., Locatelli, Y., Mermillod, P., 2004. State-of-the-art production, conservation and transfer of in-vitro-produced embryos in small ruminants. Reproduction Fertility and Development. 16, 437-445.

Costa-Neto, E.M., 2000. Knowledge and traditional uses of animal resources by an Afro-Brazilian community. Preliminary results. Interciencia. 25, 423-430.

Coyral-Castel, S., Rame, C., Monniaux, D., Freret, S., Fabre-Nys, C., Fritz, S., Monget, P., Dupont, F., Dupont, J., 2011. Ovarian parameters and fertility of dairy cows selected for one QTL located on BTA3. Theriogenology. 75, 1239-1250.

Cristofori, F., Issa, M., Yenikoye, A., Trucchi, G., Quaranta, G., Chanono, M., Semita, C., Marichatou, H., Mattoni, M., 2005. Artificial insemination using local cattle breeds in Niger. Tropical Animal Health and Production. 37, 167-172.

Custer, E.E., Berardinelli, J.G., Short, R.E., Wehrman, M., Adair, R., 1990. Postpartum interval to estrus and patterns of LH and progesterone in first-calf suckled beef cows exposed to mature bulls. Journal of Animal Science. 68, 1370-1377.

De Roest, K., 2011. Competition between production systems facing an increase of world demand for dairy and meat., In: EAAP (Ed.), 62nd Annual Meeting of the European Federation of Animal Science, Stavanger, Norway.

Delgadillo, J.A., 2011. Environmental and social cues can be used in combination to develop sustainable breeding techniques for goat reproduction in the subtropics. Animal. 5, 74-81.

Delgadillo, J.A., Gelez, H., Ungerfeld, R., Hawken, P.A.R., Martin, G.B., 2009. The 'male effect' in sheep and goats-Revisiting the dogmas. Behavioral Brain Research. 200, 304-314.

Delgadillo, J.A., Leboeuf, B., Chemineau, P., 1991. Decrease in the seasonality of sexual behavior and sperm production in bucks by exposure to short photoperiodic cycles. Theriogenology. 36, 755-770.

Delgadillo, J.A., Leboeuf, B., Chemineau, P., 1992. Abolition of seasonal variations in semen quality and maintenance of sperm fertilizing ability by photoperiodic cycles in goat bucks. Small Ruminant Research. 9, 47-59.

Delgadillo, J.A., Leboeuf, B., Chemineau, P., 1993. Maintenance of sperm production in bucks during a 3rd year of short photoperiodic cycles. Reproduction Nutrition and Development. 33, 609-617.

Delgadillo, J.A., Velez, L.I., 2010. Stimulation of reproductive activity in anovulatory Alpine goats exposed to bucks treated only with artificially long days. Animal. 4, 2012-2016.

Dobson, H., Walker, S.L., Morris, M.J., Routly, J.E., Smith, R.F., 2008. Why is it getting more difficult to successfully artificially inseminate dairy cows? Animal. 2, 1104-1111.

Escrivao, R.J.A., Webb, E.C., Garces, A.P.J.T., 2009. Effects of 12 hour calf withdrawal on conception rate and calf performance of Bos indicus cattle under extensive conditions. Tropical Animal Health and Production. 41, 135-139.

Fagundes, N.S., Nascimento, M.R.B.M., Diniz, E.G., 2006. Effect of biostimulation and shang on the pregnancy rate and postpartum oestrus manifestation of nursing beef cows. Veterinaria Noticias. 12, 123-126.

Fantinati, P., Zannoni, A., Bernardini, C., Forni, M., Tattini, A., Seren, E., Bacci, M.L., 2009. Evaluation of swine fertilisation medium (SFM) efficiency in preserving spermatozoa quality during long-term storage in comparison to four commercial swine extenders. Animal. 3, 269-274.

Fernandez, D.L., Berardinelli, J.G., Short, R.E., Adair, R., 1996. Acute and chronic changes in luteinizing hormone secretion and postpartum interval to estrus in first-calf suckled beef cows exposed continuously or intermittently to mature bulls. Journal of Animal Science 74, 1098-1103.

Fike, K.E., Bergfeld, E.G., Cupp, A.S., Kojima, F.N., Mariscal, V., Sanchez, T.S., Wehrman, M.E., Kinder, J.E., 1996. Influence of fenceline bull exposure on duration of postpartum anoestrus and pregnancy rate in beef cows. Animal Reproduction Science. 41, 161-167.

Fiol, C., Quintans, G., Ungerfeld, R., 2010. Response to biostimulation in peri-puberal beef heifers: influence of male-female proximity and heifer's initial body weight. Theriogenology. 74, 569-575.

Fischer-Tenhagen, C., Wetterholm, L., Tenhagen, B.A., Heuwieser, W., 2011. Training dogs on a scent platform for oestrus detection in cows. Applied Animal Behaviour Science. 131, 63-70.

Gamborg, C., Sandoe, P., 2005. Sustainability in farm animal breeding: a review. Livestock Production Science. 92, 221-231.

Garnsworthy, P.C., Sinclair, K.D., Webb, R., 2008. Integration of physiological mechanisms that influence fertility in dairy cows. Animal. 2, 1144-1152.

Gazal, O.S., Guzman-Vega, G.A., Williams, G.L., 1999. Effects of time of suckling during the solar day on duration of the postpartum anovulatory interval in Brahman x Hereford (F1) cows. Journal of Animal Science. 77, 1044-1047.

Getachew, T., Haile, A., Tibbo, M., Sharma, A.K., Solkner, J., Wurzinger, M., 2010. Herd management and breeding practices of sheep owners in a mixed crop-livestock and a pastoral system of Ethiopia. African Journal Agricultural Research. 5, 685-691.

Girard, L., 1813. Moyens employés avec succès, par M. Morel de Vindé, Membre de la Société d'Agriculture de Seine et Oise, pour obtenir, dans le temps le plus courts possible, la fécondation du plus grand nombre des brebis portières d'un troupeau. . Ephémérides de la Société d'Agriculture du Département de l'Indre pour l'An 1813, Séance du 5 septembre, Chateauroux, Département de l'Indre, France VII, 66-68.

Gonzalez-Stagnaro, C., 1984. Hormonal control of the oestrous cycle in small ruminants in tropical areas. Reproduction des ruminants en zone tropicale. Reunion internationale, Pointe-a-Pitre, Guadeloupe, 8-10 juin 1983., 433-471.

Gonzalez-Stagnaro, C., 2008. Benchmarking in Reproductive Control Programs of tropical dual purpose crossbred herds. Reproduction in Domestic Animals. 43, 42-42.

Goodwin, T.E., Eggert, M.S., House, S.J., Weddell, M.E., Schulte, B.A., Rasmussen, L.E.L., 2006. Insect pheromones and precursors in female african elephant urine. Journal of Chemical Ecology. 32, 1849-1853.

Hafez, E.S.E., 1952. Studies on the breeding season and reproduction of the ewe. Journal of Agricultural Science. 42, 189-265.

Hanocq, E., Bodin, L., Thimonier, J., Teyssier, J., Malpaux, B., Chemineau, P., 1999. Genetic parameters of spontaneous spring ovulatory activity in Merinos d'Arles sheep. Genetics Selection Evolution. 31, 77-90.

Hervé, V., Roy, F., Bertin, J.J., Guillou, F., Maurel, M.C., 2004. Antiequine chorionic gonadotropin (eCG) antibodies generated in goats treated with eCG for the induction of ovulation modulate the luteinizing hormone and follicle-stimulating hormone bioactivities of eCG differently. Endocrinology. 145, 294-303.

Islam, R., 2011. Synchronization of estrus in cattle: a review. Veterinary World. 4, 136-141.

Iwata, E., Kikusui, T., Takeuchi, Y., Mori, Y., 2003. Substances derived from 4-ethyl octanoic acid account for primer pheromone activity for the "male effect" in goats. Journal of Veterinary Medical Science. 65, 1019-1021.

Karl-Heinz, E., Haberl,H., Krausmann, F., Lauk, C., Plutzar, C., Steinberger, J.K., Müller, C., Bondeau, A., Waha, K., Pollack, G., 2009. Eating the Planet: Feeding and fuelling the world sustainably, fairly and humanely – a scoping study., In: Vienna, I.o.S.E.a.P.P. (Ed.), Social Ecology Working Paper No. 116.

Kawashima, C., Kida, K., Matsuhashi, M., Matsui, M., Shimizu, T., Matsunaga, N., Ishii, M., Miyake, Y.-I., Miyamoto, A., 2008. Effect of suckling on the reproductive performance and metabolic status of obese Japanese black cattle during the early postpartum period. Journal of Reproduction and Development. 54, 46-51.

Kosgey, I.S., Baker, R.L., Udo, H.M.J., Van Arendonk, J.A.M., 2006. Successes and failures of small ruminant breeding programmes in the tropics: a review. Small Ruminant Research. 61, 13-28.

Lagriffoul, G., Astruc, J.M., Barillet, F., Bouix, J., Bouffartigue, B., Francois, D., Larroque, H., Praud, J.P., Raoul, J., Rupp, R., Sidani, C., Tiphine, L., 2010. Sheep breeding programs in France using modern reproductive methods: application for genetic improvement of scrapie resistance in the national sheep flock. Proceedings of the 8th World Merino Conference, Merinoscope 2010, Rambouillet, France, 3-5 May 2010, 2-01.

Landaeta-Hernandez, A.J., Giangreco, M., Melendez, P., Bartolome, J., Bennet, F., Rae, D.O., Hernandez, J., Archbald, L.F., 2004. Effect of biostimulation on uterine involution, early ovarian activity and first postpartum estrous cycle in beef cows. Theriogenology. 61, 1521-1532.

Landaeta-Hernandez, A.J., Melendez, P., Bartolome, J., Rae, D.O., Archbald, L.F., 2006. Effect of biostimulation on the expression of estrus in postpartum Angus cows. Theriogenology. 66, 710-716.

Larson, C.L., Miller, H.L., Goehring, T.B., 1994. Effect of postpartum bull exposure on calving interval of 1st-calf heifers bred by natural service. Canadian Journal of Animal Science. 74, 153-154.

Lazar, J., Greenwood, D.R., Rasmussen, L.E.L., Prestwich, G.D., 2002. Molecular and functional characterization of an odorant binding protein of the Asian elephant, Elephas maximus: Implications for the role of lipocalins in mammalian olfaction. Biochemistry. 41, 11786-11794.

Le Bourhis, D.A., Mullaart, E., Humblot, P., Coppieters, W., Ponsart, C., 2011. 193 bovine embryo genotyping using a 50k single nucleotide polymorphism chip. Reproduction, Fertility, and Development. 23, 197.

Le Danvic C., G.O., Sellem E., Joly C., Ponsart C., Chemineau P., Humblot P. & Nagnan-Le Meillour P., 2011. Identification of oestrus chemical cues in the dairy cow, Chemical Signals in Vertebrates XII, Berlin, August 28-31.

Leboeuf, B., Delgadillo, J.A., Manfredi, E., Piacere, A., Clement, V., Martin, P., Pellicer, M., Boue, P., de Cremoux, R., 2008. Management of goat reproduction and insemination for genetic improvement in France. Reproduction in Domestic Animals. 43, 379-385.

Lee, C., Prayaga, K.C., Fisher, A.D., Henshall, J.M., 2008. Behavioral aspects of electronic bull separation and mate allocation in multiple-sire mating paddocks. Journal of Animal Science. 86, 1690-1696.

Macmillan, K.L., 2010. Recent Advances in the Synchronization of Estrus and Ovulation in Dairy Cows. Journal of Reproduction and Development. 56, S42-S47.

Martin, G.B., Milton, J.T.B., Davidson, R.H., Hunzicker, G.E.B., Lindsay, D.R., Blache, D., 2004. Natural methods for increasing reproductive efficiency in small ruminants. Animal Reproduction Science. 82-3, 231-246.

Martinat-Botte, F., Plat, M., Guillouet, P., 2009. Biotechnologies of porcine reproduction: from routine techniques to emerging methods. Produccion Animal. 22, 97-115.

Merton, S., 2011. New tools in reproduction technologies, In: EAAP (Ed.), 62nd Annual Meeting of the European Federation of Animal Science, Stavanger, Norway.

Mesa, A.R., Machado, H., 2009. Training of producers and management staff for the adoption of sustainable animal production technologies. Pastos y Forrajes. 32, 93-100.

Miller, V., Ungerfeld, R., 2008. Weekly bull exchange shortens postpartum anestrus in suckled beef cows. Theriogenology. 69, 913-917.

Molina, R., Galina, C.S., Camacho, J., Maquivar, M., Diaz, G.S., Estrada, S., Martinez, L., 2002. Effect of alternating bulls as a management tool to improve the reproductive performance of suckled Zebu cows in the humid tropics of Costa Rica. Animal Reproduction Science. 69, 159-173.

Monje, A.R., Alberio, R., Schiersmann, G., Chedrese, J., Carou, N., Callejas, S.S., 1992. Male effect on the postpartum sexual-activity of cows maintained on 2 nutritional levels. Animal Reproduction Science. 29, 145-156.

Moyo, S., McDermott, J., Herrero, M., Steeg, J.v.d., Staal, S., Baltenweck, I., 2009. Development of livestock production systems in Africa: challenges and opportunities. Animal production and animal science worldwide: WAAP book of the year 2007, 15-26.

Murage, A.W., Ilatsia, E.D., 2011. Factors that determine use of breeding services by smallholder dairy farmers in Central Kenya. Tropical Animal Health and Production. 43, 199-207.

Mwacharo, J.M., Drucker, A.G., 2005. Production objectives and management strategies of livestock keepers in South-East Kenya: Implications for a breeding programme. Tropical Animal Health and Production. 37, 635-652.

Nauta, W.J., 2009. Selective breeding in organic dairy production. Selective breeding in organic dairy production, 160 pp.

Neeteson-van Nieuwenhoven, A.M., Merks, J., Bagnato, A., Liinamo, A.E., 2006. Sustainable transparent farm animal breeding and reproduction. Livestock Science. 103, 282-291.

Nunes, J.F., Corteel, J.M., Combarnous, Y., Baril, G., 1982. Role of seminal plasma in the invitro survival of goat sperm. Reproduction Nutrition and Development. 22, 611-620.

Okamura, H., Mori, Y., 2005. Characterization of the primer pheromone molecules responsible for the 'male effect' in ruminant species. Chemical Senses. 30, I140-i141.

Paillard, S., Treyer, S., 2010. Feeding the planet: two scenarios: Agrimonde and the debate on the global future for agriculture and food. Nourrir la planete: deux scenarios: Agrimonde et le debat sur l'avenir de l'agriculture et de l'alimentation a l'echelle mondiale. Futuribles, 45-63.

Paillard S., T.S., Dorin B. (Ed.), 2010. Agrimonde: Scenarios and Challenges for Feeding the World in 2050. Éditions Quæ, Paris.

Pellicer-Rubio, M.T., Leboeuf, B., Bernelas, D., Forgerit, Y., Pougnard, J.L., Bonne, J.L., Senty, E., Breton, S., Brun, F., Chemineau, P., 2008. High fertility using artificial insemination during deep anoestrus after induction and synchronisation of ovulatory activity by the "male effect" in lactating goats subjected to treatment with artificial long days and progestagens. Animal Reproduction Science. 109, 172-188.

Rekwot, P., Ogwu, D., Oyedipe, E., Sekoni, V., 2000a. Effects of bull exposure and body growth on onset of puberty in Bunaji and Friesian x Bunaji heifers. Reproduction Nutrition and Development. 40, 359-367.

Rekwot, P.I., Akinpelumi, O.P., Sekoni, V.O., Eduvie, L.O., Oyedipe, E.O., 2004. Effects of nutritional supplementation and exposure to bulls on resumption of post-partum ovarian activity in Bunaji (Bos indicus) cattle. Veterinary Journal. 167, 67-71.

Rekwot, P.I., Ogwu, D., Oyedipe, E.O., 2000b. Influence of bull biostimulation, season and parity on resumption of ovarian activity of zebu (Bos indicus) cattle following parturition. Animal Reproduction Science. 63, 1-11.

Ronzon T., T.S., Dorin B., Caron P.., Chemineau P., Guyomard H., 2011. Feeding the world in 2050: key findings and hopes for policy making and agricultural research from the Agrimonde foresight study. Food Ethics Magazine in press.

Royal, M.D., Smith, R.F., Friggens, N.C., 2008. Fertility in dairy cows: bridging the gaps - Foreword. Animal. 2, 1101-1103.

Ruas, J.R.M., Torres, C.A.A., Silva Filho, J.M.d., Borges, L.E., Marcatti Neto, A., Machado, G.V., Borges, A.M., 2001. The effect of different types of suckling management on ovarian activity reestablishment and on plasma levels of cholesterol, glucose, urea and progesterone in anestrous Nelore cows. Arquivos de Ciencias Veterinarias e Zoologia da UNIPAR 4, 9-17.

Shelton, M., 1960. The influence of the presence of the male goat on the initiation of oestrous cycling and ovulation in Angora does. Journal of Animal Science. 19, 368-375.

Soede, N.M., Langendijk, P., Kemp, B., 2011. Reproductive cycles in pigs. Animal Reproduction Science. 124, 251-258.

Soto Belloso, E., Ramirez Iglesia, L., Guevara, L., Soto Castillo, G., 1997. Bull effect on the reproductive performance of mature and first calf-suckled Zebu cows in the tropics. Theriogenology. 48, 1185-1190.

Stumpf, T.T., Wolfe, M.W., Wolfe, P.L., Day, M.L., Kittok, R.J., Kinder, J.E., 1992. Weight changes prepartum and presence of bulls postpartum interact to affect duration of postpartum anestrus in cows Journal of Animal Science. 70, 3133-3137.

Tauck, S.A., Berardinelli, J.G., 2007. Putative urinary pheromone of bulls involved with breeding performance of primiparous beef cows in a progestin-based estrous synchronization protocol. Journal of Animal Science. 85, 1669-1674.

Tauck, S.A., Berardinelli, J.G., Geary, T.W., Johnson, N.J., 2006. Resumption of postpartum luteal function of primiparous, suckled beef cows exposed continuously to bull urine. Journal of Animal Science. 84, 2708-2713.

Tauck, S.A., Olsen, J.R., Wilkinson, J.R.C., Berardinelli, J.G., 2010. Duration of daily bull exposure on resumption of ovulatory activity in postpartum, primiparous, suckled, beef cows. Animal Reproduction Science. 118, 13-18.

Thibier, M., 2006. Biosecurity and the various types of embryos transferred. Reproduction in Domestic Animals. 41, 260-267.

Thimonier, J., Cognie, Y., Lassoued, N., Khaldi, G., 2000. The ram effect: an up-to-date method for the control of oestrus and ovulation in sheep. Produccion Animal. 13, 223-231.

Tribout, T., Caritez, J.C., Gruand, J., Bouffaud, M., Guillouet, P., Billon, Y., Pery, C., Laville, E., Bidanel, J.P., 2010. Estimation of genetic trends in French Large White pigs from 1977 to 1998 for growth and carcass traits using frozen semen. Journal of Animal Science 88, 2856-2867.

Underwood, E.J., Shier, F.L., Davenport, N. , 1944. Studies in Sheep husbandry in Western Australia. V. The breeding season of Merino crossbred and British Breed ewes in the Agricultural districts. Journal of the Department of Agriculture of Western Australia. 11, 135-143.

Vallet, J.C., Baril, G., Leboeuf, B., Perrin, J., 1992. Intrauterine insemination by laparoscopy in ewes and goats. Annales Zootechnia. 41, 305-309.

van Arendonk, J.A.M., 2011. The role of reproductive technologies in breeding schemes for livestock populations in developing countries. Livestock Science. 136, 29-37.

van Leeuwen, J.J.J., Martens, M., Jourquin, J., Driancourt, M.A., Kemp, B., Soede, N.M., 2011. Effects of altrenogest treatments before and after weaning on follicular development, farrowing rate, and litter size in sows. Journal of Animal Science. 89, 2397-2406.

Wathes, D.C., Brickell, J.S., Bourne, N.E., Swali, A., Cheng, Z., 2008. Factors influencing heifer survival and fertility on commercial dairy farms. Animal. 2, 1135-1143.

Zalesky, D.D., Day, M.L., Garciawinder, M., Imakawa, K., Kittok, R.J., Docchio, M.J., Kinder, J.E., 1984. Influence of exposure to bulls on resumption of estrous cycles following parturition in beef-cows. Journal of Animal Science. 59, 1135-1139.

Zeder, M.A., 2001. A metrical analysis of a collection of modern goats (Capra hircus aegargus and C. h. hircus) from Iran and Iraq: Implications for the study of caprine domestication. Journal of Archaeological Science. 28, 61-79.

8

DESIGN AND IMPLEMENTATION OF A COMMUNITY-BASED BREEDING PROGRAM FOR DAIRY GOATS IN NORTHERN MEXICO

Maria Wurzinger[1*], Luis Escareño [1, 2], Francisco Pastor[3], Homero Salinas[3], Luis Iñiguez[4] and Johann Sölkner[1]

[1]*BOKU - University of Natural Resources and Life Sciences, Department of Sustainable Agricultural System, Division of Livestock Sciences, Gregor – Mendel Strasse 33, A-1180 Vienna, Austria. Tel: +43-1-47654-3260, Emails: maria.wurzinger@boku.ac.at, johann.soelkner@boku.ac.at*
[2]*Universidad Autónoma de Zacatecas, Unidad Académica de Medicina Veterinaria y Zootecnia, Carretera Panamericana, Fresnillo Km 31.5 apartado postal 9 y 11 Calera de Víctor Rosales, Zacatecas, Mexico. Email: essa_lm@yahoo.com.mx*
[3]*INIFAP- Instituto Nacional de Investigaciones Forestales, Agrícolas y Pecuarias, Campo Experimental La Laguna, Blvd. José Santos Valdez No. 1200 Pte. C.P. 27440; Matamoros, Coahuila; Mexico. Emails: pastor.francisco@inifap.gob.mx, homero.salinas@inifap.gob.mx*
[4]*ICARDA-International Center for Agricultural Research in the Dry Areas, P.O. Box 5466, Aleppo, Syrian Arab Republic. Emails: l.iniguez@cgiar.org*
** Corresponding author*

SUMMARY

This paper describes the research and development process for the design and implementation of a community-based breeding program for smallholders. The emphasis of the paper is on the organisation and practical procedures of the breeding program. The selection process of young breeding bucks based on own growth performance and on mother ´s milk performance is described. The sustainability of a breeding program can only be ensured when there is a long-term commitment from researchers and extension service to collaborate with farmers and support them in the decision-making process. Ownership of farmers was created by handing-over many responsibilities and decision mechanisms to them and researchers played the role of facilitators.

Keywords: dairy goat; breeding program; participatory research; Mexico.

INTRODUCTION

In Mexico there are about 9 million goats (Montaldo *et al.* 2010), which are mostly kept in extensive production systems under harsh environmental conditions (Mellado, 2008, Escareño *et al.*, 2012). In the arid zone of Northern Mexico more than 5 million goats are kept by 320,000 families and play an important role in their livelihoods (FIRA, 2010). In this zone goats are mainly kept for milk production, the sale of kids at the age of about 45 days is a second important income source for farmers.

An intensive production system has recently emerged next to the traditional system of keeping smaller herds on communal pasture. This new system is high-input oriented and uses irrigated alfalfa as main fodder source, pure exotic dairy goats (often imported breeding stock from the USA) and high-tech management. Artificial lighting system is used to simulate an extended photoperiod in order to stimulate the libido of bucks from temperate climates. Male off-springs from these farms are sold to nearby smallholders at a very low price, highly subsidized by the regional government. The benefit of this approach is questionable as a possible Genotype x Environment-effect is ignored and the animals have never been tested under the conditions of an extensive system and farmers reported bad experiences with breeding bucks of exotic dairy breeds (Escareño *et al.*, 2011).

Therefore, the aim of this study was the development of a community-based breeding program for smallholder farmers to make them less dependent on external supply of breeding animals, which might not fit to their prevailing production system. It was planned that farmers select the best animals of their herds and set up a scheme, which ensures that animals are exchanged across herds to avoid inbreeding problems.

Other authors (Gizaw *et al.*, 2009; Haile *et al.*, 2008; Mueller, 2006; Wurzinger *et al.*, 2008), recently showed that this new community-based or village-centred approach in animal breeding can be used for smallholder systems as this ensures the full participation of the farmers and is more sustainable in the long term as it is less dependent from external inputs. These schemes are designed according to the needs of the farmers and the technical and logistical management is at the level which farmers feel comfortable with.

Detailed descriptions of the communication and consultative process between farmers and researchers and administrative and technical details for the implementation and execution of the breeding program are presented here.

MATERIAL AND METHODS

Description of study area

The project was carried out in a rural community in La Comarca Lagunera Region in Northern Mexico. The climate can be characterized as semi-arid with an annual average temperature of 19°C and an annual mean precipitation of 250 mm with a rainy season from May to July.

The selection of the study site was based on the information from INIFAP - Instituto Nacional de Investigaciones Forestales, Agrícolas y Pecuarias taking into account the importance of goat keeping for smallholder farmers and the numbers of goats in the region.

The production system can be characterized as an extensive mixed crop-livestock system, where goats are kept on pasture all year around and only supplemented with crop residuals during specific periods of the year. For further details see Escareño *et al.* (2008).

Main production aim is the sale of milk and the sale of kids (cabritos) at the age of about 45 days, directly after weaning. Milking starts immediately after weaning of the offspring.

Farmers keep a mixture of various genotypes in their herds as they practice indiscriminate crossbreeding of their local goat called Criollo with exotic breeds like Anglo-Nubian, Saanen and Alpine, depending on the availability of bucks. A phenotypic description of the population is presented in Escareño *et al* (2011).

Communication and consulting process

A series of workshops were held with farmers to design the breeding program and establish a good relationship between the different stakeholders.

A first contact between farmers and the research team was established in July 2007. A first workshop with farmers from different communities and extension workers was held at INIFAP to identify possible constraints and opportunities for dairy goat keeping in the region. Breeding practices, such as the purchase of breeding bucks, use of bucks in the herd and selection criteria were discussed. Improved milk performance of does and growth performance of young kits were identified as breeding objectives.

Based on this information a detailed questionnaire was developed and 64 farmers were interviewed. The questionnaire consisted of general information such as herd management, herd structure, use of products and socio-economic factors, hindering and encouraging factors for the establishment of a breeding program. Results of this survey are presented in Escareño *et al.* (2008).

As a next step, workshops were held in the communities, where farmers showed interest in the formation of a farmers` group and to start jointly a breeding initiative. Researchers discussed in detail the current situation and identified possible solutions for the improvement of the current scenario.

Finally, ten farmers from one community agreed to start a breeding program and developed with the support of the research team organizational and administrative rules for the breeding program. The cooperation between participants of the breeding program was also discussed in various meetings. After a few weeks three farmers decided to leave the group and the breeders' organization continued with seven members. There were different reasons for leaving the group. One person refused to send his young bucks to a common central place, another one had health problems in his herd and one farmer sold his whole herd. Dates of regular farm visits were agreed well in advance to ensure that farmers could dedicate enough time to the data collection process.

Herd monitoring

At the beginning of the project all females of the herds were registered in a herd book. At the same time, farmers identified the best females of their herds according to their own criteria. The young bucks of these best females were also registered in the herd book and were identified as the candidates for the first selection round of breeding bucks. A final selection decision was done after the completion of the growth performance test.

Individual identification was based on ear tags used by a local program for Brucellosis prevention and complemented with a collar and a medal, which includes farm and goat number. During the identification process of females, information about age, phenotype (coat colour, absence or presence of wattles, bear, horns) and number of parities were recorded. During the follow-up visits kidding dates, litter size, body weights and sex of new born kids were documented. For the young bucks, body weight was recorded monthly.

Milk recording started after weaning, which takes place on average 45 days after kidding. Young bucks, which had been identified as selection candidates, were allowed to suckle for an extended period of 90 days before they were sent to the performance testing station. This resulted in a reduced lactation period for their dams.

Data collection started in January and continued until October 2008 and took place during the morning milking before animals were guided to the pasture. The milk yield was measured every fourteen days for each animal individually with an electronic balance. A total number of 276 does from seven herds were under the milk recording scheme, the size of herds ranged from 13 to 65 animals. Milk samples for fat and protein content were taken from 227 females five times during the lactation period.

Milk samples were taken before the milking started, a preservative was added and samples were kept cold during the transport to the milk laboratory. Milk content analysis was done with Flow Cytometrics.

Selection and rearing of young bucks

In the first year of the breeding program the selection of possible candidates as breeding bucks was exclusively based on farmers' own experience and knowledge. Selection criteria were milk yield and phenotype (body size, long legs, udder conformation) of the dam and their health status. Milk yield was assessed subjectively by farmers as there was no milk recording scheme in place. Farmers were asked to select according to their own criteria the best does from their herd to be the mothers of the future breeding bucks. The only pre-conditions set by researchers were that selected females should be above herd average, healthy and at least in their second lactation. Farmers agreed that bucks from the best does were taken to a central corral of the village at the age of 3 months.

Sixteen young bucks, two per farm, were chosen and kept in a central corral after weaning. The idea of the central corral was to have a test station where the growth performance of the young animals after weaning could be tested under the same conditions. Therefore farmers agreed to construct a central corral in their village to make the monitoring of the animals easier. Although there might be also a genotype x environment interaction between farms and test station, the research team opted for this approach. This central corral was used as an educational and demonstration tool for farmers.

The young bucks were kept in the corral for 90 days and body weight was recorded weekly. The animals were provided with a diet consisting of oat hay, maize and mineral salt.

In the second year, the selection procedure was changed and a selection index, including milk performance of the dam and own growth performance of the young bucks, was established. This selection index is based on phenotypic information. Total milk yield of the dam was used, fat and protein content was not used as there was not enough information available. As there is no information on economic values available, both traits were given equal importance and the ratio was 1:1. Milk yield was recorded twice per month and body weight of young bucks was measured on a monthly basis.

A total number of 8 young bucks after weaning at the age of three months were brought to the central corral for evaluation of their growth performance and kept there for duration of five months.

Before the young breeding bucks were re-distributed to herds, a sanitary check was done by the veterinarian of the research team. Only animals free of parasites and brucellosis were allowed to re-enter a herd.

Data management and analysis

Finally, all records from the field were entered into a data base and the information was used for further analysis. The following data was recorded: milk yield of all females was recorded every 15 days, milk fat and milk protein from 5 milkings throughout the lactation period, body weight of young bucks in the central corral was recorded once per month.

Total milk yield was calculated according to ICAR standards (ICAR, 2009). Lactation length was defined as the milking period after weaning of the kids until drying off.

Total milk yield (TMY) was calculated by using the milk yield of one test day, multiplied by the days of the interval until the next test day and the sum was made over the whole period.

For presentation of the milk performance, a cut-off point of at least 90 days in milk was used as suggested by Valencia *et al.* (2007).

The selection index was calculated based on the total milk yield of the dam and the own growth performance of the young bucks. In a first step, dams were ranked according to their milk performance from rank 1(=best) to rank 8 (=worst). Independently a ranking was done for the young bucks based on their daily gain from best to worst. In a next step, the final index was then calculated by dividing the sum of the two ranks by two. This information was used to build the final rank of the young selection candidates. All data analysis was performed using SAS 9.1 (SAS, 2003).

RESULTS AND DISCUSSION

Constraints and opportunities for dairy goat keeping

During the initial workshop different problems were highlighted by farmers. Limited accesses to credit system, feed shortage and volatile prices for products were frequently mentioned as hindering factors for the development of the dairy goat sector. Farmers were also concerned that breeding bucks were not easily available and they often had to buy animals without any information about dam's performance or in some cases they bought sick animals and infested their whole herd with brucellosis. Farmers also mentioned that they choose, depending on availability, bucks from different exotic breeds. Saanen and Alpine goats were seen as animals with high potential for high milk production, but one of the disadvantages mentioned was the relatively low fat content. Some farmers even reported that milk traders refused to buy milk from Saanen crossbred goats as they thought that the milk was diluted with water.

Farmers also explained that Saanen animals could not stand the hot weather, could not walk for long distances in the sunshine and got sick more frequently.
Anglo-Nubian was another breed which was used by farmers. This breed was associated with a lower milk yield, but much better milk fat content than the other exotic dairy breeds.

Artificial insemination was not seen as a viable way to improve production under the current situation. It was also noted that a missing recording system is another limitation for the establishment of a breeding program.

Farmers agreed that a breeding program, which will make them less dependent from external actors, would be a good opportunity for improving the goat management and also the household income. They even identified the sale of their own breeding bucks to nearby farmers as a possible new source of income.

Herd monitoring

The results of the milk recording data are presented in Table 1. The total milk yield showed a very large variation with a minimum of 52 kg and a maximum of 406 kg. This result can partly be explained by the length of the lactation period, which ranged between 90 and 248 days. It also has to be taken into account that the milk yield was only recorded after kids were weaned. In comparison, Valencia et al. (2007) reported for Sannen breed in Mexico a milk yield of 800 kg in a lactation period of 285 days, whereas Mellado *et al.* (1991) reported a total milk yield of 140 kg over 180 days in local goats. Montaldo *et al.* (1995) reported for a local Mexican goat breed a total milk yield of 299 ± 34 kg and a lactation length of 288 ± 11 days. Another study from Southern Mexico identified for local Creollo goats a production of 856 ± 28.1 kg in 100 days (Sánchez de la Rosa *et al.*,

2006). Mellado (2008) concluded that for the selection of local goats under rangeland conditions, these must produce at least 60 kg / 6 months which means 300 g/day.

The milk fat and protein content is shown in Table 2. The measured fat content was between 4.18 % and 4.96 % and the milk protein varied between 3.48 and 4%. There was no literature available from Mexico to compare with.

Selection of dams, young bucks, selection index

As a first attempt to include the important information of the dams´ milk performance the total milk yield was used. For further refinement of the breeding program it would be much better to use energy corrected milk. This would help to make better informed decisions, but this would also mean additional costs for laboratory analysis of the milk. As most of the milk is sold for processing, information on fat and protein content are essential for farmers.

Using this information might result in a re-ranking of the first two dams, but would not influence the final ranking of the young bucks (Table 3).

The daily gain of the young bucks for selection was also very heterogeneous and varied between 40 g and 168 g. Mellado (2008) reported daily gain after weaning of 75 g during the dry season, but during the rainy season daily gain can go up to 100 g.

The current used selection index is a first attempt to use recorded data instead of only subjectively assessed performance. As a next step economic values of selection criteria have to be calculated.

Table 1. Total milk yield and lactation length of all animals under recording.

			TMY			TDM	
Farm	n	Mean (kg)	SD (kg)	Min-Max (kg)	Mean (days)	SD (days)	Min-Max (days)
1	13	136	61.5	52-210	142	42.7	92-193
2	57	214	84.2	74-406	170	46.0	98-242
3	31	168	66.3	78-343	150	39.5	98-220
4	47	223	60.1	122-421	188	30.0	123-248
5	32	152	67.9	52-299	156	48.2	90-237
6	19	124	45.6	63-217	149	37.5	93-203
7	17	136	43.3	36-198	157	35.3	104-208

TMY = Total milk yield, TDM=Total days in milk, SD=Standard deviation

Table 2. Average content of milk fat (%) and milk protein (%).

			FAT			PROT	
Farm	n	Mean (%)	SD (%)	Min-Max (%)	Mean (%)	SD (%)	Min-Max (%)
1	11	4.24	0.56	3.47-4.97	3.48	0.28	2.80-3.90
2	50	4.96	0.57	2.99-8.67	3.63	0.55	1.44-4.93
3	33	4.72	0.72	3.58-6.01	3.83	0.30	3.17-5.06
4	56	4.18	1.00	2.64-8.98	3.73	0.57	2.66-5.72
5	37	4.91	1.40	2.20-8.83	4.00	0.68	2.88-6.39
6	24	4.54	0.93	3.08-6.64	3.57	0.49	2.90-4.70
7	16	4.71	0.75	3.60-5.83	3.61	0.39	2.92-4.27

FAT= average fat content (%), PROT= average protein content (%), SD=Standard Deviation

Table 3. Rank of young bucks based on selection index (milk performance of dam and own performance on growth from birth to 6 months).

Farm	Number of dam	TDM (days)	TMY (kg)	Ranking of dam	Daily gain (g)	Rank of young buck	Index	Overall rank
4	10	168	249	1	142	2	1.5	1
5	2	167	219	5	168	1	3.0	2
4	9	168	228	3	133	5	4.0	3
3	27	145	247	2	101	7	4.5	4
5	9	145	225	4	126	6	5.0	5.5
3	43	167	202	6	134	4	5.0	5.5
6	7	126	76	8	138	3	5.5	7
6	5	126	84	7	40	8	7.5	8

TDM= Total days in milk, TMY=Total Milk Yield

Organisation of breeding program

Farmers set up their own rules for their breeding program. In principal, they agreed that they would allow new members in their association, but they have to agree to follow their rules and procedures. One important point was that the mating period should not be too long as this would lead to kiddings over a longer period and as a further consequence milking would be shifted towards the dryer months with more problems of feed shortage. This was one criterion for a farmer to be allowed to enter the group.

The members also agreed that a person, who has left once the association, was not allowed to join again. The members decided to take turns in taking care of the young bucks in the central corral. At the initial stage feed costs were met by the research team, but farmers thought that a given percentage of income from sale of bucks could be kept in the future to cover the costs themselves.

Farmers also developed a simple mechanism for the selection of the young bucks. Firstly, a person was not allowed to take back his own buck to his herd. Farmers, who had provided a buck to the central corral, had the right to choose first. By drawing lots, the winner of the first round was allowed to choose an animal. The winner of the second round was then allowed to select and this procedure was repeated until everybody had a breeding buck. The remaining breeding bucks were put for sale to non-members.

After the first experiences farmers convinced INIFAP to take the young bucks and raise them on the experimental station. This was more convenient for them as they could "out-source" the daily care and fodder supply, which would mean also additional costs for them. The partners of INIFAP agreed to take over this responsibility in order not to jeopardize the initiative. The only condition was that they only take bucks from does with milk records. This was seen as an option to stimulate the recording scheme.

Communication process

Interaction between farmers and investigators was easy, because there were no language barriers. This ensured a free and direct exchange of ideas and suggestions, which is usually more difficult when communication has to be done through a translator. Regular visits of the research team in the study site ensured that farmers gained trust. The role of the researchers was more to act as facilitators in the process and to give scientific back-up support when important decisions had to be made. This constant communication process ensured that farmers gained ownership in the breeding program. The low number of only seven farmers at the beginning of the program made the communication and decision process easier as everybody had the possibility to express concerns or bring forward new ideas. As the group will continue to grow in the near future, a new form of organisation might be necessary like the election of board members of the organisation etc.

Interventions like milk recording were organized in such a way that the daily routine of farmers was not disturbed.

The research team started simultaneously a communication with policy makers to create more awareness of the needs of smallholder farmers.

Way forward

At the moment farmers are paid by the volume of milk sold. It might happen that the dairy factories change their price policy and pay farmers based on energy corrected milk. In this case it would be important that farmers include in their milk recording scheme regular analysis of milk fat and protein content. A drawback at the moment is that there is no laboratory nearby that offer this service on a routine basis.

New organisational structures have to be developed as new farmers are interested to join.

CONCLUSIONS

Farmers were open-minded, creative in problem-solving and interested to participate in a community-based breeding program. To ensure that farmers develop ownership of the program, the management of the program has to be designed according to the level of expertise and willingness of farmers to cooperate.

A community-based breeding program can work, but needs some technical support to get started and a continuous follow-up by technicians, which can decrease over time, is essential.

Acknowledgement

The authors thank the farmers for their willingness to share their knowledge with the research team. The Austrian Exchange Service is also acknowledged for financing the scholarship of L. Escareño.

REFERENCES

Escareño, L., Pastor, F., Wurzinger, M., Iñiguez, L., Salinas, H., Sölkner, J. 2008. Description of a low-input goat production system in Northern Mexico. 9th International Conference on Goats, Queretaro, Mexico, August 31- 4 September 2008.

Escareño, L., Wurzinger, M., Pastor, F., Salinas, H., Sölkner, J., Iñiguez, L. 2011. The goat and goat production systems of small-scale producers of the Comarca Lagunera in Northern Mexico, Revista Chapingo Serie Ciencias Forestales y del Ambiente. XVII (Esp): 235-246.

Escareño, L., Salinas-Gonzalez, H., Wurzinger, M., Iñiguez, L., Sölkner, J., Meza-Herrera, C. 2012. Dairy goat production systems: Status quo, perspectives and challenges. Tropical Animal Health and Production, vol. 45: 17-34.

FIRA-Fideicomisos Instituidos en Relación con la agricultura http://portal.fira.gob.mx (accesed 22/04/10)

Gizaw, S., Komen H., van Arendonk J. 2009. Optimal village breeding schemes under smallholder sheep farming systems. Livestock Science 124: 82-88.

Haile, A., Tibbo, M., Duguma, G., Mirkena, T., Wurzinger, M., Sölkner, J., 2008. Community-based sheep breeding: a new approach to genetic improvement. 16th Annual Conference of the Ethiopian Society of Animal Production, Addis Ababa, Ethiopia, October, 8-10, 2008.

ICAR. 2009. Rules, Section 2.3.Standards and guidelines for milk recording in goats. http://www.icar.org [accessed 29 April 2009]

Mellado, M., 2008. Técnicas para el manejo reproductivo de las cabras en agostadero (Goat reproductive management under rangeland conditions). Tropical and Subtropical Agroecosystems 9:47-63.

Mellado, M., Foote, R., Borrego, E., 1991. Lactational performance, prolificacy and relationship to parity and body weight in crossbred native goats in northern Mexico. Small Ruminant Research 6:167-174.

Montaldo, H., Torres, G., Valencia, M., 2010. Goat breeding research in Mexico. Small Ruminant Research 89, 2:155-163.

Montaldo, H., Juárez, A., Berruecos, J., Sánchez, F. 1995. Performance of local goats and their backcrosses with several breeds in Mexico. Small Ruminant Research 16:97-105.

Mueller, J.P., 2006. Breeding and conservation programs with local communities. FAO-WAAP workshop "Sustainable utilization as a support to the further development of the global strategy for animal genetic resources management", Ferentillo, Italia, June 28 – July 6, 2006.

Sánchez de la Rosa, I., Martínez, R., Torres, G., Becerril, C., Mastache, A., Suárez, J., Rubio, M., 2006. Milk production and lactation curves in three goat breeds in the dry tropic of Mexico. Veterinaria México. 37 (4): 493-502.

SAS. 2003. Statistical Analysis System. SAS for Windows, Release 9.1 (2002–2003). SAS Institute Inc., Cary, NC, USA.

Valencia, M., Dobler, J., Montaldo, H., 2007. Genetic and phenotypic parameters for lactation traits in a flock of Saanen goats in Mexico. Small Ruminant Research. 68: 318-322.

Wurzinger, M.,Willam, A., Delgado, J., Nürnberg, M., Valle Zarate, A., Stemmer, A., Ugarte, G., Sölkner, J., 2008. Design of a village breeding programme for a llama population in the High Andes of Bolivia. Journal of Animal Breeding and Genetics 125, (5), 311-319.

9

QUANTITATIVE TRAIT LOCI FOR FLEECE TRAITS IN BALUCHI SHEEP

G.R. Dashab[a,c *], A. Aslaminejad[a], M. Nassiri[a], A.K. Esmailizadeh [b] and D.A. Saghi [d]

[a] *Dept. Animal Science, Fac. Agriculture, Ferdowsi Univ. of Mashhad, Mashhad, Iran*
[b] *Dept. Animal Science, Fac. Agriculture, Shahid Bahonar Univ. of Kerman, Kerman, Iran*
[c] *Dept. Animal Science, Univ. of Zabol, Iran*
[d] *Agricultural and Neutral Resource Center, Mashhad, Iran*

E-mail address: dashab5@yahoo.co.in
[*]Corresponding author

SUMMARY

Three sheep chromosome regions (OAR1, OAR5 and OAR25) were selected to study quantitative trait loci (QTL) segregating for wool traits in Baluchi sheep, an indigenous sheep breed in Iran. A total of 503 progeny from 13 half-sib families were genotyped for 15 microsatellite markers. The data were collected from Research Centre on Baluchi breed from 2009- 2010. Wool traits were corrected for fixed effects of birth-year, sex, herd, and litter size. The average number of progeny per sire was 38 and ranged between 16 and 59. The QTL analyses were performed using regression-based interval mapping. The results revealed 7 QTL on OAR1 for type fiber traits (hetero-type, kemp and true wool percent in fleece), coefficient of variation of fiber diameter, greasy and clean fleece weight and staple length, 2 QTL on OAR5 for hetero-type wool percent and clean fleece, and 6 QTL on OAR25 for type wool traits, greasy and clean fleece weight and coefficient of variation of average fiber diameter.

Key words: DNA markers; QTL mapping;Baluchi sheep; fleece traits.

INTRODUCTION

Genomic research and the detection of quantitative trait loci (QTL) provide an opportunity to investigate the underlying genetic mechanism and improve the understanding of the biology of economically important traits. Although, some of wool traits have moderate to high heritability (Fogarty, 1995), because of antagonistic correlations among wool traits and also between wool and growth traits, there is a great interest in the study of genes of major effect and identifying QTLs to improve wool quality and hence, production efficiency.

In recent years, the construction of comprehensive marker maps in different species has been feasible through using techniques of molecular biology (Vaiman et al., 1996; Schibler et al., 1998; Maddox et al., 2001). These techniques allow mapping of quantitative trait loci (QTL) influencing economically important traits, including fleece and wool traits. Microsatellite makers from these maps are being used to identify inheritance patterns of linked segments of the genome in structured pedigree populations. Associations of marker allele with phenotype of interest indicate the presence of quantitative traits loci (QTL). The development of efficient and robust linear regression methods to detect and mapping of QTL in simple and complex pedigree (Haley and Knott, 1992; Knott et al., 1998; de Koning et al., 2001) have made it possible to locate the chromosomal regions influencing traits of interest.

Identifying genes affecting quantitative trait loci (QTL) of economically importance traits in livestock species has the potential to significantly increase the accuracy of predicted breeding values and to reduce the generation interval which consequently would increase the rate of genetic progress (Spelman and Bovenhuis, 1998; Dekkers and Hospital, 2002). QTL mapping also opens the possibility of studying genes and causal polymorphisms for traits of agricultural importance (Seaton et al., 2006).

The total sheep population of Iran is 54 million heads which consisted of 27 breeds and ecotypes (ASRI, 2004). The Baluchi sheep is one of the most frequent breeds, which constitutes about 30 percent of the total sheep population of Iran. It plays an important role in meat and wool production (648000 tons meat and 81000 tons wools) in the area (ASRI, 2004). The wool yield is being used for carpet production. The coat color in this breed is white and has been adapted to the harsh local climate condition.

The objective of the present study was to identify QTL affecting sheep wool production and wool quality in a half-sib design in Baluchi sheep, through the analysis of specific regions of chromosomes 1, 5 and 25.

MATERIALS AND METHODS

Animal and phenotype traits

The experimental population for the QTL study was 503 individuals belonging to 13 paternal half-sib families of Baluchi sheep from two flocks at Research Center of Baluchi sheep located in north-east of Mashhad, Iran. The average family size was 38 offspring per sire, ranging from 16 to 59. The data were collected in two consecutive years (2009 and 2010).

Wool production and wool quality traits were recorded on sire's progeny. Mid-side fleece samples were taken from individuals at 15-17 months of age and staple length (SL, mm) and greasy fleece weight (GFW, g) were measured. Then, after washing, wool samples were analyzed in a laboratory (based on textile standard) for determining clean fleece weight (CFW, g), clean wool yield (YLD, %), average fiber diameter (AFD; μm), coefficient of variation of AFD (CVAFD; %), percentage of Kemp fiber and continuous medullated fibers (KEMP; %), percentage of discontinuous medullated fibers (HET; %) and percentage of true wool (TRUE, %).

Wool trait records were analyzed by least squares analysis of variance (SAS, 2009) to identify known fixed effects including birth-year, sex, litter size, flock, shearing date and birth weight (as covariate) with significance ($p<0.05$) and such effects were included in the mixed model for variance component estimation. Further, (co)variance components were estimated with mixed model by DMU 5.0 (Madsen et al., 2011) from single-trait analyses using animal model:

$$y = X\beta + Za + e$$

In this model, y was a vector of observations for different wool traits, β was vector of fixed effects influencing traits, α was the direct additive effect with the normal distribution $N(0, A\sigma_{\alpha}^{2})$, where **A** is the additive relationship matrix and σ_{α}^{2} is the additive genetic variance and e was a randomly distributed vector of residuals. X and Z were the corresponding incidence matrices relating observation to the respective fixed effects and random animal effects. Because most of traits had not normal distribution their data were transformed (arcsine was used for YLD and natural logarithm for CVAFD, TRUE, HET and KEMP) for improvement of residual distribution.

Microsatellite genotyping

Fifteen microsatellite markers from the available web-based sheep genetic map (www.thearkdb.org/arkdb/) on chromosomes 1, 5 and 25 were used. Chromosomal region studied in the present study is given in Figure 1. Our linkage analysis covered specific regions on OAR1 (8 microsatellites from 80.8 to 234.6 cM), OAR5 (4 markers from 12.8 to 95.7 cM) and OAR25 (3 markers from 0 to 52.6), with an average distance between markers of 19.2, 20.7 and 17.5 cM, respectively.

Blood samples of thirteen sires and their progeny were collected from the jugular vein in tubes with k3-EDTA 5% as anticoagulant. The blood samples were frozen at -20 °C. The DNAs was extracted from refrigerated blood using a DNA extraction kit (CinnaGen Co., Iran). Purity of all extracted DNA was assessed by calculating the 260/280 nm ratios using Nano drop machine. PCR amplifications were carried out in 25μl reactions with 20-50 ng genomic DNA as template. Reaction mixtures contained *Taq* DNA Polymerase, dNTP, Tris-HCL, KCL and MgCl2. The cycling protocol was as follows: An initial denaturation step at 95 ^{0}C for 10 min, followed by 35 cycles with the following steps: 94^{0}C for 30s, 48-62^{0}C for 55s and 72^{0}C for 30s. The reactions were terminated by a final extension step at 72^{0}C for 10 min. The primer, amplification conditions and other information were obtained from the Roslin Institute (University of Edinburg) website (www.roslin.ed.ac.uk/). Amplification products were electrophoresed on 6 and 8% denaturing polyacrylamide gels and the DNA bands were visualized by silver staining.

In the first step, all sires were genotyped for all microsatellite markers and then the offspring were genotyped for those markers heterozygous in their sire. Dams were not genotyped. Allele size was calculated using 20 bp DNA ladder (In the rage Of 100-200bp). The scanned gels were visually scored independently by two researchers and the results incorporated in a database containing maker genotype and phenotypic data.

QTL analysis

Single trait QTL analysis was performed by multi-marker regression (Knott et al., 1996; Elsen et al., 1999) using GridQTL software (Seaton et al., 2006). Each of the adjusted phenotypes was regressed on the inheritance probabilities, at each location along each chromosome for each family. For each regression an F-ratio ($p<0.05$) of the model including the phase probability versus the same model without the phase probability was calculated. The best estimated position for a QTL in each family, for each trait, was taken to be the location with the largest F-ratio. Appropriate F-statistic thresholds for chromosome-wise type 1 error rate were generated by permutation test of 10,000 iterations (Churchill and Doerge, 1994). Confidence interval for the location of QTL was calculated by bootstrapping (Visscher et al., 1996) with 10,000 resamples.

RESULTS

Family phenotypic means and their standard deviations are shown in Table 1. Phenotypic and genetic estimates of wool traits in the overall Baluchi sheep population are presented in Table 2. Some of wool traits showed medium to high heritability.

This study detected QTL for wool traits on three chromosomes (Table 3 and Fig. 2-6). Chromosome 1 showed segregation of seven putative QTL at position 91.8, 92.8, 93.8, 93.8, 95.8, 151.8 and 227.8 affecting CFW, GFW, CVAFD, SL, TRUE, KEMP and HET traits, respectively ($p<0.01$ chromosome-wise significance level for GFW and KEMP and $p<0.05$ chromosome-wise significance level for CFW, CVAFD, TRUE, HET and SL).

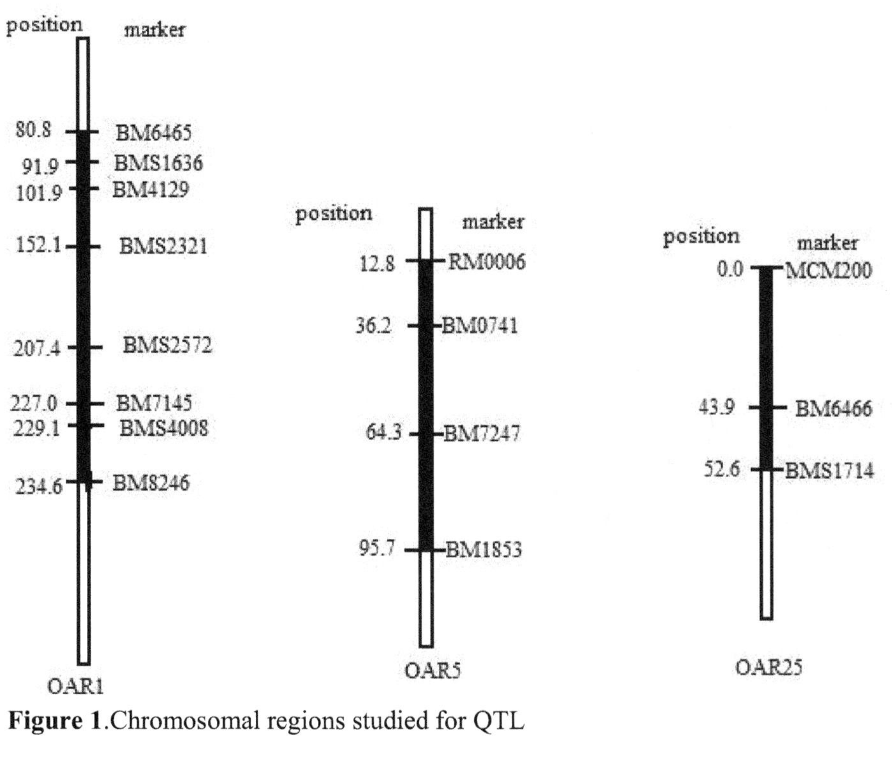

Figure 1.Chromosomal regions studied for QTL

Table 1. Family phenotypic mean ± standard deviation (SD) for wool traits in Baluchi sheep.

Family (size)	GFW (g)	YLD (%)	CFW (g)	SL (cm)	FD20 (%)	FD30 (%)	FD40 (%)	FD50 (%)	FDUP50 (%)	AFD (μm)	CVAFD (%)	HET (%)	KEMP (%)	TRUE (%)
1001 (59)	1245.6	73.7	991.7	6.69	0.60	0.24	0.09	0.05	0.02	24.8	41.9	0.03	0.02	0.95
SD	5.20	1.34	4.92	0.40	0.12	0.08	0.05	0.03	0.02	0.78	1.01	0.02	0.02	0.15
1062 (51)	1410.5	70.5	1050.9	6.54	0.56	0.26	0.07	0.04	0.06	25.8	47.2	0.03	0.03	0.94
SD	6.09	1.62	6.24	0.49	0.14	0.10	0.05	0.04	0.05	0.98	1.32	0.03	0.03	0.19
1314 (43)	1492.9	71.7	1150.5	7.14	0.63	0.23	0.08	0.03	0.02	23.9	43.6	0.03	0.01	0.96
SD	7.30	1.81	7.23	0.57	0.17	0.10	0.06	0.04	0.03	1.04	1.41	0.04	0.03	0.21
1369 (38)	1325.0	72.0	1041.4	6.83	0.61	0.24	0.09	0.05	0.02	24.2	41.5	0.03	0.02	0.96
SD	7.43	1.90	7.22	0.58	0.17	0.11	0.07	0.05	0.03	1.1	1.44	0.04	0.03	0.22
1414 (37)	1313.8	71.6	993.2	6.92	0.52	0.29	0.10	0.05	0.04	26.4	40.9	0.03	0.03	0.95
SD	6.73	1.76	6.57	0.55	0.15	0.11	0.07	0.05	0.04	1.07	1.33	0.03	0.03	0.20
1433 (45)	1421.1	73.4	1073.4	6.81	0.62	0.23	0.08	0.04	0.03	24.0	43.4	0.03	0.01	0.96
SD	6.12	1.49	5.7	0.45	0.14	0.08	0.05	0.04	0.03	0.85	1.15	0.03	0.02	0.17
1584 (16)	1241.2	73.5	900.8	6.21	0.64	0.23	0.08	0.04	0.02	23.7	41.1	0.02	0.01	0.97
SD	8.54	2.08	7.28	0.60	0.19	0.12	0.07	0.05	0.03	1.18	1.55	0.03	0.03	0.24
2310 (34)	1474.1	74.3	1083.4	7.92	0.53	0.26	0.11	0.06	0.04	26.6	43.5	0.03	0.03	0.94
SD	7.39	1.88	7.18	0.61	0.16	0.11	0.07	0.05	0.04	1.13	1.44	0.04	0.04	0.21
2338 (29)	1278.3	71.1	927.8	6.75	0.55	0.28	0.08	0.05	0.04	25.8	41.1	0.03	0.02	0.96
SD	7.45	1.84	6.65	0.57	0.16	0.12	0.06	0.05	0.04	1.11	1.40	0.03	0.03	0.21
2348 (30)	1458.3	73.5	1136.4	6.95	0.61	0.22	0.08	0.06	0.03	24.7	41.2	0.03	0.02	0.95
SD	7.8	1.87	7.36	0.58	0.17	±0.10	±0.06	±0.05	±0.03	±1.08	±1.40	±0.04	±0.03	±0.21
2365 (34)	1558.6	71.5	1130.7	7.26	0.58	0.25	0.07	0.04	0.06	25.1	45.1	0.03	0.02	0.95
SD	7.33	1.73	6.86	0.55	0.16	0.10	0.05	0.04	0.05	1.02	1.37	0.03	0.03	0.20
2382 (36)	1808.3	68.2	1256.5	7.26	0.60	0.25	0.09	0.04	0.02	24.3	41.9	0.02	0.01	0.97
SD	8.68	1.76	7.56	0.57	0.17	0.11	0.06	0.04	0.03	1.05	1.38	0.03	0.03	0.21
2413 (38)	1373.3	70.2	997.9	6.83	0.59	0.24	0.09	0.05	0.02	24.9	41.4	0.03	0.02	0.95

Family (size)	GFW (g)	YLD (%)	CFW (g)	SL (cm)	FD20 (%)	FD30 (%)	FD40 (%)	FD50 (%)	FDUP50 (%)	AFD (µm)	CVAFD (%)	HET (%)	KEMP (%)	TRUE (%)
SD	6.76	1.67	6.31	0.52	0.15	0.15	0.06	0.04	0.03	0.99	1.28	0.03	0.03	0.19
Mean (490)	1410.3	72.0	1055.5	6.92	0.59	0.25	0.09	0.05	0.03	24.9	42.7	0.03	0.02	0.95
SD	24.62	0.43	19.37	0.08	0.008	0.01	0.00	0.00	0.00	0.19	0.45	0.00	0.00	0.00

Greasy fleece weight, GFW; Clean fleece weight, CFW; Clean wool yield, YLD; Average fiber diameter, AFD; Coefficient of variation of AFD, CVAFD; The percentage of fiber with diameter under 20 µm, FD20; The percentage of fiber with diameter between 20 to 30 µm, FD30; The percentage of fiber with diameter between 30 to 40 µm, FD40; The percentage of fiber with diameter between 40 to 50 µm, FD50; The percentage of fiber with diameter up 50 µm, FDUP50; The percentage of hetero type wool, HET; The percentage of medullated wool, KEMP; The percentage of true wool, TRUE.

QTL for different traits were mapped to similar regions of the chromosome and this may suggest that the same gene(s) might be involved in the control of the traits. In chromosome 1, QTL for GFW, CFW in the small distance and QTL for CVAFD and SL were mapped in similar regions on OAR1 (Figure 2)

The two QTL segregation on chromosome 5 were related to HET and CFW at 87.8 cM with effects from 0.67 to 3.8 phenotypic standard deviations for HET in 2348 and 2382 families and an effect of 0.97 phenotypic standard deviation for CFW in 2382 families ($p<0.05$).

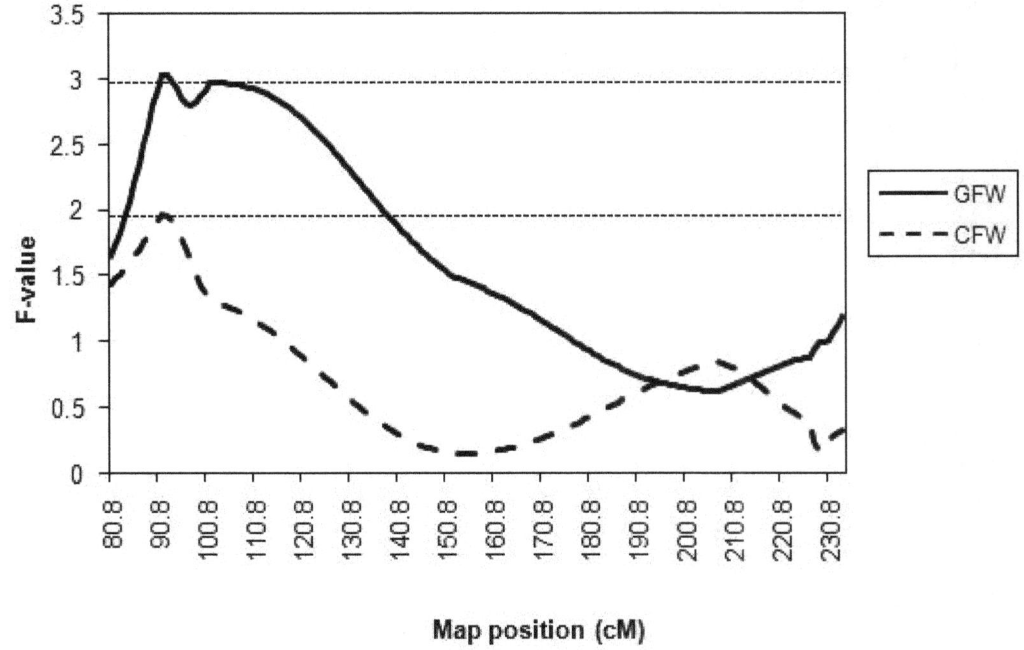

Figure 2. The F-value profile obtained from across family analysis for greasy fleece weight (the solid line) and clean fleece weight (the dash line) on sheep chromosome 1. The lower and upper horizontal dotted lines represent 5% and 1% chromosome-wide significant levels of linkage, respectively.

Table 2. Summary of phenotypic and genetic parameters of fleece traits in Baluchi sheep.

Traits	Average	SD	MIN	MAX	δ^2a	R	h^2	S.E.(h^2)
GFW	1519	0.544	100.0	4400	10634	43836	0.195	0.08
YLD	72.35	7.795	45.19	91.95	6.0303	19.647	0.235	0.01
CFW	1103	0.432	72.00	3549	6973.6	28146	0.198	0.01
SL	6.998	1.550	3.375	15.00	0.9999	1.0742	0.482	0.18
AFD	25.11	3.555	17.43	38.98	1.8313	3.8434	0.322	0.07
CVAFD	42.87	8.280	20.80	83.85	9.1300	59.300	0.133	0.11
HET	0.026	0.028	0.000	0.228	0.8332	0.43515	0.650	0.23
KEMP	0.021	0.025	0.000	0.230	0.8332	0.43510	0.650	0.25
TRUE	0.954	0.042	0.708	1.000	0.8682	0.4632	0.650	0.44

Minimum, MIN; Maximum, MAX; Additive genetic variation, δ^2a; Residual error, R; Heritability, h^2 Standard error of heritability, S.E.

There was a highly significant QTL on OAR25 affecting HET, KEMP, TRUE and GFW. Six QTL related to wool traits were detected on this chromosome, i.e. GFW at 23cM with effects from 0.75 to 0.57 phenotypic standard deviations for 1414 and 1314 families ($p<0.01$), TRUE at 28 cM with effects from 0.91 to 1.35 phenotypic standard deviations for 1001, 1414 and 1062 families ($p<0.01$), KEMP at 32 cM with effects from 1.06 to 1.96 phenotypic standard deviations for 1414 and 1062 families ($p<0.01$), HET at 42 cM with effects from 0.33 to 4.17 phenotypic standard deviations for 2413, 1001 and 1414 families, CFW at 46cM with effects from 0.55 to 0.79 phenotypic standard deviations for 2310 and 1414 families ($p<0.05$) and CVAFD at 52cM with effects from 0.92 to 1.21 phenotypic standard deviations for 1001 and 2310 families ($p<0.05$) (Figure 3, 5 and 6).

Significant results for the two-QTL models are presented in Table 4. For GFW two QTL were detected but they were located very close to each other and could be one QTL. Also two-QTL model for CVAFD and HET were significant.

Confidence interval for all QTL situations included the complete region studying on chromosome. The effect/standard deviation ranged from 0.33 to 5.34 phenotypic standard deviations and is shown in the last column of Table 5.

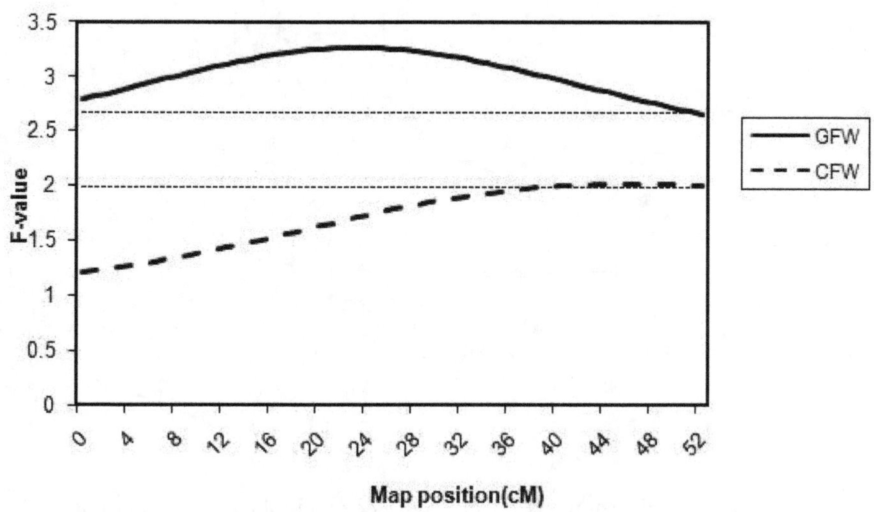

Figure 3. The F-value profile obtained from across family analysis for greasy fleece weight (the solid line) and clean fleece weight (the dash line) on sheep chromosome 25. The lower and upper horizontal dotted lines represent 5% and 1% chromosome-wide significant levels of linkage, respectively.

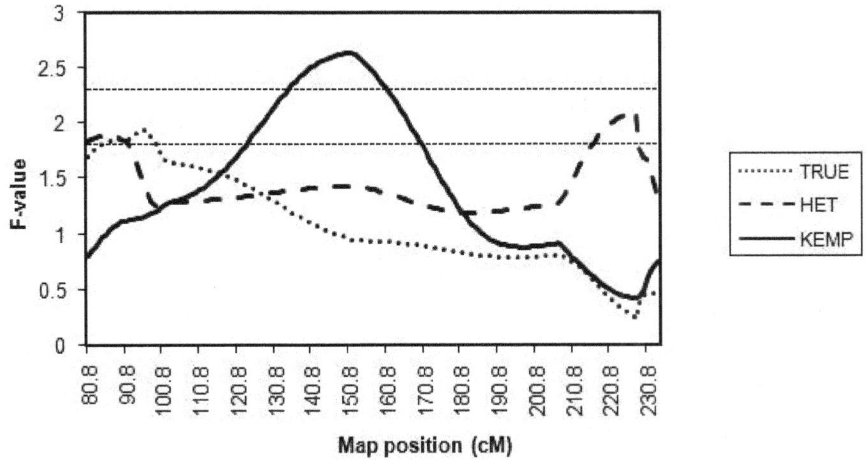

Figure 4. The F-value profile obtained from across family analysis for KEMP (the solid line), TRUE (the square dot line) and HET (the dash line) on sheep chromosome 1. The lower and upper horizontal dotted lines represent 5% and 1% chromosome-wide significant levels of linkage, respectively.

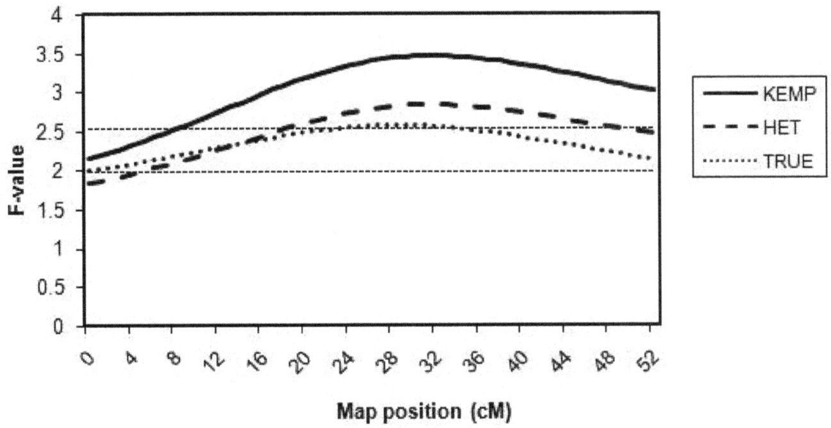

Figure 5. The F-value profile obtained from across family analysis for KEMP (the solid line), TRUE (the square dot line) and HET (the dash line) on sheep chromosome 25. The lower and upper horizontal dotted lines represent 5% and 1% chromosome-wide significant levels of linkage, respectively.

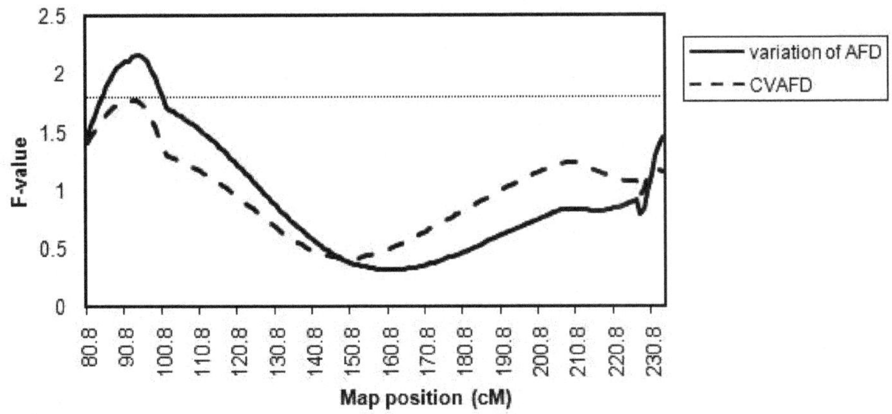

Figure 6. The F-value profile obtained from across family analysis for variation of fiber diameter (AFD), and coefficent of variation of fiber diameter (CVAFD) on sheep chromosome 1. The horizontal dotted line represent 5% chromosome-wide significant level of linkage.

Table 3. Summary of QTL segregation for wool traits in Baluchi sheep

OAR	Position	Trait	F-value	Family	Effects± S.E.	Nearest marker	Effect/SDp[a]
1	227.8	HET	2.11**	2310	0.71±0.34	BM7145	3.69
				2382	0.78±0.35		4.33
				1433	0.78±0.29		1.56
	151.8	KEMP	2.64**	1433	1.97±0.76	BMS2321	5.34
				1001	0.87±0.26		1.47
	95.8	TRUE	1.95*	2348	0.56±0.31	BMS1636	0.59
				2413	0.46±0.21		0.48
				1314	0.42±0.20		0.44
	93.8	CVAFD	1.76*	1062	23.52±11.25	BMS1636	0.5
				1314	22.12±9.92		0.51
				2310	18.29±10.29		0.42
	92.8	GFW	3.11**	1314	841.97±218.55	BMS1636	0.56
				1414	540.86±262.84		0.41
	91.8	CFW	1.95*	1314	567.60±260.34	BMS1636	0.49
				2413	483.08±221.91		0.48
				2310	468.52±240.19		0.43
	93.8	SL	1.71*	2413	3.34±1.54	BMS1636	1.28
				2310	3.016±1.69		1.26
				1314	3.29±1.63		1.23
5	87.8	HET	1.97*	2348	1.14±0.36	BM1853	3.8
				2382	1.21±0.54		0.67
	87.8	CFW	2.09*	2382	1215.78±402.1	BM1853	0.97
25	42	HET	2.33**	2413	0.075±0.03	BM6466	4.17
				1001	0.056±0.025		2.0
				1414	0.024±0.01		0.33
	32	KEMP	3.46**	1414	0.051±0.01	BM6466	1.96
				1062	0.034±0.01		1.06
	28	TRUE	2.57**	1001	1.29±0.61	BM6466	1.35
				1414	0.86±0.42		0.91
				1062	0.86±0.40		0.91
	52	CVAFD	2.18*	2310	39.91±17.97	BMS1714	0.92
				1001	50.73±30.11		1.21
	23	GFW	3.26**	1414	989.13±272.28	BM6466	0.75
				1314	849.42±335.25		0.57
	46	CFW	2.01*	1414	547.12±230.11	BM6466	0.55
				2310	857.43±473.14		0.79

* and ** = Genome-wise significant level ($p<0.05$ and $p<0.01$, respectively).
a = QTL effect in phenotypic deviation units.

DISCUSSION

In the present study, we reported several putative QTL for wool traits on chromosome 1 (seven QTL related to HET, KEMP, TRUE, GFW, CFW, CVAFD and SL), chromosome 5 (two QTL related to CFW and HET) and chromosome 25 (Six QTL related to HET, KEMP, TRUE, GFW, CFW, and CVAFD) in Baluchi sheep.

The putative QTLs affecting hetero-type (HET), kemp and true wool percentage were observed on chromosome 1. The position of the putative QTLs related to these traits was far from others. The results suggested that these traits could be controlled by different set of genes. Two QTL were detected for GFW and CFW on chromosome 1, i.e. close to BMS1636 marker. The short distance between these QTLs, and the high genetic and phenotypic correlation between GFW and CFW traits in the present study (r_g= 0.89 and r_p=0.91), suggested that these QTLs could be a single QTL with the same effect on both traits.

A QTL influencing coefficient of variation of fiber diameter on chromosome 1 was also detected. Cano et al. (2007) detected a putative QTL for CVAFD at position of 96 cM on chromosome 1 in Angora goat. Due to homology of sheep and goat (Maddox, 2005) it is confirmed that a QTL related to fiber diameter is at this position of chromosome. Also, a putative QTL related to staple length was detected close to the QTL responsible for wool yield. Beh et al. (2001) also detected a putative QTL on chromosome 1 related to staple length.

In sheep, there are several reports available for the linkage between genes and QTL in wool production traits (Ponz et al., 2001, Bray et al., 2002, Purvis and Franklin, 2005). The putative QTL for wool traits found in Baluchi sheep on chromosome 1 could be related to keratin (KRT) and keratin-associated protein (KRTAP) family genes as pointed out by McLaren et al. (1997).

The Kemp fibers are large modulated fibers with a latticed deficient in sulphur. OAR1 harbors one of the keratin-associated-protein clusters (KAP, KAP7 and KAP8) involved in the formation of matrix surrounding cortex wool micro fibrils (McLaren et al., 1997; Ede et al., 1995; Parsons et al., 1993). Another important wool follicle protein, i.e. trichohyalin (THH), encoded by a single gene, was also mapped in chromosome 1 (McLaren et al., 1997). A linkage between this keratin cluster gene and fiber diameter in a merino half-sib was observed (Parsons et al., 1994) confirming that KAP cluster genes are strong candidates for the wool quality traits studied in Baluchi sheep.

In this study, two QTL on chromosome 5 were detected which are related to clean fleece weight and hetero-type wool percent traits. In Angora goat a QTL for Kemp wool percent trait on the position of 20 cM was detected (Cano et al., 2007). Two keratin family genes (KRT8 and KRT1B) have been assigned to chromosome 5 in sheep, goat and cattle (Fries et al., 1991; Schibler et al., 1998; Pinton et al., 2000), all of them could be related to those QTLs found here on chromosome 5 of Baluchi sheep.

Significant QTLs affecting several wool traits were observed on chromosome 25 suggesting the evidence of a gene with major effect on wool traits. Ponze et al. (2001) in INRA401 breed have described three QTLs related to AFD, CVAFD and staple length traits. Allain et al. (2006) in a Lacaune × Sarda backcross detected four QTLs for fleece characteristics on OAR25, including greasy fleece weight at position of 17 cM, mean diameter at position of 2 cM, coefficient of variation of diameter fiber at position of 4 cM and medullated fiber percent at position of 7 cM. In this study, putative QTLs for CVAFD, GFW, CFW, and Kemp were detected at positions of 52, 23, 46, 35 and 32 cM on this chromosome, respectively. Therefore, these results and others confirm that there is a gene or genes with major effect on chromosome 25 which are responsible for wool production and wool quality traits.

CONCLUSION

The study has been successful in detecting QTL in Baluchi sheep populations, and contributes to the understanding of chromosomal region that explain part of the variability of wool quality traits in Baluchi sheep especially in chromosome 1 and 25. The QTL effects raged from 0.33 to 5.34 in units of phenotypic standard deviation, thus the experiment had only a power to detect medium to large QTL. The QTL involved in wool traits, might allow considerable improvement in the response to selection.

ACKNOWLEDGMENTS

This study was funded and supported by Department of Animal Science, Ferdowsi University of Mashhad, Iran. The authors also acknowledge the staff of the Research Centre of Baluchi sheep in north-east of Iran.

REFERENCES

Allain, D., Schibler, L., Mura, L., Barillet, F., Sechi, T., Rupp, R., Casu, S., Cribiu, E., Carta, A. 2006. QTL detection with DNA markers for wool traits in a sheep backcross Sarda* Lacaune resource population. 8th Word Congress on Genetics Applied to Livestock Production, Belo Horizonte, MG, Brazil, August 13-18.

Animal Science Research Institute of IRAN Tehran 2004. Draft Iran's Country Report on Farm Animal Genetic Resources: Ministry of Jihad-e-Agriculture.

Bray, A.R., O'Connell, D., Saville, D.J. 2002. Genes with major effects on wool traits detected in Finn cross sheep. Proc. N. Z. Doc. Animal Production. 62:62-68.

Cano, E.M., Marrube, G., Roldan, D.L., Bibinost, F., Abad, M., Allain, D., Vaiman, D., Taddeo, H., Poli, M.A. 2007. QTL affecting fleece traits in Angora goats. Small Ruminant Research. 71:158-164.

Churchill, G.A., Doerge, R.W. 1994. Empirical threshold values for quantitative trait mapping. Genetics. 138:963-971.

de Koning, D.J., Schulmant, N.F., Elo, K., Moisio, S., Kinos, R., Vilkki, J., Maki-Tanila, A. 2001. Mapping of multiple quantitative trait loci by simple regression in half-sib design. Journal of Animal Science. 79:616-622.

Dekkers, J.C., Hospital, M. 2002. The use of molecular genetics in the improvement of agricultural populations. Nature Reviews Genetics. 3: 22-32.

Ede, A.J., Crawford, A.M. 1995. Mutations in the sequence flanking the microsatellite at the KAP8 locus prevent the amplification of some alleles. Animal Genetic. 26:43-44.

Elsen, J.M., Mangin, B., Goffinet, B., Boichard, D., LeRoy, P. 1999. Alternative models for QTL detection in livestock. I. General introduction. Genetic Selection Evolution. 31: 213-224.

Fogarty N. M. 1995. Genetic parameters for live weight, fat and muscle measurements, wool production and reproduction in sheep. a review. Animal Breeding Abstract. 63:101.

Fries, R., Threadgill, D.W., Hediger, R., Gunawardana, A., Blessing, M., Jorcano, J.L., Stranzinger, G., Womack, J.E. 1991. Mapping of bovine cytokeratin sequences to four different sites on three chromosomes, Cytogenetic Cell Genetic. 57:135-141.

Haley, C.S., Knott, S.A. 1992. A simple regression method for mapping quantitative trait loci in line crosses using flanking markers. Heredity. 69: 315-375.

Knott, S.A., Elsen, J.M., Haley, C.S. 1996. Methods for multiple marker mapping of quantitative trait loci in half-sibs population. Theoretical Applied Genetic. 93:71-80.

Knott, S.A., Marklund, L., Haley, C.S., Andersson, K., Davies, W., Ellegren, H., Fredholm, M., Hannsson, I., Lundstrom, K., Moller, M., Andersson, L. 1998. Multiple marker mapping of quantitative trait loci in an outbred cross between wild boar and large white pigs. Genetics, 149: 1069-1080.

Maddox, J.F. 2005. A presentation of the differences between the sheep and goat genetic maps, Genetic Selection Evolution. 37 (Suppl. 1):S1-S10.

Maddox, J.F., Davies, K.P., Crawford, A.M., Hulme, D.J., Vaiman, D., Cribiu, E.P., Freking, B.A., Beh, K.J., Cockett, N.E., Kang, N., Riffkin, C.D., Drinkwater, R., Moore, S.S., Dodds, K.G., Lumsden, J.M., van Stijin, T.C., Phua, S.H., Adelson, D.L., Burkin, H.R., Broom, J.E., Buitkamp, J., Cambridge, L., Cushwa, W.T., Gerard, E., Galloway, S.M., Harrison, B., Hawken, R.J., Hiendleder, S., Henry, H.M., Medrano, J.F., Paterson, K.A., Schibler, L., Stone, R.T., van Hest, B. 2001. An enhanced linkage map of the sheep genome comprising more than 1000 loci. Genome Research. 7:1275-1289.

Madsen, P., Jensen, J. 2011. A package for analyzing multivariate mixed models, Version 6, Release 5.

McLaren, R.J., Roger, G.R., Davies, K.P., Maddox, J.F., Montgomery, G.W. 1997. Linkage mapping of wool keratin and keratin-associated protein genes in sheep. Mammalian Genome. 8:938-940.

Parsons, Y.M., Cooper, D.W., Piper, L.R. 1994. Evidence of linkage between high-glycine-tyrosine keratin gene loci and wool fiber diameter in a Merino half-sib family. Animal Genetic. 25:105-108.

Parsons, Y.M., Cooper, D.W., Piper, L.R., Powell, B.C. 1993. A BamHI polymorphism in ovine glycine-tyrosine-rich type II (KATAP6) keratin sequences. Animal Genetic. 24:218.

Pinton, P., Schibler, L., Cribiu, E., Gellin, J., Yerle, M. 2000. Localization of 113 anchor loci in pigs: improvement of the comparative map for humans, pigs and goats, Mammalian Genome. 11:306-315.

Ponz, R., Moreno, C., Allain, D., Elsen, J.M., Lantier, F., Lantier, L., Brunel, J.C., Perez-Enciso, M. 2001. Assessment of genetic variation explained by markers for wool traits in sheep via a segment mapping approach. Mammalian Genome. 7(12):569-572.

Purvis, I.W., Franklin, I.R. 2005. Major genes and QTL influencing wool production and quality: a review. Genetic Selection Evolution. 37:S97-S107.

SAS User Guide: statistics (2009), SAS institute (version 9.2) Cary: NC.

Schibler, L., Vaiman, D., Oustry, A., Giraud-Delville, C., Cribiu, E.P. 1998. Comparative gene mapping: a fine-scale survey of chromosome rearrangements between ruminants and humans. Genome Research. 9: 901-915.

Seaton, G., Hernandez, J., Grunchec, J.A., White, I., Allen, J., de Koning, D.J., Wei, W., Berry, D., Haley, C., Knott, S. 2006. GridQTL: a grid portal for QTL mapping of compute intensive datasets, In: Proceedings of the 8th Word Congress on Genetics Applied to Livestock Production, Belo Horizonte, Brazil, August 13-18.

Vaiman, D., Schibler, L., Bourgeois, F., Oustry, A., Amigues, Y., Cribiu, E. 1996. A genetic linkage map of the male goat genome. Genetics. 144:279-305.

Visscher, P., Thompson,B. R., Haley, C. 1996. Confidence intervals in QTL mapping by bootstrapping. Genetics. 143:1013-1020

10

SMALLHOLDER SHEEP FEEDING BASED ON DEFOLIATED CASSAVA AND MAIZE LEAVES

O.A. Fasae[+*] , I.F. Adu and A.B.J. Aina

Department of Animal Production and Health, University of Agriculture, P.M.B. 2240, Abeokuta, Nigeria.

[+] Email: animalexp@yahoo.co.uk

**Corresponding author*

SUMMARY

The recent surge in the production of cassava and maize in Nigeria has generated a surplus of leaves which if adequately utilized could alleviate the feed shortage problem encountered by smallholder in sheep production as a result of seasonal reduction in feed supply and reduced pasture quality. An experiment was conducted with West African Dwarf yearling rams to investigate the feeding value of defoliated cassava leaves (CL) and maize leaves (ML) at different ratios (CL/ML) in the diets; 100:0 (solely ML); 25: 75; 50: 50; 75: 25; and 0:100 (solely CL). The experiment that lasted for 105 days involved twenty (20) rams in a completely randomized design used to evaluate the dry matter (DM) intake, weight gain, nutrient digestibility and nitrogen utilization as well as the effect of the diets on the haematological profile of the rams. Results showed that DM intake and digestibility of nutrients were similar in the mixed forage diets and significantly ($P<0.05$) higher than that of the sole forage diets of cassava and maize. Live weight gain (g/day) peaked at the ratio 75%CL/25%ML diet with rams fed sole forage diets having significantly lower gain in weight. The intake of the CP increased significantly as the component of CL in the diet increased. The N balance (g/day) and retention (%) were also highest in the mixed forage diets and lowest in the solely ML diets. Blood parameters monitored were not affected by the experimental diets. It was concluded that DM intake, weight gain, nutrient digestibility and nitrogen utilization of sheep can be enhanced by feeding 75:25 (CL/ML) in the diet.

Keywords: Cassava leaves; digestibility; maize leaves; sheep; smallholder; weight gain

INTRODUCTION

Sheep production in Nigeria is a growing enterprise among smallholders that combine sheep husbandry with other agricultural and non-agricultural activities. The production of sheep is however limited among other factors by inadequacy of year round feed availability. The use of concentrate rations based on cereal grains or root crops as energy feeds and grain legumes have led to direct competition with man and industries for the same resources (Fasae and Alokan, 2006). This coupled with high cost of concentrate has guided to the search for non- competitive alternative feedstuffs (Olorunsomo, 2008).

The recent surge in the production of cassava and maize in Nigeria has generated a lot of leaves which are currently underutilized. These leaves are considered waste products in settled rural areas and they are readily available for smallholder and their use in agro-pastoral systems of the region. Their availability and abundance makes them a cheap source of nutrients especially dietary energy, protein and fermentable products for sheep production. This is because sheep can conveniently handle these fibrous wastes through microbial fermentation (Adebowale and Taiwo, 1996). Cassava and maize leaves have been found to be a valuable feed material for ruminants, meeting the nutritional requirement for growth, reproduction and maintenance of ruminants' year round (Semenye et al. 1994; Wanapat et al. 2000). In a previous study, Fasae et al. (2009a, b) assessed the effect of defoliation time on leaf yield of cassava (TMS 30572 variety) and maize (DMR-SR-L-Y variety) plants, they found that cassava and maize leaves can be defoliated after 5 and 3 months planting, respectively and fed to ruminants without adverse effect on the crop yield. Also, earlier reports presented by Muller (1977) and Wanapat et al. (1997) showed that cassava foliage could be harvested after 4 months planting to ensure high nutrient content and to avoid reduction in tuber yield while Abate and Abate (1994) reported that maize defoliated after 79 days can be a source of quality feed for ruminants. This however, shows the potential of defoliated cassava and maize leaves as available sources of nutrients for sheep production thereby addressing the major problem facing stockowners in Nigeria as a result of provision of adequate nutrition for sheep especially during the dry season.

This study therefore investigates the feed intake, weight gain, nutrient digestibility, nitrogen utilization and haematological profile of West African Dwarf sheep fed varying levels of defoliated cassava and maize leaves diets which are readily available feedstuffs among smallholders in Nigeria.

MATERIALS AND METHODS

Animals and management

The study was conducted in the small ruminant experimental unit of the Teaching and Research Farms, University of Agriculture, Abeokuta, Nigeria. Twenty West African Dwarf (WAD) yearling rams with average body weight of 15.25 kg were purchased from smallholder farmers in villages near to the University. Prior to the experiment, the rams were kept under inspection for 30 days. During this period they were vaccinated with broad spectrum Terramycin L.A. (1 ml/10 kg body weight) and dewormed with albendazole tablets (12.5 g/kg body weight) and diazintol solution (6 ml/Lt).

On arrival at the experimental site, the rams were housed individually in open sided well ventilated disinfected pens, with slatted floors to prevent direct contact with their faeces and urine. They were randomly allocated to 5 treatments of 4 rams each balanced for body weights. Rams were allowed to adapt to the experimental diets for two weeks before the beginning of the experiment.

Experimental diets preparation

Cassava leaves (CL) TMS30572 and maize leaves (ML) DMR-SR-L-Y were randomly defoliated during the late rainy season (October to November) from an established plot of an hectare each at 5 months and 12 weeks after planting, respectively as in Fasae et al., (2009a, b). Leaves were chopped by a cutlass into smaller parts of about 4cm and sun dried for 5 days. Animal feeding lasted 105days and 4% of the rams' body weight was fed and they had clean water *ad libitum*. . The rams were randomly assigned to the treatments of dehydrated CL and ML at different ratio (CL/ML) in the diets; 100:0 (solely ML); 25: 75; 50: 50; 75: 25; and 0:100 (solely CL). Each treatment was supplemented with 100g of concentrate composed of 40% wheat bran, 35% corn bran, 22% palm kernel cake, 2% bone meal, 0.5% common salt and 0.5% vitamin premix balanced for protein and energy requirements (NRC, 1975). Individual troughs were used for placing the feed and water.

Apparent digestibility and nitrogen balance trials

The rams were transferred immediately after the growth trial into individual metabolic crates which allowed collection of total faecal and urine outputs. Each experiment consists of a 7-day dietary adjustment period followed by a 7-day balance trial. The feed was offered to the animals twice daily at 9 a.m. and 2 p.m., respectively. The quantity of feed offered and refusal, faeces and urine were determined

for each sheep in the last 7 days of the trial. Nitrogen loss by volatilization was prevented by introducing 15 ml of 10% H_2SO_4 into urine collection container. Samples collected were pooled over a 7- day period, and then frozen until subsequent laboratory analysis.
5 ml of blood samples were collected in ethylenediamine tetracetic acid bottles from each ram used at the beginning and end of the experiment. Blood samples were obtained from jugular vein (Shewati and Dagash, 1994) to determine the effect of the diets on the blood cells. Blood samples were taken to the laboratory to determine the packed cell volume, red blood cells, White blood cells, glucose, blood urea nitrogen, total protein (Lamb 1991; Davice and Lewis, 1991).

Chemical Analyses

The chemical compositions of the leaves and faecal samples as well as nitrogen in urine were determined following the methods described in AOAC (1990). The DM was determined by oven drying at 100^0C for 24 hours, Crude protein by Kjedhal method and fat by Soxhlet fat analysis. The concentration of neutral detergent fibre (NDF), acid detergent fibre (ADF) and lignin both in feed and faecal samples were also determined by the method of Van Soest et al. (1991). The hydrocyanic (HCN) content of cassava leaves in feed and faeces were also determined as described by Bradbury et al. (1999). Analyses of tannins were carried out by methods described in Makkar et al. (1993).

Data Analysis

Data collected from parameters investigated were based on completely randomized design and subjected to one way analysis of variance using the statistical package (SAS, 1999). Significant means were separated using Duncan Multiple Range Test (Duncan, 1955).

RESULTS

The DM content of the sole and mixed forages of cassava and maize leaves was similar across the treatments. The CP content of the diets differs across the treatments with an increase with the inclusion of cassava leaves in the diets (Table 1). The NDF content was higher in ML relative to CL while ADF and ADL contents were lower in ML compared the CL. HCN and tannins were not detected in ML sole diet. The CL sole diet was richer in calcium compared to ML while phosphorus content was vice versa.

Table 1. Chemical composition (g/Kg DM) of cassava and maize leaves based diets fed to West African Dwarf sheep.

			Diets			
Composition	0%CL	25%CL	50%CL	75%CL	100%CL	Conc.
Dry matter	908.2	903.4	910.0	912.1	909.1	84.24
Crude protein	118.5	147.0	160.8	181.4	205.0	167.2
Neutral Detergent Fibre	645.6	609.1	571.1	550.7	527.6	296.2
Acid Detergent Fibre	404.0	394.4	381.1	357.3	344.0	412.4
Acid Detergent Lignin	72.0	70.3	68.3	65.1	62.4	31.4
Energy (MJ/kg DM)	14.9	15.6	14.9	16.2	17.2	17.9
Hydrocyanic acid (Mg/kg)	0.00	11.1	16.3	19.8	22.4	0.00
Tannin	0.00	1.0	1.4	1.6	1.8	nd
Calcium	5.0	6.3	7.8	10.1	13.4	nd
Phosphorus	4.0	3.9	3.3	2.9	2.3	nd

[MH] Maize hay [CH] Cassava hay [Conc.] Concentrate [nd] Not determined

The DM intake was highest ($P < 0.05$) in 75:25 (CL/ML) diets which ranks the same statistically with other mixed forage diets, compared the sole forage diets of cassava and maize (Table 2). The intake of the CP increased ($P < 0.05$) with an increase in the level of CL in the diet while the average DM intake for concentrate diets which was all consumed was 84.2g/day in all the treatments. The live weight gain (g/day) in rams fed mixed forage diets was significantly ($P < 0.05$) higher compared to those fed sole cassava and maize forage. The feed conversion ratio was statistically similar ($P > 0.05$) in all the treatments.

The DM and CP digestibilities varied significantly ($P < 0.05$) across the treatments with the mixed forages having the highest ($P < 0.05$) compared to the sole diets (Table 3). The DM digestibility was least in 0%CL diet and highest ($P < 0.05$) in the mixed forage diets. The NDF digestibility coefficient varied ($P < 0.05$) across the treatments. Significant differences ($P > 0.05$) was not observed in the ADF and ADL digestibilities.

Table 2. Performance indices of West African Dwarf sheep fed cassava and maize leaves based diets.

	Diets					
Parameters	0%CL	25%CL	50%CL	75%CL	100%CL	SEM
DM Intake (g/Kg DM/day)						
Cassava leaves	-	1244.1	2382.2	3868.0	4599.1	
Maize leaves	4265.2	3577.0	2589.1	1342.3	-	
Concentrate	842.4	842.4	842.4	842.4	842.4	
Total	5107.6^{b}	5663.5^{ab}	5813.7^{ab}	6052.7^{a}	5441.5^{b}	33.50
Live weight gain (g/day)	38.8^{c}	44.5^{b}	46.4^{ab}	47.9^{ab}	41.6^{c}	1.51
LW gain (g DM/$W^{0.75}$/day)	62.5^{b}	68.4^{a}	68.3^{a}	67.2^{a}	61.0^{c}	1.89
Feed conversion ratio	13.1	12.7	12.5	12.6	13.1	0.82

Table 3. Digestibility coefficients (g/Kg DM) of major nutrients of the experimental diets fed to West African Dwarf sheep.

	Diets					
Parameters	0%CL	25%CL	50%CL	75%CL	100%CL	SEM
Dry matter	614.4^{c}	675.6^{ab}	716.4^{a}	742.2^{a}	649.1^{b}	19.8
Crude protein	634.4^{b}	691.4^{a}	712.4^{a}	718.1^{a}	649.7^{b}	16.3
Neutral detergent fibre	635.5^{b}	665.0^{ab}	696.1^{a}	723.5^{a}	633.2^{b}	6.79
Acid detergent fibre	649.1	673.5	698.1	719.5	659.0	6.17
Acid detergent lignin	496.3	531.8	540.4	565.7	531.5	8.32

Mean values in the same row with the same superscripts are not significantly ($P > 0.05$) different.

The nitrogen intake in rams increased ($P<0.05$) with increase in the level of CL in the diets (Table 4). The faecal and urinary N excretions varied among treatments with the excretions from sheep fed sole CL and 75:25 (CL/ML) having significantly ($P < 0.05$) higher levels compared to the other treatments. The N balance was least ($P < 0.05$) in 0%CL diets and highest ($P < 0.05$) in sole CL and 75:25 (CL/ML) diets. However, the N retention values were highest ($P < 0.05$) for mixed forage diets than the sole diets of cassava and maize.

The ranges and mean values for the hematological parameters of the experimental sheep prior and after the experiment are presented in Table 5. The results indicate that the experimental diets did not have significant ($P > 0.05$) effect on the blood parameters of the animals.

DISCUSSION

The CP contents of CL and ML suggests that they contain nutritional characteristics that can support sheep production as they were higher than the minimum range of 11% of CP recommended for growing sheep by Gatenby, (2002). They are therefore adequate for meeting the protein requirement for growing sheep thereby generating a high level of ammonia in the rumen to ensure an efficient digestion process (Orskov, 1995). The NDF contents were close to 60% which is considered safe for acceptable intakes of forage in ruminants (Messiner et al., 1991). The ADF and lignin contents were however lower than the majority of the Nigerian browse as reported by Onwuka et al. (1989).

The HCN and tannin contents of CL are lower than 50mg/kgDM and 4%, respectively recommended as a safety level for animals (Becker, 1997 and Phuc et al., 2000). Tannins above 4% of DM inhibit microbial activity and depress voluntary intake (Neptana et al., 2001). The calcium and phosphorus contents of both CL and ML appeared adequate for ruminant animals based on recommended by Puls (1994).

Table 5. Hematological parameters of West African Dwarf sheep prior and after feeding cassava and maize leaves based diets.

	Diets						
Parameters	RPE	0%CL	25%CL	50%CL	75%CL	100%CL	SEM
Packed cell volume (%)	22.4-36.7	29.5	27.1	30.5	32.0	31.5	2.20
Red blood cells ($x10^2L$)	2.3- 4.5	2.8	2.8	3.0	3.1	2.9	0.83
White blood cells ($x10^9L$)	5.9-6.7	6.4	6.2	6.6	6.6	6.4	1.10
Glucose (mmol/L)	2.1-3.3	2.9	2.6	3.1	3.2	2.9	0.68
Blood urea N (mmol/L)	3.4-5.7	3.9	3.7	4.2	4.3	3.7	0.94
Total protein (g/dl)	35.3-57.8	44.3	41.0	45.6	45.5	44.3	5.11

Mean values in the same row are not significantly different ($P > 0.05$)
[RPE] Range prior to experiment.

The range for sheep intake compares favorably with 556.4 to 582.1g/day reported by Olateru and Adegbola (2001) for Yankasa sheep with an average weight of 18 kg fed sorghum stover. The concentrates fed to the rams would have increased the feeding value of the entire diet by addition of energy over and above that supplied by the leaves. Mpairwe et al. (2003) observed that providing supplements with adequate CP to ruminants has promoted DMI, rumen degradation and nutrient flow to the small intestine and culminated in higher animal performance.

The higher feed intake observed in rams fed mixed forage diets compared to the sole forage diets of CL and ML could be as a result of interaction within the diets. This supports the findings of Lamidi et al. (1998) who reported higher feed intake by sheep fed mixture of *Ficus thonningia* leaves with *Hyparrhenia spp* leaves compared to sole leaves. Moreover, the low feed intake observed for rams on 100%CL with sole cassava leaves despite the highest dietary CP content however, confirms the reports of Adu and Osinowo (1985) that DMI was significantly lower in diets containing 20% CP compared to those containing lower values when fed to lambs. The least DMI reported for rams fed with diet containing zero level of cassava leaves (100%ML) could be attributable to the lower CP level of maize leaves compared to the other diets with the inclusion of varying levels of cassava leaves which supports the findings of Muniga *et al.* (1992) where inclusion of browse with high protein content in sheep's diet increased DM intake.

In the present study, all the rams fed the experimental diets gained weight, which confirms the reports of Devendra and Mcleroy (1982) that a CP of 11% is adequate for normal weight gain of sheep and goats. The variation observed in the daily gain in weight (g/day) of the rams could be attributed to the variation in the nutrient utilization by each animal from the diets. The observed weight gain of the experimental animals across the treatments could suggest the ability of the leaves coupled with the concentrate supplement to provide sufficient nitrogen and energy which ensure optimal microbial growth in the rumen.

The digestibility of nutrients in this study was generally high but highest in CL and ML forage mixtures which may imply that weight gain in rams was boosted by good DMI. The DM and CP digestibility seemed to be better digested at high levels of CL supplementation to ML. The NDF and ADF digestibility values fall within the range reported by Oduguwa et al. (2006) reported for the same breed of sheep fed forages with concentrate. The tannin content in CL seems not to have any adverse effect on the performance of the animals; it may even be responsible for the enhanced efficiency of utilizing the CP of the CL by sheep.

Moreover, the non-significant difference observed for FCR across the treatments implies equal efficient utilization of the animals in body tissue formation. The higher values obtained for FCR in this study could be as a result of the effect of the use of by products on these animals, which may not have allowed a very high growth performance. This is in consonance with the reports of Alkoriet et al. (2007) for Djallonke sheep fed cotton seed husks with *Panicum maximum* and cotton seed with 40 and 50% animal feed supplementation. However, lower values for FCR of 8 to 11 were observed by Wildeus et al. (2007) with hair sheep fed 100 % alfalfa hay and Hue et al. (2008) with lambs fed urea treated rice straw and molasses with different forage species.

The intake, faecal and urinary excretions of nitrogen was significantly affected by the dietary CP. This contradicts the reports of Rodehutscord and Preffer (1996) that faecal and urinary excretion of N in sheep was not significantly affected by dietary CP concentration. All animals were in positive nitrogen balance, an indication that the rams received adequate amounts of nitrogen from the diets fed which is a reflection of the potential of the leaves to enhance nitrogen utilization (Osakwe, 2006) and high digestibility of nutrients in the diets (Kawas et al. 2007).

The concentration of blood components of sheep was used in this study to monitor nutrient status between the treatments. The range of the hematological constituents of the rams before the commencement of the feeding trials were taken to serve as reference range for animals fed the experimental diets and to distinguish between the levels of the effect of each treatment on the animals. Values observed were within the normal range for sheep (Meyer and Harvey, 1998). These shows there were no appreciable effects of feeding this leaves on the blood profile of the experimental sheep. This reflects an efficient feed utilization that resulted in a stable energy and protein status of the animals.

CONCLUSION

Cassava leaves can be fed in mixture with maize leaves and can be used in diet mixtures of up to 75% of DM fed. This will improve feed intake, nutrient digestibility and N utilization leading to a better weight gain of sheep. Therefore, cassava and maize leaves which are otherwise waste materials can be converted to hay and used in seasons when grass forage supply is scarce especially in smallholder sheep production system.

REFERENCES

Abate, A.N and Abate, A. 1994. Performance of weaner lambs fed maize leaves and napier grass. In : S.H.B. Lebbie, B. Rey and E.K. Irungu (eds.), Small ruminant research and development in Africa . Proceedings of the 2nd biennial conference of Africa Research Network, AICC, Arusha, Tanzania. 7-11 December, 1992.

Adebowale, E.A and Taiwo, A.A.1996. Utilization of crop residues and agro-industrial by-products as complete diets for West African Dwarf sheep and goats. Nigerian Journal for Animal Production 23 (2), 153-160.

Adu, I.F. 1985. Feedlot fattening of sheep and goats. The Shika experience. National Conference on Small Ruminant Production, Zaria, 6th – 10th October. Pp. 202 – 210.

Adu, I.F. and Osinowo, O.A. 1985. Effects of dietary protein concentration and feeding level on the performance of weaned lambs. Journal of Animal Production and Research 5, 45-58.

Alkoiret T.I., Manne A.A.S., Gbangboche A.B., Attakpa, E.Y. 2007. Fattening performance of Djallonke sheep supplemented with cotton seed husks in Benin. Livestock Research for Rural Development (19) 10 , www.lrrd.org/lrrd19/10/alko19141.htm

AOAC (Association of Analytical Chemist). 1995. Official Methods of Analysis, 15th edition. Washington, D.C., USA. pp. 69-88.

Becker, K. 1997. Tannin-animal interactions. Aquaculture-Systems and Animal Nutrition in the tropics and subtropics, Annual reports, University of Hohenheim. No. 4.

Bradbury, M.G., Egan, S.V., and Bradbury, J.H.1999. Determination of all forms of cyanogens in cassava roots and cassava products using picrate paper kits. Journal of the Science, Food and Agriculture, 79, 593-601.

Davice J.U. and Lewis S.M. 1991. Practical haematology. 8th edition, Longman Ltd., London. pp. 187-188.

Devendra, C. and McLeroy, G.B. 1982. Goat and sheep production in the tropics. Longman Singapore publishers. B.V. Amsterdam. pp. 9-15.

Fasae, O.A. and Alokan, J.A. 2006. Growth performance of weaner Yankasa sheep fed varying levels of *Leucaena leucocephala* leaf residues. Assets Series A. Vol. 6, (2), 323-328.

Fasae O.A., Adu I.F; Aina, A.B.J. and Elemo K.A. 2009a. Effects of defoliation time of maize on leaf yield, quality and storage of maize leaves as dry season forage for ruminant production. Brazilian Journal of Agricultural Science 4 (3), 358-362.

Fasae, O.A., Adu I.F., Aina A.B.J. and Elemo, K.A. 2009b. Production, defoliation and storage of cassava leaves as dry season forage for small ruminants in smallholder crop – livestock production system. Agricultura Tropical et Subtropica 42, 15-19.

Gatenby, R.M. 2002. Sheep. Revised edition. Tropical Agriculturalist. Macmillan Educational Ltd., London, U.K.pp.8.

Hue, K.T., Van, D.T.T. and Ledin, I., 2008. Effect of supplementing urea treated rice straw and molasses with different forage species on the performance of lambs. Small Ruminant Research, 78, 134-143.

Kawas J.R., Garc´ıa-Castillo R. , Fimbres-Durazo H., Garza-Cazares F., Hern´andez-Vidal J.F.G., Olivares-S´aenz E. and Lu C.D. 2007. Effects of sodium bicarbonate and yeast on nutrient intake, digestibility, and ruminal fermentation of light-weight lambs fed finishing diets, Small Ruminant Research 67, 149–156.

Lamb, G.N. 1991. Manual of Veterinary Laboratory Technique. CIBA – GEIGY, Kenya, pp. 98 – 99.

Lamidi, O.S., Osinowo, O.A; Adamu, A.M. and Afolayan, R.A. 1998. Performance and nutrient utilization by sheep fed Ficus leaves. Nigerian Journal for Animal Production, 25 (1), 63 -67.

Makkar, H. P. S.; Blummel, M; Borowy, and Becker, K. 1993. Gravimetric determination of tannins and their correlation's with chemical and protein precipitation methods. Journal of the Science of Food and Agriculture. 61, 161 – 165.

Meissner, H.H., Koster, H.H., Nieuwoudt, S.H. and Coertze, R.J. 1991. Effect of energy supplementation on intake and digestion of early and mid season rye grass and Panicum/smuts finger hay on *in sacco* disappearance of various forage species. South African Journal of Animal Science, 21,33-42.

Meyer, D.J. and Harvey, J.W. 1998. Veterinary laboratory Medicine, Interpretation and diagnosis. 2nd edition. W.B. Saunders Company. An imprint of Elsevier Science, Philadelphia, Pennsylvania.

Mpairwe, D.R., Sabiti E.N., Umana, N.N., Tegegne, A. and Osuji, P. 2003. Integration of forage legumes with cereal crops. I. Effects of supplementation on graded levels of Lablab hay on voluntary intake, digestibility and milk composition of cross bred cows fed maize Lablab stover or oats vetch hay *ad libitum*. Livestock Production Science, 79, 193-212.

Muniga, R.W.; Thorpe, W., Topps, J.H. and Murethi, J.G. 1992. Responses of a Pennisetum (Napier grass) based diets harvested at two different heights and fed with three levels of Leuceana forage to cross bred dairy cow in the sub humid tropics. In, The complementarity's of feed resources for Animal Production in Africa. Proc. joint feed resources network held in Gaborore, Bostwana, 4 - 8th March. 1991. (John, E.S, Said, N. and Jackson A, Kategile, A.) African Feed Resources Network, ILCA, Addis Ababa, Ethiopia, pp. 173-180.

Muller, Z.1977. Improving the quality of cassava root and leaf product technology. Pp 120-126 in Cassava as animal as animal feed. Proceedings, Cassava as Animal Feed Workshop, edited by B. Nestel and M. Graham, 18-20 April 1977. University of Guelph, Ontario, Canada, IDRC, Ottawa.

Neptana, N., Wanapat, M., Poungchompu, O. and Toburan, W. 2001. Effect of condensed tannins in cassava hay on feacal parasitic egg counts in swamp buffaloes and cattle. In, Proceedings of the International Workshop Current Research and Development of Cassava as Animal Feed (Editors, T.R. Preston, B Ogle and M, Wanapat). July 23-24, 2001.Khon Kean University, Thailand.

NRC, 1975. Nutrient requirements of domestic animals. No. 5. Nutrient requirement of sheep. National Research Council, Washington D.C.

Oduguwa, B.O.; Babayemi, J.O.; Jolaosho, A.O., Aina, A.B.J., Ozoje, M.O. and Adu, I.F. 2006. Utilization of malted sorghum sprout, shrimp waste meal and Tephrosia bracteolate hay by West African Dwarf Sheep fed soyabean stover based diets. Nigerian Journal for Animal Production 33(2), 254-267.

Olorunsomo, O.A. 2008. Sweet potato as a ruminant feed. Performance of sheep fed mixtures of forage and root. Nigerian Journal for Animal Production 3(2), 242-251.

Onwuka, C.F.I., Akinsoyinu, A.O. and Tewe O.O. 1989. Feed value of some Nigeria browse plants, Chemical Composition and in-vitro digestibility of leaves. East African Agricultural Journal, 54(3), 157 – 163.

Ørskov, E. R. 1995. Optimizing rumen environment for cellulose digestion. In Rumen Ecology Research Planning (eds. R. J. Wallace and A. Lahlou-Kassi). Proceedings of a Workshop held at ILRI, Addis Ababa, Ethiopia, 13 -18 March 1995. pp. 177-182.

Osakwe, I.I. 2006. Effects of *Leucaena leucocephala* supplementation to basal hay on energy and protein metabolism in West African Dwarf sheep. Nigerian Journal for Animal Production 33(1), 94-101.

Phuc B.H.N., Ogle, R.B. and Lindberg, J.E. 2000. Effect of replacing soya protein with cassava leaf protein in cassava root meal based diets for growing pigs on digestibility and nitrogen retention. Animal Feed Science and Technology, 83, 233-235.

Puls, R. 1994. Mineral levels in animal health, diagnostic data. Normal levels summary sheets. pp. 302-308.

Rodehutscord, M and Pfeffer, E. 1996. Bilanzstudien zur N-Ausscheidung Laktierender Ziegen bei stark variierten Gehalten an Rohprotein im futter. Proc.Soc. Nutr. Physiology 5,44.

SAS (Statistical Analysis system) 1999. SAS User's guide, SAS Institute Inc., Cary North Carolina, USA.

Semenye, P.P.; Shsya, M. and Getz W.R. 1994. Overcoming the constraints of Dry matter intake on dual purpose goat production by feeding defoliated maize leaves. Small Ruminant Research Network Workshop held in Lilongwe, Malawi. 5-9 December, 1988. B.H. Dzowela, A.N. Said and J.A. Kategile (eds). pp 302-303.

Shewati, M.M and Dagash, H.A. 1994. Effects of pregnancy and lactation on some biochemical components in the blood of Egyptian coarse wool ewes, Assist Veterinary Medical Journal, 30(59), 64- 73.

Van Soest, P.J. and Robertson, J.B. and Lewis B.A. 1991. Methods of dietary fibre, neutal detergent fibre and non starch polysaccharides in relation to animal nutrition. Journal Dairy Science, 74, 3583 – 3583.

Wanapat, M., Pimpa O., Petlum, A. and Boontao U. 1997. Cassava hay, A new strategic feed for ruminants during the dry season. Livestock Research for Rural Development 9 (2).

Wanapat M; Puramangkan T; Siphuak, W. 2000. Feeding of cassava hay to lactating dairy cows. Asian Australian Journal of Animal Science 13 (4), 478-482.

Wildeus S, Turner K.E. and Collins J.R. 2007. Growth, intake, diet digestibility, and nitrogen use in three hair sheep breeds fed alfalfa hay. Small Ruminant Research 69, 221-227.

11

EFFECT OF BODY CONDITION SCORE ON ESTRUS AND OVARIAN FUNCTION CHARACTERISTICS OF SYNCHRONIZED BEEF-MASTER COWS

Fernando Centurión-Castro, Juan Orihuela-Porcayo, Ricardo J. Aké-López, Juan G. Magaña-Monforte, Rubén C. Montes-Pérez and José C. Segura-Correa

Campus de Ciencias Biológicas y Agropecuarias,
Universidad Autónoma de Yucatán, km 15.5 carretera Mérida-Xmatkuil.
Apartado postal 4-116, Mérida, Yucatán, México
E-mail: centuri@uady.mx
**Corresponding Author*

SUMMARY

The effect of body condition score (BCS) on estrus characteristics and ovarian function was evaluated in 44 synchronized Beef-Master cows. Weaned adult cows in anestrous were grouped according to their BCS as BCS1-2 (n = 13), BCS3-4 (n = 17) and BCS5-6 (n = 14). All cows were synchronized using norgestomet implants plus estradiol valerate for 9 days. The estrus detection began 24 hours after implant removal and it was detected every 6 hours for 72 hours. The time from implant removal to the onset of estrus (TIR) and the duration of estrus were registered. All animals were examined 4 times with a transrectal ultrasound kit to determine the size of the follicles (6 to 9 mm or > 9 mm), the diameter of the largest follicle (DLF), ovulation rate, ovulation rate relative to estrus detection (ORE) and ovulation rate relative to the follicle size (ORF). The effect of BCS on TIR, the duration of the estrus, the number of follicles of size 6 to 9 mm or > 9 mm, and DLF was determined using analysis of variance. The effect of BCS on ovulation rate, ORE and ORF, were determined using Chi-square tests. There were no differences (P> 0.05) between TIR, duration of estrus and follicle sizes between BCS groups. The DLF increased from 9.8 to 12.9 mm, the ovulation rate from 7.7 to 85.7%, the ORE from 0 to 100% and the ORF from 16.7 to 92.9%. In conclusion, under the range of BCS of the synchronized cows, evaluated in this study, the BCS of Beef Master cows did not affect the estrus characteristics; which agree with results observed in other breeds of cattle in the tropics. However, the BCS affect follicle development and ovulation rate.

Keywords: body condition; Beef Master; estrus; follicular development; ovulation.

INTRODUCTION

Annual calf production in the tropics is low due to the late sexual maturity of heifers, prolonged postpartum anestrous (>200 days) and long calving intervals (>450 days) of cows (Nogueira, 2004; Abeygunawardena and Dematawewa, 2004). This situation is probably caused by high temperature and humidity, the presence of parasites, high variation in the quantity and quality of pasture and loss of cow body condition score during lactation (Delgado et al. 2004; Aban *et al.*, 2008).

In the tropics, season of calving, body condition score and suckling of calves are among the main factors that influence the length of the anestrous period. Under the extensive management conditions of Southeastern Mexico, Aban *et al.* (2008) in Zebu and crossbred cows reported 40% pregnancy rate at 120 days postpartum. Delgado *et al.* (2004) in the same breed type of cows found a pregnancy rate of 37%, also at 120 days postpartum, and 63% of cows in anestrous. In both studies, cows were lactating with a calf at foot and more than 50% of them had low body condition score. So it was likely that cows were in estrus and ovulated after weaning.

In the tropics, one strategy to decrease the length of anestrous is the use of hormones to induce and/or synchronize estrus in cows (Hattab, 2000). However, results may vary depending on the breed or body condition score (Silva-Mena *et al.*, 2002). The reduction of food consumption pattern and nutritional status of the cows in estrus synchronization programs, affect the expression of estrus, due to the lack of growth and maturation of ovarian follicles caused by a negative energy balance. Cow body condition score (BCS) reflects the nutritional status of the herd and is used as a criterion to measure the response to hormonal treatment to stimulate the resumption of ovarian activity in weaned cows when the negative effects of suckling has been suppressed.

Some studies on reproductive events surrounding the induction and/or synchronization of estrus in *Bos indicus* cattle have been developed in the tropics (Ross et al., 2004; Kim *et al.*, 2005; Colazo *et al.*, 2007). However there is little information on body condition and follicle development and its relationship with ovulation rate in *Bos taurus* beef cattle under tropical conditions. The objective of this study was to evaluate the effect of BCS on the onset of estrus, follicular development and ovulation rate in synchronized Beef Master cows in Southeastern Mexico.

MATERIALS AND METHODS

The study was carried out during April to June (dry season) in a Beef-Master herd located in eastern Yucatán, Mexico. The climate is classified as AW_0, with summer rains, with a mean annual temperature of 26.3 °C, rainfall of about 469 mm and annual average relative humidity ranging from 66% in March to 89% in December (INEGI, 2005).

Forty-four adult weaned Beef-Master cows with an average of 243 ± 21 days after calving were used. Cows were managed on paddocks of Guinea grass (*Panicum maximum*) under rotational grazing from 17:00 to 7:00 hours. The remaining hours of a day the animals were housed in a "corral" for watering, minerals, salts supplementation and general management.

Ovarian inactivity of cows was measured by the absence of a corpus luteum, checked twice with a 15 days interval, using a real time ultrasound kit (Pie Medical Falco 100, Maastricht, Netherlands).

Body condition score

To assess cow BCS, a visual scheme proposed by Ayala *et al.* (1995) was used, which is based on a 9-point scale, where 1 corresponded to an emaciated and 9 to an obese cow. Thereafter, according to their BCS, cows were grouped in three categories: cows with BCS1-2 (n = 13), cows with BCS3-4 (n = 17) and cows with BCS5-6 (n = 14).

Timing and estrus detection

For estrus induction and synchronization, a subcutaneous implant of Norgestomate (Crestar, Intervet, Mexico) was used in each cow for 9 days. Cows were also injected 3 mg of Norgestomet plus 5 mg estradiol valerate intramuscularly. After 24 hours of the implant removal, estrus was visually detected every 6 hours for 72 hours (40 minutes each time). The criterion to identify a cow in estrus was the observation of the homosexual behavior of cows. The onset of estrus was determined when a cow allowed to be mounted the first time, and the end of estrus when the cow did not accept to be mounted. The duration of estrus was determined based on the mean time of two consecutive observations.

Identification of ovarian structures

An ultrasound kit performed the identification of ovarian structures, via rectal. The first diagnosis was made the day of implant removal and then every 24 hours, until one day after the end of estrus. Sheet records were used for each cow, recording the number and diameter of the follicles. Follicles were classified in two categories: those with a diameter of 6 to 9 mm, and those with a diameter > 9 mm (Guzeloglu *et al.*, 2001). The ovulation was identified by the presence of a corpora luteum, by ultrasound, on day 11 after implant removal, and confirmed when the plasma progesterone concentration of the cow was > 1 ng / ml (Diaz *et al.*, 2002).

Plasma concentrations of progesterone were measured in blood samples taken from the coccygeal vein. Blood samples were collected in vacutainer tubes containing EDTA and then, centrifuged at 1500 x g for 15 minutes to obtain the plasma (Burke *et al.*, 2001), which was decanted in 1.5 ml Eppendorf tubes and stored at -20° C until processed. Serum progesterone was determined by radioimmunoassay solid phase, using a L125 marker (Coat-A-Count Procedure, Diagnostic Products Corporation, Los Angeles, CA, USA). The method was highly specific for progesterone with approximately 2.0% cross-reactivity with 20 α dihydroxyprogesterone, and 0.3% with 17 α-hydroxyprogesterone. The detection limit was 0.05 ng/ml and the intra-and inter-assay coefficients of variation were 7% and 8.8%, respectively.

Statistical analysis

Analyses of variance were used to evaluate the effect of BCS group on the time from implant removal to onset of estrus, duration of estrus, the number of follicles and the diameter of the largest follicle. To determine the effect of BCS on ovulation rate, ovulation rate with respect to the cows in estrus and with respect to the cows with follicles greater than 9 mm, Chi-square tests were used. All statistical analyses were carried out using SAS (2002).

RESULTS

The time from implant removal to estrus detection, estrus duration and number of follicles of size 6-9 mm or > 9 mm was similar for the three BCS groups (Table 1). However, the diameter of the largest follicle was largest in the cows with BCS5-6. The overall estrus rate was 61.4%. As BCS improved the percentage of cows that ovulated, the ovulation rate in relation to the cows that showed estrus and those that ovulate in relation to the follicles > 9 mm improved (Table 2). Cows with BCS5-6 had the highest ovulation rate, ovulation rate relative to cows in estrus and to those with follicles >9 mm, followed by BCS3-4 group cows. However, only the differences between the BCS1-2 group cows and the other two BCS groups were significant (P< 0.05).

Table 1. Means (±SD) by body condition score for time from implant removal to onset of estrus, estrus duration, number of follicles by size and diameter of the largest follicle in Beef Master synchronized cows.

	n	Implant removal to estrus detection (hours)	Estrus duration (hours)	Follicle size		Diameter of largest follicle (mm)
				6-9 mm	>9 mm	
BCS1-2	13	36±9.5	10.2±1.4	2.8±1.6	2.7±0.8	9.8±0.9^{a}
BCS3-4	17	36±9.0	8.1±1.0	2.3±1.9	2.0±1.1	10.8±1.3^{ab}
BCS5-6	14	33±6.3	10.3±1.3	4.0±4.6	2.6±1.6	12.9±1.0^{b}

^{a,b} Different literals between rows, means statistical differences (P<0.05).

Table 2. Number and rates by body condition score groups of ovulated cows, ovulation rate in relation to estrus and to follicle size >9 mm in synchronized cows.

	n	Ovulated cows (%)	Ovulations in relation to estrus (%)	Ovulations in relation to follicles >9 mm (%)
BCS1-2	13	1 (7.7)^{a}	0/5 (0)^{a}	1/6 (16.7)^{a}
BCS3-4	17	11(64.7)^{b}	8/10 (80)^{b}	9/13 (69.2)^{b}
BCS5-6	14	13(85.7)^{b}	12/12 (100)^{b}	13/14 (92.9)^{b}

^{a,b,c} Different literals between rows, means statistical differences (P<0.05).

DISCUSSION

The rate of cows in estrus (61%) after implant removal, in this study, was higher than the 21% obtained by Velázquez *et al.* (2011) in crossbred *Bos taurus* x *Bos indicus* cows. However, the values here obtained were lower than those reported in Zebu or crossbred cows in anestrous, where the response rate to estrus after progesterone implant removal was greater than 70% (Peralta-Torres *et al.*, 2010). The low response to estrus may be due to the effect of the warm months of the year (April-June) when the present experiment was carried out. It is known that the stress caused by high temperatures may affect the sexual behavior of animals (White *et al.*, 2002).

The results of this study show that the time from implant removal to the onset of estrus was similar in the three BCS groups. This agree with Soto-Camargo *et al.* (1999), who did not found differences (p> 0.05) between the time of implant removal and the onset of estrus in Brahman heifers with different BCS (3.0, 3.5 and 4.0, on a scale of 1-5). Soto-Camargo *et al.* (1999) means were 42.5 ± 10.3, 41.0 and 39.3 ± 8.0 ± 3.0 hours in the three BCS groups, respectively. In this study the mean time from implant removal to the onset of estrus was 35 ± 1.7 hours, a result that is lower (43%) than the obtained by Peralta-Torres *et al.* (2010) in Zebu cattle. The fact that no differences were found among BCS groups for the time from implant removal to the onset of estrus, suggest that BCS of cows did not affect estrus response, probably because cows responded to the effect of exogenous estrogen administered with the implant; and because estrogen influenced the expression of estrus and ovulation (Lemaster *et al.*, 1999). However, the estrus in BCS 1-2 cows were anovulatory. BCS had also no effect on duration of estrus (P> 0.05), result that agree with those reported by Soto-Camargo *et al.* (1999). The latter authors found no differences in Brahman heifers with different BCS (the estrus duration means were 6.5, 8.3 and 4.4 hours). However, the overall estrus duration mean (9.5 hours) was similar to that observed in *Bos indicus* cows (10.2 ± 6.8 hours) with BCS of 2.5 (Medrano, 1996). This indicates that *Bos taurus* in the tropics shows no difference in the estrus expression compared with *Bos indicus* cows, whose estrus is of short duration. The effect of BCS group on the number of follicles of size 6-9 mm or > 9 mm, agrees with the results of Burke *et al.* (1998). They found no difference (P> 0.05) in the number of follicles of three different sizes in cows with two BCS: low (2.3 follicles) and high (4.4 follicles) using a scale of 1 to 5 points.

In this study it was expected that the BCS5-6 group cows had the largest follicles; however, there were not differences in their diameters among the three groups; this probably due to the postpartum time when this study was carried out (243 days on average). According to Lucy *et al.* (1991), 243 days after calving, the body condition no longer influences the number or size of the follicle; however, other factors might also be involved. The mean of the diameter of the largest follicles was higher in the BCS5-6 group (12.9 mm) as compared to the follicles for the cows with BCS1-2 (9.8 mm). These results differ from those reported by Burke *et al.* (1998), who found no difference in the diameter of the follicles of cows with low body condition (16.3 mm) and high body condition (14.2 mm). However, the diameter of the largest follicles for BCS5-6, in this study, is similar (11 mm and 12 mm) to those reported in *Bos indicus* cows (Burke *et al.*, 2001; Calvalho, 2008; Peralta-Torres *et al.,* 2010). The diameter of the largest follicles for the BCS5-6 group cows suggests that the body condition, and therefore nutrition, might have a specific effect on the growth of follicles, because the diameter of preovulatory follicles is positively correlated with the weight of *Bos indicus* cattle (Rhodes *et al.*, 1995). It has also been observed in dairy cattle that negative energy balance influences follicular growth (Bean and Butler, 1997; Wiltbank *et al.*, 2002) and the size of the ovulated follicle (Armstrong *et al.*, 2001). This situation was evident in this study, because cows with poor BCS had follicles with small diameters. Thus, a nutritional deficiency can affect the diameter of the dominant follicle and therefore ovulation. Diskin *et al.* (2003) suggest that the nutritional status of the cow affects follicular growth, maturation and ovulation.

The highest ovulation rate, ovulation rate relative to estrus and to follicles > 9 mm are similar to those reported by Sepulveda *et al.* (2001), who assessed the body condition of cows at 60 days postpartum, and found significant difference (P <0.05) in ovulation; the ovulation rates were 53% and 100% in cow with BCS of 2.5 and 2.9 in a scale of 1 to 5.

The lowest ovarian activity in the group with the lowest BCS was probably due to the small follicle sizes reached by the cows of this group. Sartori *et al.* (2001) reported that follicles smaller than 10 mm in diameter, in lactating dairy cows, are unable to ovulate due to the lack of effect of the LH hormone, but the response begins to increase when the diameter of the follicles exceeds 10 mm.

It has been shown that the nutritional status of the female affects growth and follicular diameter (Rhodes *et al.*, 1995; Armstrong *et al.*, 2001; Diskin *et al.*, 2003), follicle maturation and ovulation (Diskin *et al.*, 2003). This suggests that body condition, and therefore nutrition and follicle size, had a specific effect on the proportion of cows that ovulate (Sá Filho *et al.*, 2010). It has been established that the nutrition acts at various levels of the system that controls reproduction (Butler, 2000) and that the main pathway by which nutrition affects follicular development is through the hypothalamic-pituitary-ovarian axis (Diskin *et al.*, 2003). It is also known that the central nervous system is capable to detect subtle changes in the availability of several metabolic compounds, and as a result, the pattern of secretion of GnRH and the gonadotropins can decrease or increase depending on the change of metabolism. The administration of a diet with low protein and energy content influence negatively the LH concentration, and therefore affect the growth, persistence and ovulation of the dominant follicle (Rasby *et al.*, 1992; Rhodes *et al.*, 1995; Rhodes *et al.*, 1996).In conclusion, under the conditions of this study, there was no effect of the body condition of Beef Master cows on estrus traits, results that agree to those observed in other cattle breeds in the tropics. However, body condition affected follicle development and ovulation rate.

REFERENCES

Abán, J., Delgado, R., Magaña, J.G., Segura J.C. 2008. Factores que afectan el porcentaje de gestación a 120 días posparto en vacas Cebú y cruzadas con europeo en el sureste de México. Avances en Investigación Agropecuaria. 12: 45-56.

Abeygunawardena H., Dematawewa C.M.B. 2004. Pre-puberal and postpartum anestrus in tropical Zebu cattle. Animal Reproduction Science 83-83: 373-387.

Ayala, A., Delgado, R.; Honhold, N.; Magaña, J. 1995. A visual condition scoring scheme for Bos indicus and crossbred cattle. In: Dual Purpose Cattle Production Research. Eds. S. Anderson and J. Wadsworth. IFS (International Foundation For Science) FMVZ-UADY 1995, Proceedings of an International Workshop, March 1992, Merida, Mexico, 119-128.

Armstrong, D.G., McEvoy, T.G., Baxter, G., Robinson, J.J., Hogg C.O., Woad, K.J., Webb, R. 2001. Effect of dietary energy and protein on bovine follicular dynamics and embryo production in vitro: Associations with the ovarian insulin-like growth factor system. Biology of Reproduction 64: 1624-1632.

Burke, S.M., Hampton, J.H., Staples, C.R., Thatcher, W.W.. 1998. Body condition maintenance of a persistent Influences first wave dominant follicle in dairy cattle. Theriogenology 49: 751-760.

Burke, C.R.; Mussard, M.L., Grum, D.E., Day, M.L. 2001. Effects of maturity of the ovulatory follicle on induction potential of oestrus and ovulation in cattle with oestradiol benzoate. Animal Reproduction Science 66: 161-174.

Beam, S.W., Butler, W.R. 1997. Energy balance and ovarian follicle prior to the first Development ovulation postpartum in dairy cows receiving three levels of dietary fat. Biology of Reproduction 56: 133-142.

Butler, W.R. 2000. Nutritional interactions with reproductive performance in dairy cattle. Animal Reproduction Science 60-61: 449-457.

Colazo, M.G.,. Kastelic, J.P., Small, J.A., Wilde, R.E., Ward, D.R., Mapletoft, R.J. 2007. Resynchronization of estrus in beef cattle: Ovarian function, estrus and fertility following progestin treatment and treatments to synchronize ovarian follicular development and estrus. Canadian Veterinary Journal 48: 49-56.

Calvalho, J.B.P., Carvalho N.A.T., Reis, E.L., Nichi, M., Souza, A.H., Baruselli, P.S. 2008. Effect of early luteolysis in progesterone-base timed AI protocols in Bos indicus x Bos taurus, and Bos taurus heifers. Theriogenology 69: 167-175.

Diskin, M.G., Mackey, D.R., Roche, J.F., Sreenan, J.M. 2003. Effects of nutrition and metabolic status on circulating hormone and ovarian follicle development in cattle. Animal Reproduction Science 78: 345-370.

Delgado, R., Magana, J.G., Galina, C., Segura, J. 2004. Effect of body condition at calving and its changes during early lactation on postpartum reproductive performance of Zebu cows in a tropical environment. Journal of Applied Animal Research 26: 23-28.

Díaz, G.S., Galina, C.S., Basurto, C.H., Ochoa, G.P. 2002. Effect of natural progesterone with or without the addition of estradiol benzoate on the occurrence of oestrus, ovulation and pregnancy in *Bos indicus* animals in tropical Mexico. Archives of Veterinary Medicine 34: 283-286

Guzeloglu, A., Ambrose, J.D., Kassa, T., Diaz, T., Thatcher, M.J., Thatcher, W.W. 2001. Long-term follicular dynamics and biochemical characteristics of dairy cows in dominant follicles subject to acute heat stress. Animal Reproduction Science 66: 15-34.

Hattab, I.N.C., Kadoom, A.K., Palme, R., Bamberg, E. 2000. Effect of estrus synchronization and Crestar on the relationship between concentrations of plasma and fecal progestagens in buffalo cows. Theriogenology. 54: 1007-1017.

INEGI. 2005. Encyclopedia of the municipalities of Mexico, State of Yucatan. http://www.e-local.gob.mx/work/templates/enciclo/yucatan/municipios/31006a.htm.

Kim, U.H., Suh, G.H., Nam, H.W., Kang, H.G., Kim, I.H. 2005. Follicular wave emergence, luteal function and synchrony of ovulation or estradiol benzoate following GnRH in CIDR-Treated, lactating Holstein cows. Theriogenology 63:260-268.

Lucy, M.C., Staples, C.R., Michel, F.M., Thatcher, W.W. 1991. Energy balance and size and number of ovarian follicles detected by ultrasonography in early postpartum dairy cows. Journal of Dairy Science 74: 473-482.

Lemaster, J.W., Yelich, J.V., Kempfer, J.R., Schrick, F.N. 1999. Ovulation and estrus characteristics in crossbred Brahman heifers treated with in an intravaginal progesterone-releasing insert in combination in prostaglandin F2α and estradiol benzoate. Journal of Animal Science 77: 1860-1868.

Medrano, E., Hernandez, O., Lamothe, C., Galina, C. 1996. Evidence of asynchrony in the onset of signs of estrous in Zebu cattle treated with a progestogen ear implant. Research in Veterinary Sciences 60: 51-54.

Nogueira, G.P. 2004. Puberty in southamerican *Bos indicus* (Zebu) cattle. Animal Reproduction Science 82-83: 361-372.

Peralta-Torres, J., Ake-Lopez, R., Centurion-Castro, F., Magaña-Monforte, J. 2010. Comparison of estradiol cypionate and estradiol benzoate effects on ovaric activity, estrus and ovulation on Bos indicus cows anestrus. Journal of Animal and Veterinary Advances 9: 466-470.

Rasby, R.J., Wettemann, R.P., Harms, P.G., Lusby, K.S., Wagner, J.J. 1992. GnRH in the infundibular stalk-median eminence is related to percentage body fat in carcasses of beef cows. Domestic Animals. Endocrinology 9: 71-76.

Ross, P.J., Aller, J.F., Callejas, S.S., Butler, H., Alberio, R.H. 2004. Estradiol benzoate given 0 or 24 h after the end of progestagen in postpartum suckled beef Treatment cows. Theriogenology, 62: 265-273.

Rhodes, F.M., De´ath, G., Entwistle, K.W. 1995. Animal and temporal effects on ovarian follicular dynamics in Brahman heifers. Animal Reproduction Science. 38: 265-277.

Rhodes, F.M., Entwistle, K.W., Kinder, J.E. 1996. Changes in ovarian function and gonadotropin secretion preceding the onset of nutritionally induced anestrus in Bos indicus heifers. Biology of Reproduction 55:1437-43.

SAS. User's Guide. 2002. Statistics 8th ed. Cary (NC): SAS Institute Inc.

Sartori, R., Fricke, P.M., Ferreira, J.C., Guinther, O.J, Wiltbank. 2001. Follicular deviation and adquisition of ovulatory capacity in bovine follicles. Biology of Reproduction 65: 1403-1409.

Soto-Camargo, R., Galina, C.S., Rubio, G.I., Castille, G.E., Basurto, C.H. 1999. Effect of nutritional supplementation, body condition on estrus synchronization and mounting activity of Brahman heifers on pasture. Revista de la Facultad de Agronomía (Luz). 16: 663-676.

Sepúlveda, N.G.; Inostrosa, M.A., Peña, P., Risopatrón, J., Rodero, E. 2001. The onset of postpartum ovarian function in primiparous and multiparous dairy cows. Archivos de Zootecnia 50: 399-402.

Sá Filho, M.F., Crespilho, A.M., Santos, J.E., Perri, G.A., Baruselli, P.S. 2010. Ovarian follicle diameter at timed insemination and estrous response Influence of ovulation and pregnancy likelihood after estrous synchronization with progesterone and progestin-based protocols in suckled Bos indicus cows. Animal Reproduction Science 120: 23-30

Silva-Mena, C., Guzmán-Casas, R., Delgado-Leon, R., Ake-Lopez. R. 2002. Brahman heifers response to estrus synchronization with progestagen, sexual behavior and pregnancy rate. Revista Biomédica 13: 265-271.

Velázquez, R.L., Fregoso, A.C., Lopez, R.W., Harris, C.J. 2011. Estrous response and conception rate in Bos taurus-Bos indicus pospartum cows, treated with bovine somatotropin ovulation induction in a program with progestogens and eCG. Veterinaria México 42: 245-251.

Wiltbank, M.C.; Gumen, A., Sartori, R. 2002. Physiological classification of anovulatory conditions in cattle. Theriogenology 57: 21-52.

White, F.J., Wettemann, R.P., Looper, M.L., Prado, T.M., Morgan, G.L. 2002. Seasonal effects on estrous and time of ovulation behavior in nonlactating beef cows. Journal of Animal Science 80: 3053-3059.

12

SENSORIAL AND PHYSIOLOGICAL CONTROL OF MATERNAL BEHAVIOR IN SMALL RUMINANTS: SHEEP AND GOATS

H. Hernández [1*], A. Terrazas [2], P. Poindron[1], S. Ramírez-Vera[1], J. A. Flores[1], J. A. Delgadillo[1], J. Vielma[1], G. Duarte[1], I. G. Fernández[1], G. Fitz-Rodríguez[1], S. Retana-Márquez[3], M. Muñoz-Gutierrez[3] and N. Serafín[4]

[1] *Centro de Investigación en Reproducción Caprina, Universidad Autónoma Agraria Antonio Narro, Periférico Raúl López Sánchez y Carretera a Santa Fe, CP 27054, Torreón, Coahuila, México. E-mail: hernandezhoracio@hotmail.com*
[2] *Departamento de Ciencias Pecuarias, Facultad de Estudios Superiores Cuautitlán, Universidad Nacional Autónoma de México. Km 2.5 Cuautitlán-Teoloyucan, San Sebastián Xhala, Cuautitlán Izcalli, México. E-mail: angeterr_2000@yahoo.com*
[3] *Departamento de Biología de la Reproducción, Universidad Autónoma Metropolitana Iztapalapa. Edificio de Ciencias Biológicas y de la Salud, Av. San Rafael Atlixco No. 186, Colonia Vicentina, 09340 Iztapalapa México D.F.*
[4] *Departamento de Neurobiología Conductual y Cognitiva, Instituto de Neurobiología, Universidad Nacional Autónoma de México, Juriquilla, Querétaro, México.*
**Corresponding author*

SUMMARY

Contrary to rodents in which maternal behaviors is characterized by the nest formation and give birth to altricial offsprings, maternal behavior in sheep and goats is characterized by the establishment of a selective bond between the mother and their progeny during the first postpartum hours. In both species, maternal behavior ethogram at parturition consists of a series of behaviors that initiates with the prepartum isolation of the female from their coespecifics and that culminates with the successful exclusive nursing behavior toward their newborn during postpartum. The sensory and physiological factors that control the expression of maternal behavior are very similar in both sheep and goats, although in the goats there is little information generated about this aspect. The increased peripherical concentrations of oestradiol at the end of gestation and the vaginocervical stimulation are two primary physiological events that are very important to the expression of the maternal behavior. However, olfactory cues from the offspring also are involved in the maintenance of maternal responsiveness once the birth takes place. Although mother-young spatial relationships during postpartum are different between sheep (followers) and goats (hiders) the maternal behavior had many similarities and their sensorial and physiological control are basically identical for many aspects.

Keywords: Maternal care; sheep; goats; oestradiol; olfactory cues; suckling.

INTRODUCTION

Maternal behavior presents a wide variety of behavioral patterns among the mammals. Its ethological characteristics depend mainly on the degree of precocity of the young at birth, but also on the social nature of the specie. Thus, while mothers of altricial young build a nest (e.g. rodents), this is rarely the case in ungulates, whose maternal behavior is rather characterized by a rapid development of the inter-individual recognition and exclusive care (maternal selective behavior).

In this respect, they represent an interesting alternative to rodents for the study of mother-young interactions. Additionally, interest to study this aspect in small ruminants is furthermore enhanced by the fact that the onset of maternal behavior is strictly associated with the process of parturition by contrast to what observed in rodents. In fact, in this last specie, maternal behavior can be elicited by the only presence of the young (Numan, 1994).

In the fist part of this review will be focus on the characteristics of maternal behavior before, during and after parturition in sheep and to a lesser extent in goats. In the second part of this review concerns to the respective roles of physiological and sensorial roles on the maternal behavior studied in these species.

Prepartum Maternal Behavior

Sheep and goats are social animals and show strong gregarious tendencies. In both species when remain apart from conspecifics (social isolation) is highly stressful with very strong behavioral responses indicative of agitation (running, rearing against the sides of the open, escape attempts) and frequent, high-pitched vocalization (Price and Thos, 1980; Carbonaro *et al*., 1992; Le Neindre *et al*., 1993; Dwyer, 2009). However, pregnant ewes and goats show social isolation from the rest of the flock and a reduction of their behavioral response to social isolation (i.e social agitation index) between 24 h and about 1 h prepartum (Sheep: Arnold and Morgan, 1975; Echeverri *et al*., 1992; Lécrivain and Janeau, 1987; Shackleton and Haywood, 1985; goat: Das and Tomer, 1997; Lickliter, 1985; O´Brien, 1983, 1984; Ramirez *et al*., 1995; Rudge, 1970). Therefore, during this state of gestation the females show a clear a reduction in the gregariousness behavior (Poindron *et al*., 1997, 1998a,b).

The imminent labour of parturition in sheep and goats is characterized by an initial increase in restlessness, pawing, lip licking, frequency of vocalizations, and then straining in both standing, lying positions and the intolerance from conspecifics also increase (Collias, 1956; Herscher *et al*., 1963a; Smith, 1965; Das and Tomer, 1997; Dwyer, 2009). In ewes and goats, isolation from the rest of her conspecific could be beneficial to the early relationship with their young (Val-Laillet and Nowak, 2006; Dwyer, 2009). Furthermore, mother select a birth site and offspring survival is increased with increased duration spent on the birth site (Dwyer, 2009). This isolation improves the establishment of the mother-young bond by preventing separations of newborn young from their mothers and reducing interference or newborns stealing by other mothers or predators (Gonyou and Stookey, 1983, 1985; Poindron *et al*., 1993). Therefore, this isolation without interference from others ewes is indeed of the utmost importance for the onset of the mother-young relationship and this behavior may facilitate mutual recognition and rapid access to the udder at birth, improving the chances of newborn survival (Val-Laillet *et al*., 2006). Ewes can display full maternal behavior at three hours before parturition in the presence of the others newborn lambs (Arnold and Morgan, 1975).

Maternal and Neonatal Behavior at Birth

Maternal behavior towards the neonate is characterized by a number of very typical components; these behaviors are similar in sheep and goats (Hersher *et al*., 1963a,b). Immediately after birth the mothers show intense and focused interest in the newborn and in the amniotic fluids in its coat and on the ground. In this regard, amniotic fluids are important in experienced ewes not only for the establishment of maternal responsiveness as in primiparous, but also these fluids carries some chemosensory information to facilitate exclusive bonding (Poindron *et al*., 2010). A few minutes after expulsion of the fetus the mother rises and starts to lick her kid very vigorously, generally beginning with its head and neck (Collias, 1956; Shillito and Hoyland, 1971; Lickliter, 1985). Licking is accompanied by numerous low and high pitched bleats and occasionally by flehmen behavior, especially when consuming urine from the neonate (Sambraus and Wittmann, 1989; Dwyer *et al*., 1998). These behaviors serve for two purposes: first, the mothers help to dry (facilities the remove of the fetal membranes from the face that could lead to suffocation), and stimulate the lamb (the teat-seeking activity of the young), and secondly they form an exclusive and selective attachment to her own offspring (Nowak *et al*., 2000; Dwyer, 2009). Furthermore, the mother must be coordinated with the movements of the neonate lamb which has to be able to stand before it can orient and reach the mammary gland (Nowak *et al*., 2007). The newborn stands on its feet, usually at 10 to 30 min after birth, it start to nuzzle its mother´s body and rapidly reaches the udder, to successfully suck in 30 to 60 min after birth

(Allan *et al.*, 1991; Poindron *et al.*, 1993). In sheep and does a selective attachment is formed within an hour or less of birth, this mother-young bonding is important for the rapidly learns to mutual recognition at distance (Nowak *et al.*, 2000; Terrazas *et al.*, 2003; Poindron *et al.*, 2003). A disturbance in the maternal care, maternal selectivity or in the mutual recognition may have deleterious effects on the survival of the young during the first week of life (Nowak *et al.*, 1997; Nowak *et al.*, 2000; Poindron *et al.*, 2007a,b; Nowak *et al.*, 2007).

Formation of an exclusive mother-young bond and maternal behavior during the postpartum

The survival of the newborn is dependent on the coordinated expression of appropriate behaviors between both the ewe and the lamb that lead to the formation of a strong relationship between both partners (González-Mariscal and Poindron, 2002). In sheep and goats, expulsion of the fetus facilitates the formation of the maternal bond. Also, the vaginocervical stimulation (VCS) performed after the establishment of maternal selectivity allows mothers to form a new bond with an alien newborn (sheep: Kendrick *et al.*, 1991; goat: Romeyer *et al.*, 1994b).
Whether estrogen may participate in the establishment of maternal selectivity remains an open question. Some results in sheep suggest that they may modulate the establishment of bonding (Alexander *et al.*, 1986, 1989) or allow bonding in the absence of VCS (Le Neindre *et al.*, 1979). However, in studies of the sensitive period in which ewes were induced to lamb using high doses of estradiol, was found no evidence for an effect of estradiol on selectivity (Poindron and Le Neindre, 1980). Immediately after parturition mothers groom and nurse any kid, but then rapidly learn to discriminate and nurse only their own young (Collias, 1956; Herscher *et al.*, 1963b). In sheep and goats, the maternal bond is established within the first 2 h leading to the later rejection of any infant that tries to suckle from a dam other than its own (Poindron *et al.*, 1993). In some conditions, 5 to 10 min of contact with the neonate are sufficient for does to become selective (Gubernick, 1981) and mothers reject alien kids at 2 to 4 h postpartum, regardless of their age or the similarity of their coat to that of their own young (Romeyer and Poindron, 1992; Romeyer *et al.*, 1994a; Poindron *et al.*, 2007b). Previous studies have demonstrated that 30–60 min of immediate postpartum contact with a neonate may be sufficient for the ewe to become familiar with that lamb's unique signature and discriminate her lamb from alien ones (Keller *et al.*, 2003). Maternal olfaction plays an important role; this full access to olfactory cues through licking facilitates the establishment of maternal selectivity. Of the mothers that remain maternal despite 4 h of licking deprivation, half fail to be selective and still accept alien kids, in contrast to mothers that have fully interacted with their neonate during the same period (Poindron *et al.*, 2007a,b). It could be argued that the inability of mothers to display maternal care after deprivation of the neonate at birth is due to the fact that they have not had the opportunity to bond to their young. These in turn have higher incidence of separations and higher mortality of the offspring (Murphy *et al.*, 1994; Alexander *et al.*, 1983; Putu *et al.*, 1988). Therefore, the first hours postpartum represent a critical time for the establishment of the maternal bond towards her young (Poindron *et al.*, 2007b).

One way to assess whether the establishment of maternal selectivity depends on the sensitive period is to study the rapidity with which selectivity develops at various times during this period (Poindron *et al.*, 2007b). The sensitive period is generally considered a transition from a control of maternal responsiveness by physiological factors internal in the mother, to a neurosensory control by sensory cues from the young (Poindron and Le Neindre, 1980; Rosenblatt *et al.*, 1979). Hence, presenting or suppressing cues of the neonate during the sensitive period will provide valuable information about their importance for the maintenance of maternal responsiveness and selectivity beyond the sensitive period (Poindron *et al.*, 2007a,b). The results of such studies will be discussed in the section "The sensitive period and sensorial control of maternal behavior" of the present review. Nonetheless, some results suggest that the sensitive period may be shorter in goats than in sheep (Ramírez *et al.*, 1996).

Maintenance of maternal behavior relies on behavioral cues from the neonate, particularly sucking interactions, and mothers whose offspring die lose interest in the body of their young within a few hours (Dwyer, 2009). Thus, the ability of the neonate to attach rapidly to its mothers (as much as the reverse) is a primary condition for the maintain access to food and maintain maternal behavior in the postpartum, consequently increase the opportunity of survival of kid (Poindron *et al.*, 1993; Nowak *et al.*, 2000). Thus, 50% of mothers do not show maternal interest in their young after 4 h of separation beginning at birth and 75% of them reject their own young after 12 h of privation in sheep and goats (Poindron *et al.*, 1979; Lickliter, 1982). Mother-young interactions at birth are important for the establishment of its mutual recognition after parturition. Thus, the establishment of early recognition of the dam may be vital for the lamb (Poindron *et al.*, 1993; Nowak *et al.*, 2007; Poindron *et al.*, 2007a,b). However, the maintenance of maternal behavior during postpartum depends very much on the events immediately after parturition (Poindron *et al.*, 2007b).

During the first week of life the mother allows the offspring to access the udder and suck whenever it approaches. Therefore, during first month of life the mothers will actively seek out her lamb when they are separated, and mother-young distances are generally no more than a few meters in lamb, and distances higher in kids (Nowak *et al.*, 2000; Dwyer, 2009).

In ungulates, two main types of mother-young spatial relationships have been described, which depends on the behavior of the neonate. In some species as in sheep at birth the neonate start to follow their mothers as soon as she moves from the birth site classified as "followers" (Lent, 1974). This "follower" species spend less time in close physical proximity to their dams as they mature (Herscher *et al.*, 1963b). Others species, for example goat neonates, do not follow their mothers when she goes foraging. Rather, for the first few days they lie down and hide, waiting for the mother to return for nursing, thus it mother-young spatial relationship is classified as "hiders" (Lickliter, 1987). This hider species gradually come to spend more time in close physical proximity to their mothers as they mature (Lickliter, 1984). However, contrary to a greater quantity of published studies on the onset of maternal behavior, mechanisms that contribute to their spontaneous disappearance in the mothers and the spontaneous weaning of the young in small ruminants are also little studied.

Physiological control of maternal behavior

Contrary to rodents, in sheep the activation of maternal motivation is completely controlled by the physiological events that accompany parturition. Thus, full maternal responsiveness is strictly related to the event of parturition, but fades after few hours in the absence of the new-born lamb (Poindron and Le Neindre, 1980). Thus, 4 h of separation beginning at birth induced disturbances of maternal responsiveness in half of the mothers and after 12 h of deprivation 75% of them rejected the lamb (Poindron *et al.*, 1979). On the other hand, when 4 or 24 h of immediate postpartum contact is allowed, it was sufficient to maintain maternal responsiveness for the following 24-h period of separation (Poindron *et at.*, 1979; Lévy *et al.*, 1991). If immediately separated from their neonate, parturient goats display maternal behavior spontaneously only for a short time after giving birth, although the duration of this period of initial responsiveness towards the neonate is longer in multiparous mothers (Lickliter, 1982). This indicates the existence of a sensitive period for the establishment of the relationship between the newborn kid and its mother (Herscher *et al.*, 1963b; Poindron *et al.*, 2007b) and suggests that it is controlled by the physiological factors associated with parturition.

The physiological events around parturition

In sheep and goats, as in many others mammals, one of the physiological events that characterizes the parturient female is the rapid and important changes in steroid hormones (progesterone and oestrogens) that take place in the days or hours just before birth. Both preparturient ewes and does experience a fall in the peripheral plasma concentrations of progesterone few days prior to parturition, while oestradiol rises one day before parturition to reach a peak at parturition and return to basal levels within the first 4 h postpartum (Thorburn *et al.*, 1972; Chamley *et al.*, 1973; Shipka and Ford, 1991). As a result of the shift in balance in steroid exposure, being oestradiol dominant, serum prolactin (PRL) levels increase dramatically around one day before parturition in sheep (Chamley *et al.*, 1973) and during the last 3 days in does (Currie *et al.*, 1988), and are maintained at high levels during postpartum because of suckling stimulation (Davis *et al.*, 1971).

The changes of concentration in estrogen and progesterone in the goat during late pregnancy and parturition are very similar to those found in sheep, in which these hormones play an important role in the display of maternal behavior (Poindron *et al.*, 2007a). However, an exogenous treatment of estradiol and progesterone that induces lactation and facilitates maternal behavior in non-pregnant ewes is ineffective in the goat (Rosenblatt and Siegel, 1981). The alternative of inducing maternal behavior with treatments mimicking more closely the hormonal changes occurring at parturition has not been investigated.

Fetus expulsion and vaginocervical stimulation (vcs)

The importance of VCS for the activation of maternal behavior has been most extensively studied in sheep. The induction of maternal behavior in nonpregnant animals is achieved with some success only by using pharmacological doses of oestradiol (Poindron and Le Neindre, 1980; Kendrick *et al.*, 1992). In addition, the type of behaviors induced, their latency and sequence of appearance, are quite variable, thus suggesting that additional factors are required for the activation of an adequate maternal behavior. Although a large number of physiological changes occur during parturition, some strong evidences indicate that VCS is the initial step that triggers the immediate onset of maternal behavior. The actual process of expulsion of the fetus appears of utmost importance. For example, the acceptance of a second neonate presented to a ewe 30 min after the birth of her own single lamb, it is not accepted as readily as an alien presented to a twin-bearing ewe at the time of expulsion of her second born (Poindron *et al.*, 1980). The application of artificial VCS for 5 min, to nonpregnant multiparous females, induced the maternal behavior in 80% of them within 30 min (Keverne *et al.*, 1983). This facilitatory effect of VCS is evident only in multiparous ewes, primed with ovarian esteroids (Poindron *et al.*, 1988; Kendrick and Keverne, 1991). In parturient ewes, expulsion of the second lamb enhances licking behavior toward any neonate (Poindron *et al.*, 1980) as does additional vaginal stimulation applied one hour postpartum (Keverne *et al.*, 1983). Conversely, the suppression of the sensory information associated with labor (by means of applying peridural anesthesia or performing a Cesarean section) results in the failure of parturient mothers to display maternal behavior, especially if they are primiparous (Alexander *et al.*, 1988; Krehbiel *et al.*, 1987). Similar results are obtained in goats (Poindron *et al.*, 2007b), although the effects of peridural anesthesia are more variable, possibly due to interindividual variations of genital innervation in this species (Labussiere, 1999). Although the nervous pathways that convey the transmission of sensory information from the vagina and uterus to the brain are not known in detail, it is well established that one way by which VCS facilitates maternal behavior in sheep is through the intracerebral release of oxytocin (OT).

Main neurobiological events

The neurobiological role of oestradiol has been studied mainly in the rat, in which this hormone had a more facilitatory effect than in the ewe (Bridges, 1996; Rosenblatt *et al.*, 1979; Rosenblatt *et al.*, 1988). Thus, Numan (1994) had established that medial preoptic area (MPOA) is the target structure to facilitate the maternal behavior by the oestradiol. In fact, this structure showed an increase in the number of receptor for oestradiol (ER) at the end of pregnancy (Giordano *et al.*, 1989). Furthermore, oestradiol implants in the MPOA facilitates the maternal behavior in virgin females (Numan, 1994), whereas applications of antiestrogens inhibit it (Ahdieh *et al.*, 1987). In the ewe, the presence of ER during oestrus in the MPOA (between some others structures) is well established (Lehmann *et al.*, 1993; Blache *et al.*, 1994). In the sheep, contrary to mice, at parturition the number of cells with ER is low and this number does not depend on the previous maternal experience (Meurisse *et al.*, 2005; Ehret and Buckenmaier, 1994). However, more recently Meurisse *et al.* (2005) indicated that in sheep ER expression is influenced by previous physiological and/or maternal experience at specific times of the reproductive cycle. They are also congruent with the higher ability of multiparous than nulliparous ewes to show maternal behavior several days prepartum.

The mechanical stimulation caused by the expulsion of fetus is associated with a short, but important, liberation of OT during the expulsion phase in the peripheral circulation, due to the Ferguson reflex (Kendrick *et al.*, 1991). As with PRL, the later release of OT is also provoked by suckling. However, the physiological events associated with VCS, caused by the passage of the fetus through the birth canal, also have neurobiological correlates. A rise of OT is also found in the cerebrospinal fluid (CSF) and the olfactory bulbs (Kendrick *et al.*, 1986, 1988). Furthermore, during parturition significant increases of the noradrenaline metabolite MHPG have been found at the CSF levels, and of noradrenaline (NA) in the median preoptic area, the bed nucleus of the stria terminalis, and the olfactory bulbs (Kendrick *et al.*, 1992), which are structures involved in the manifestation of maternal behavior in rodents (Numan, 1990).

Besides the role of the steroids hormones and VCS, some other peripheral and central physiological parameters also participate in the activation of the maternal behavior at partum, even though their role may not be as determinant. For example, the decrease in the peripheral concentrations of progesterone facilitates the effects of VCS (Kendrick and Keverne, 1991; Poindron, 2005). This is the case for the central release of opiates, and also maybe for CRH, although for this factor, its intracerebral releasing at the moment of parturition has not yet been shown in the ewe. These physiological factors and probably others that have not yet been studied, result in a rapid increase of maternal motivation in the 3 or 4 h preceding parturition, culminating at delivery. Then, maintaining this motivation will depend on the possibility of the mother to interact with the newborn. In the following section, the participation of some sensory cues on the maintenance of maternal responsiveness is examined.

The sensitive period and sensorial control of maternal behavior

If a mother is not allowed to interact with her young at the time of internally driven maternal responsiveness, she losses her ability to care for her progeny until the next parturition. This phenomenon evidences the existence of a so called "sensitive period" presented in most mammals studied. The removal of the neonates at parturition results in the rapid fading of maternal responsiveness, in contrast to what occurs if the parturient dam is allowed to have some initial contact with her young. It is, internal factors render the mother sensitive to specific stimuli from the neonate and the perception of such stimuli allows her to gain experience necessary to maintain her maternal responsiveness once those internal factors have disappeared (González-Mariscal and Poindron, 2002). These effects of early mother-young separation have been confirmed in many studies and extended to other species (rats: Bridges, 1975; Fleming and Sarker, 1990; sheep: Poindron *et al.*, 1979; goats: Klopfer *et al.*, 1964; Lickliter, 1982; Ramírez *et al.*, 1996; cattle: Hudson and Mullord, 1977; Le Neindre and Garel, 1976; rabbits: González-Mariscal *et al.*, 1998). In sheep and goats, it was initially thought that existence of a sensitive period pertained to the establishment of selective bonding to the lamb. However, the fact that anosmic ewes deprived of their lamb show a fading of maternal responsiveness similar to the intact (deprived) mothers indicates that the sensitive period concerns maternal responsiveness, as it does in the rat (Poindron and Le Neidre, 1980; Rosenblatt and Siegel, 1981).

The sensory cues, essential for the consolidation of maternal responsiveness have been investigated in rats and sheep by depriving mothers of some of the sensory stimulation provided by the young. In sheep and goats, the critical sensory modality for maintaining maternal responsiveness appears to be olfactory. For example, it is important that the mother has access to olfactory cues provided by licking their kid. About half of does prevented from licking their kid during first 4 h postpartum fail to show adequate maternal care when reunited with her young (Bordi *et al.*, 1994; Romeyer *et al.*, 1993). Additionally, the full access to olfactory cues through licking facilitates the establishment of maternal selectivity. Indeed, when selectively preventing the perception of stimuli from the lamb, postparturient ewes lose their responsiveness only if deprived of the neonate's olfactory cues (Poindron and Le Neindre, 1980). In fact, the single deprivation of olfactory stimulation provoked by placing the lamb in a smell-proof box leads to the fading of maternal behavior in a way similar to that caused by total separation (González-Mariscal and Poindron, 2002; Fig. 1). In contrast, prepartum anosmia does not provoke such effect (Baldwin and Shillito, 1974; Bouissou, 1968; Lévy *et al.*, 1995b; Poindron, 1976a,b) and only maternal selectivity is affected. As a whole, this illustrates the possibility that compensation in the sensory regulation of maternal responsiveness during the sensitive period may occur, but only if maternal olfactory system is impaired and not if it is the perception of the olfactory cues that is impeded in an intact mother (González-Mariscal and Poindron, 2002).

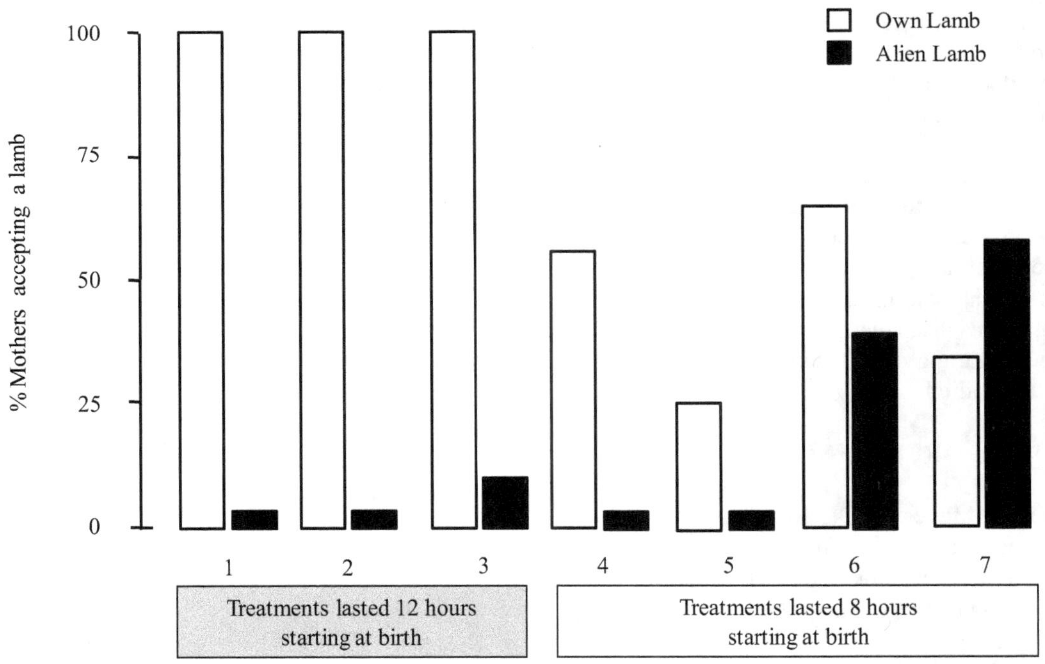

Figure 1. Percentage of mothers accepting their lamb or an alien one when a partial contact with own lamb is allowed in postparturient ewes. Groups were 1, full contact; 2, full contact except suckling; 3, lamb separated from dam by a 5cm thick double wire mesh; 4, lamb in a transparent smellproof box with ventilation toward the dam; 5, lamb in a transparent smell-proof box with ventilation away from the dam; 6, lamb in an opaque smell-proof box, with ventilation toward the dam; 7, ewe totally separated from her lamb. Visual and acoustic cues alone (group 5) did not allow maternal responsiveness to be maintained significantly better than when the mother was totally deprived of the lamb (group 7). Ewes with unrestricted access to olfactory cues from the lamb, even without direct physical contact (groups 3, 4 and 6), generally developed maternal selectivity (but note also group 5). Grey and white bars denote that treatments lasted 12 (groups 1, 2 and 3) and 8 (groups 4, 5, 6 and 7) hours, respectively starting at birth (Adapted from González-Mariscal and Poindron, 2002).

CONCLUSIONS AND PERPECTIVES

At the present, a wide body in the literature concerning to the characteristics of maternal behavior and their control mechanisms in sheep and goats has been generated. In both species, parturient females show of a series of behaviors that initiates with the prepartum isolation of the female from their coespecifics and that culminates with the successful exclusive nursing behavior toward their newborns during postpartum. Some studies underlie the role of the changes in peripheral circulating of both steroids hormones, progesterone and oestradiol. However, the sensory stimulation caused by fetus expulsion or by VCS induces an intracerebral OT release that is implicated in the onset of maternal responsiveness. Overall, estrogen and VCS induce a temporary state of maternal responsiveness that last for several hours, and interacting with the neonates allows its consolidation. While most of the studies on the characteristics and the control factors of maternal behavior in small ruminants have been conducted under experimental conditions, there is little literature concerning the possible complications on maternal behavior in animals maintained under natural grazing conditions, which could suffer an under nutrition. Because an inadequate maternal behavior at parturition as a result from under nutrition can be a common cause of neonatal mortality in flocks of sheep and goats; then, studying the effects of a short term nutritional supplementation before parturition on the mother-young interactions could help to apply some feeding strategies that reduce the productive lost in these flocks. Furthermore, contrary to a greater quantity of published studies on the onset of maternal behavior, mechanisms that contribute to their spontaneous disappearance in the mothers and the spontaneous weaning of the young in small ruminants are also little studied.

REFERENCES

Ahdieh, H.B., Mayer, A.D., Rosenblatt, J.S. 1987. Effects of brain antiestrogen inplants on maternal behavior and on postpartum estrus in pregnant rats. Neuroendocrinology. 46:522-531.

Alexander, G., Stevens, D., Kilgour, R., De Langen, H., Mottershead, B.E., Lynch, J.J. 1983. Separation of ewes from twin lambs: Incidence in several sheep breeds. Applied Animal Ethology. 10:301-317.

Alexander, G., Poindron, P., Le Neindre, P., Stevens, D., Lévy, F., Bradley, L. 1986. Importance of the first hour postpartum for exclusive maternal bonding in sheep. Applied Animal Behaviour Science. 16:295-300.

Alexander, G., Stevens, D., Bradley, L.R. 1988. Maternal behaviour in ewes following caesarian section. Applied Animal Behaviour Science. 19:273-277.

Alexander, G., Stevens, D., Bradley, L.R. 1989. Maternal acceptance of alien lambs in ewes treated and untreated with oestrogen at birth. Australian Journal of Experimental Agriculture. 29:173-178.

Allan, C.J., Holst, P.J., Hinch, G.N. 1991. Behaviour of parturient Australian bush goats I. Doe behaviour and kid vigour. Applied Animal Behaviour Science. 32:44-64.

Arnold, G.W., Morgan, P.D. 1975. Behaviour of the ewe and lamb at lambing and its relation to lamb mortality. Applied Animal Ethology. 2:25-46.

Baldwin, B.A., Shillito, E.E. 1974. The effects of ablation of the olfactory bulbs on parturition and maternal behavior in Soay sheep. Animal Behaviour. 22:220-223.

Blache, D., Batailler, M., FabreNys, C. 1994. Oestrogen receptors in the preopticohypothalamic continuum: Immunohistochemical study of the distribution and cell density during induced oestrous cycle in ovariectomized ewe. Journal of Neuroendocrinology. 6:329-339.

Bordi, A., De Rosa, G., Napolitano, F., Litterio, M., Marino, V., Rubino, R. 1994. Postpartum development of the mother-young relationship in goats. Applied Animal Behaviour Science. 42:145-152.

Bouissou, M.F. 1968. Effet de l'ablation des bulbes olfactifs sur la reconnaissance du jeune par sa mère chez les Ovins. Revue de Comportement Animal. 2:77-83.

Bridges, R.S. 1975. Long-term effects of pregnancy and parturition upon maternal responsiveness in the rat. Physiology and Behavior. 14:245-249.

Bridges, R. 1996. Biochemicals basis of parental behavior in the rat. Advances in the Study of Behavior. 25:215-242.

Carbonaro, D.A., Friend, T.H., Dellmeier, G.R. 1992. Behavioral and physiological response of dairy goats to isolation. Physiology and Behavior. 51:297-301.

Chamley, W.A., Buckmaster, J.M., Cerini, M.E. Cumming, I.A., Goding, J.R., Obst, J.M., Williams, A., Winfield, C. 1973. Changes in the levels of progesterone, corticosteroids, estrone, estradiol-17 β, luteinizing hormone, and prolactin in the peripheral plasma of the ewe during late pregnancy and at parturition. Biology of Reproduction. 9:30-35.

Collias, N.E. 1956. The analysis of socialization in sheep and goats. Ecology. 37:228-239.

Currie, W.B., Gorewit, R.C., Michel, F.J. 1988. Endocrine changes, with special emphasis on oestradiol-17β, prolactin and oxytocin, before and during labour and delivery in goats. Journal of Reproduction and Fertility. 82:299-308.

Das, N., Tomer, O.S. 1997. Time pattern on parturition sequences in Beetal goats and crosses: Comparison between primiparous and multiparous does. Small Ruminant Research. 26:157-161.

Davis, S.L., Reichert, JR L.E., Niswender, G.D. 1971. Serum levels of prolactin in sheep as measured by radioimmunoassy. Biology of Reproduction. 4:145-153.

Dwyer, C.M., McLean, K.A., Deans, L.A., Chirnside, J., Calvert, S.K., Lawrence, A.B. 1998. Vocalisations between mother and young in the sheep: Effects of breed and maternal experience. Applied Animal Behaviour Science. 58:105-119.

Dwyer, C.M. 2009. The ethology of domestic animals. In: Jensen, P. (ed.). The behavior of sheep and goats. CAB International, London UK. pp. 161-176.

Echeverri, A.C., Gonyou, H.W., Ghent, A.W. 1992. Preparturient behavior of confined ewe: Time budgets, frequencies, spatial distribution and sequential analysis. Applied Animal Behaviour Science. 34:329-344.

Ehret, G., Buckenmaier, J. 1994. Estrogen-receptor occurrence in the female mouse brain: Effects of maternal experience, ovariectomy, estrogen and anosmia. Journal of Physiology-Paris. 88:315-329.

Fleming, A.S., Sarker, J. 1990. Experience-hormone interactions and maternal behavior in rats. Physiology and Behavior. 47:1165-1173.

Giordano, A.L., Siegel, H.I., Rosenblatt, J.S. 1989. Nuclear estrogen receptor binding in the preoptic area and hypothalamus of pregnancy-terminated rats: Correlation with the onset of maternal behavior. Neuroendocrinology. 50:248-258.

Gonyou, H.W., Stookey, J.M. 1983. Use of lambing cubicles and the behavior of ewes at parturition. Journal of Animal Science. 56:787-791.

Gonyou, H.W., Stookey, J.M. 1985. Behavior of parturient ewes in group-lambing pens with and without cubicles. Applied Animal Behaviour Science. 14:163-171.

González-Mariscal, G., Melo, A.I., Beyer, C. 1998. Importance of mother/young contact at parturition and across lactation for the expression of maternal behavior in rabbits. Developmental Psychobiology. 32:101-111.

González-Mariscal, G., Poindron, P. 2002. Parental care in mammals: Immediate internal and sensory factors of control. In: Pfaff, D.W., Arnold, A.P., Etgen, A.M., Fahrfbach, S.E. and Rubin, R.T. (Eds.). Hormones, Brain and Behavior. Academic Press, New York. pp. 215-298.

Gubernick, D.J. 1981. Parent and infant attachment in mammals. In: Gubernick, D.J. and Klopfer, P.H. (Eds.), Parental Care in Mammals. Plenum Press, New York. pp. 243-305.

Hersher, L., Richmond, J.B., Moore, A.U. 1963a. Maternal behavior in sheep and goats. In: Rheingold, H.L. (Ed.), Maternal behavior in mammals. John Wiley and Sons Inc, New-York. pp. 203-232.

Hersher, L., Richmond, J.B., Moore, A.U. 1963b. Modifiability of the critical period for the development of maternal behavior in sheep and goats. Behaviour. 20:311-319.

Hudson, S.J., Mullord, M.M. 1977. Investigations on maternal bonding in dairy cattle. Applied Animal Ethology. 3:271-276.

Keller, M., Meurisse, M., Poindron, P., Nowak, R., Ferreira, G., Shayit, M., Levy, F. 2003. Maternal experience influences the establishment of visual/auditory, but not olfatory recognition of the newborn lamb by ewes at parturition. Developmental Psychobiology. 43:167-176.

Kendrick, K.M., Keverne, E.B., Baldwin, B.A., Sharman, D.F. 1986. Cerebrospinal fluid levels of acetylcholinesterase, monoamines and oxytocin during labour, parturition, vaginocervical stimulation, lamb separation and suckling in sheep. Neuroendocrinology. 44:149-156.

Kendrick, K.M., Keverne, E.B., Chapman, C., Baldwin, B.A. 1988. Microdialysis measurement of oxytocin, aspartate, γ-aminobutyric acid and glutamate release from the olfactory bulb of the sheep during vaginocervical stimulation. Brain Research. 442:171-174.

Kendrick, M.K., Keverne, E.B., Hinton, M.R., Goode, J.A. 1991. Cerebrospinal fluid and plasma concentrations of oxytocin and vasopressin during parturition and vaginocervical stimulation in the sheep. Brain Research Bulletin. 26:803-807.

Kendrick, K.M., Keverne, E.B. 1991. Importance of progesterone and estrogen priming for the induction of maternal behavior by vaginocervical stimulation in sheep: Effects of maternal experience. Physiology and Behavior. 49:745-750.

Kendrick, M.K., Keverne, E.B., Hinton, M.R., Goode, J.A. 1992. Oxytocin, amino acid and monoamine release on the region of the medial preoptic area and bed nucleus of the stria terminalis of the sheep during parturition and suckling. Brain Research. 569:199-209.

Keverne, E.B., Lévy, F., Poindron, P., Lindsay, D.R. 1983. Vaginal stimulation: an important determinant of maternal bonding in sheep. Science. 219:81-83.

Klopfer, P.H., Adams, D.K., Klopfer, M.S. 1964. Maternal imprinting in goats. Proceedings of the National Academy of Sciences of the United States of America. 52:911-914.

Krehbiel, D., Poindron, P., Lévy, F., Prud`Homme, M.J. 1987. Peridural anesthesia disturbs maternal behavior in primiparous and multiparous parturient ewes. Physiology and Behavior. 40:463-472.

Labussière, J. 1999. The physiology of milk ejection: consequences on milking techniques. In: Martinet, J., Houdebine, L.M. and Head, H.H. (eds.). Biology of Lactation. INRA Editions, Paris. pp. 307-343.

Le Neindre, P., Garel, J.P. 1976. Existence d`une période sensible pour l`établissement du comportement maternel chez la vache après la mise-bas. Biology of Behavior. 1:217-221.

Le Neindre, P., Poindron, P., Delouis, C. 1979. Hormonal induction of maternal behavior in non-pregnant ewes. Physiology and Behavior. 22:731-734.

Le Neindre, P., Poindron, P., Trillat, G., Orgeur, P. 1993. Influence of breed on reactivity of sheep to humans. Genetics Selection Evolution. 25:447-458.

Lécrivain, E., Janeau, G. 1987. Comportement d'isolement et de recherche d'abri de brebis agnelant en plein air dans un système d'élevage à caractère extensif. Biology of Behavior. 12:127-148.

Lehmann, M.N., Ebling, J.P., Moenter, S.M., Karsh, F.J. 1993. Distribution of estrogen receptor-inmunoreactive cells in the sheep brain. Endocrinology. 133:876-886.

Lehmann, J., Pryce, C.R., Bettschen, D., Feldon, J. 1999. The maternal separation paradigm and adult emotionality and cognition in male and female wistar rats. Pharmacology Biochemistry and Behavior. 64:705-715.

Lent, P. 1974. Mother infant relationships in ungulates. In: Geist, V. and Walther, F. (eds.). Behaviour of ungulates and its relation to management. International Union for Conservation of Nature and Natural Resources, Morges, Switzerland. pp.14-55.

Lévy, F., Gervais, R., Kinderman, U., Litterio, M., Poindron, P., Porter, R. 1991. Effects of early postpartum separation on maternal selectivity and responsiveness in parturient ewes. Applied Animal Behaviour Science. 31:101-110.

Lévy, F., Kendrick, K.M., Goode, J.A., Guevara-Guzman, R., Keverne, E.B. 1995a. Oxytocin and vasopressin release in the olfactory bulb of parturient ewes: Changes with maternal experience and effects on acetylcholine, gammaaminobutyric acid, glutamate and noradrenaline release. Brain Research. 669:197-206.

Lévy, F., Locatelli, A., Piketty, V., Tillet, Y., Poindron, P. 1995b. Involvement of the main but not the accessory olfactory system in maternal behavior of primiparous and multiparous ewes. Physiology and Behavior. 57:97-104.

Lickliter, R.E. 1982. Effects of a post-partum separation on maternal responsiveness in primiparous and multiparous domestic goats. Applied Animal Ethology. 8:537-542.

Lickliter, R.E. 1984. Mother-infant spatial relationships in domestic goats. Applied Animal Ethology. 13:93-100.

Lickliter, R.E. 1985. Behavior associated with parturition in the domestic goat. Applied Animal Behaviour Science. 13:335-345.

Lickliter, R.E. 1987. Activity patterns and companion preferences of domestic goat kids. Applied Animal Behaviour Science. 19:137-145.

Meurisse, M., Gonzalez, A., Delsol, G., Caba, M., Lévy, F., Poindron, P. 2005. Estradiol receptor-alpha expression in hypothalamic and limbic regions of ewes is influenced by physiological state and maternal experience. Hormones and Behavior. 48:34-43.

Murphy, P.M., Lindsay, D.R., Purvis, I.W. 1994. The importance of the birth site on the survival of Merino lambs. Proceedings of the Australian Society for Animal Production. 20:251-254.

Nowak, R., Murphy, T.M., Lindsay, D.R., Alster, P., Andersson, R., Uvnäs-Morberg, K. 1997. Development of a preferential relationship with the mother: Importance of the sucking activity. Physiology and Behavior. 62:681-688.

Nowak, R., Porter, R.H., Levy, F., Orgeur, P., Schaal, B. 2000. Role of mother-young interactions in the survival of offspring in domestic mammals. Reviews of Reproduction. 5:153-163.

Nowak, R., Keller, M., Val-Laillet, D., Lévy, F. 2007. Perinatal visceral events and brain mechanisms involved in the development of mother-young bonding in sheep. Hormones and Behavior. 52:92-98.

Numan, M. 1990. Long-term effects of preoptic area knife cuts on the maternal behavior of postpartum rats. Behavioral and Neural Biology. 53:284-290.

Numan, M. 1994. Maternal behavior. In: Knobil, E. and Neill, J.D. (eds.). The Physiology of Reproduction. Raven Press, New York. pp. 221-302.

O'Brien, P.H. 1983. Feral goat parturition and lying-out sites: Spatial, physical and meteorological characteristics. Applied Animal Behaviour Science. 10:325-339.

O'Brien, P.H. 1984. Leavers and stayers: Maternal postpartum strategies in feral goats. Applied Animal Behaviour Science. 12:233-243.

Poindron, P. 1976a. Effets de la suppression de l'odorat, sans lésion des bulbes olfactifs, sur la sélectivité′ du comportement maternel de la Brebis. Comptes Rendus Hebdomadaires Des Séances de l'Académie des Sciences de Paris. Série D: Sciences Naturelles. 282:489-491.

Poindron, P. 1976b. Mother-young relationships in intact or anosmic ewes at the time of suckling. Biology of Behaviour. 2:161-177.

Poindron, P., Le Neindre, P. 1979. Hormonal and behavioral basis for establishing maternal behavior in sheep. In: Zichella, M.L. and Pancheri, P. (eds.). Psychoneuroendocrinology in Reproduction. Elsevier, Amsterdam. pp. 121-128.

Poindron, P., Martin, G.B., Hooley, R.D. 1979. Effects of lambing induction on the sensitive period fro the establishment of maternal behavior in sheep. Physiology and Behavior. 23:1082-1087.

Poindron, P., Le Neindre, P., Raksanyi, I., Trillat, G., Orgeur, P. 1980. Importance of the characteristics of the young in the manifestation and establishment of maternal behaviour in sheep. Reproduction Nutrition and Development. 20:817-826.

Poindron, P., Le Neindre, P. 1980. Endocrine and sensory regulation of maternal behavior in the ewe. Advances in the Study of Behavior. 11:75-119.

Poindron, P., Lévy, F., Krehbiel, D. 1988. Genital, olfactory, and endocrine interactions in the development of maternal behaviour in the parturient ewe. Psychoneuroendocrinology. 13:99-125.

Poindron, P., Nowak, R., Lévy, F., Porter, H.R., Schaal, B. 1993. Development of exclusive mother-young bonding in sheep and goats. In: Milligan, S. R. (ed.). Oxford Reviews of

Reproductive Biology. Oxford University Press, New York. pp. 311-363.

Poindron, P., Soto, R., Romeyer, A. 1997. Decrease of response to social separation in preparturient ewes. Behavioural Processes. 40:45-51.

Poindron, P., Hernandez, H., Gonzalez, F., Navarro, M.L., Delgadillo, J.A. 1998a. Mother-young relationships in goats: Mechanisms of control and possible implications for production. In: Veissier, I. and Boissy, A. (eds.). Proceedings of the 32nd Congress of the International Society for Applied Ethology. INRA, Clermont-Ferrand, France. pp. 85.

Poindron, P., Hernandez, H., Navarro, M.L., Gonzalez, F., Delgadillo, J.A., Garcia, S. 1998b. Relaciones madre-cria en cabras. Proceedings of the XIII Reunion Nacional Sobre Caprinocultura. Universidad autónoma de San Luis Potosí, San Luis Potosi, Mexico. pp. 48-66.

Poindron, P., Gilling, G., Hernandez, H., Serafín, N., Terrazas, A. 2003. Early recognition of newborn goat kids by their mother: I. Nonolfactory discrimination. Developmental Psychobiology. 43:82-89.

Poindron, P. 2005. Mechanisms of activation of maternal behaviour in mammals. Reproduction Nutrition and Development. 45:341-351.

Poindron, P., Keller, M., Lévy, F. 2007a. Maternal responsiveness and maternal selectivity in domestic sheep and goats: The two facets of maternal attachment. Developmental Psychobiology. 49:54-70.

Poindron, P., Terrazas, A., Navarro-Montes de Oca M. de la L., Serafín, N., Hernández, H. 2007b. Sensory and physiological determinants of maternal behavior in the goat (Capra hircus). Hormones and Behavior. 52:99-105.

Poindron, P., Otal, J., Ferreira, G., Keller, M., Guesdon, V., NowakR., Lévy, F. 2010. Amniotic fluid is important for the maintenance of maternal responsiveness and the establishment of maternal selectivity in sheep. Animal. 4:2057-2064.

Price, E.G., Thos, J. 1980. Behavioral response to short-term isolation in sheep and goat. Applied Animal Ethology. 6:331-339.

Putu, I.G., Poindron, P., Lindsay, D.R. 1988. Early disturbance of Merino ewes from the birth site increases lamb separations and mortality. Proceedings of the Australian Society for Animal Production. 17:298-301.

Ramírez, A., Quiles, A., Hevia, M., Sotillo, F. 1995. Behavior of the Murciano-Granadina goat in the hour before parturition. Applied Animal Behaviour Science. 44:29-35.

Ramírez, A., Quiles, A., Hevia, M.L., Sotillo, F., Ramírez, M.C. 1996. Effects of immediate and early postpartum separation on maintenance of maternal responsiveness in parturient multiparous goats. Applied Animal Behaviour Science. 48:215-224.

Romeyer, A., Poindron, P. 1992. Early maternal discrimination of alien kids by post-parturient goats. Behavioural Processes. 26:103-112.

Romeyer, A., Porter, R.H., Lévy, F., Nowak, R., Orgeur, P., Poindron, P. 1993. Maternal labelling is not necessary for the establishment of discrimination between kids by recently parturient goats. Animal Behaviour. 46:705-712.

Romeyer, A., Poindron, P., Orgeur, P. 1994a. Olfaction mediates the establishment of selective bonding in goats. Physiology and Behaviour. 56:693-700.

Romeyer, A., Poindron, P., Porter, R.H., Lévy, F., Orgeur, P. 1994b. Establishment of maternal bonding and its mediation by vaginocervical stimulation in goats. Physiology and Behaviour. 55:395-400.

Rosenblatt, J.S., Siegel, H.I., Mayer, A.D. 1979. Progress in the study of maternal behavior in the rat: Hormonal, nonhormonal, sensory, and developmental aspects. Advances in the Study of Behavior. 10:225-311.

Rosenblatt, J.S., Siegel, H.I. 1981. Factors governing the onset and maintenance of maternal behavior among nonprimate mammals. In: Gubernick, D.J. and Klopfer, P.H. (Eds.). Parental care in mammals. Plenum Press, New York. pp. 13-76.

Rosenblatt, J.S., Mayer, A.D., Giordano, A.L. 1988. Hormonal basis during pregnancy for the onset of maternal behavior in the rat. Psychoneuroendocrinology. 13:29-46.

Rudge, M.R. 1970. Mother and kid behaviour in feral goats (Capra hircus L.). Zeitschrift für Tierpsychologie. 27:687-692.

Sambraus, H.H., Wittmann, M. 1989. Observations of the birth and suckling behavior of goats. Tierärztliche Praxis. 17:359-365.

Shackleton, D.M., Haywood, J. 1985. Early mother-young interactions in California bighorn sheep.Ovis Canadensis californiana. Canadian Journal of Zoology. 63:868-875.

Shillito, E.E., Hoyland, V.J. 1971. Observation on parturition and maternal care in Soay sheep. Journal of Zoology, London. 165:509-512.

Shipka, M.P., Ford, S.P. 1991. Relationship of circulating estrogen and progesterone concentrations during late pregnancy and the onset phase of maternal behavior in the ewe. Applied Animal Behaviour Science. 31:91-

Smith, F.V. 1965. Instinct and learning in the attachment of lamb and ewe. Animal Behaviour. 13:84-86.

Terrazas, A., Serafin, N., Hernández, H., Nowak, R., Poindron, P. 2003. Early recognition of newborn goat kids by their mother: II. Auditory recognition and evidence of an individual acoustic signature in the neonate. Developmental Psychobiology. 43:311-320.

Thorburn, G.D., Nicol, D.H., Basset, J.M., Shutt, D.A., Cox, R.I. 1972. Parturition in the goat and sheep changes in corticosteroids, progesterone, oestrogens and prostaglandin F. Journal of Reproductive Physiology. 16:61-84.

Val-Laillet, D., Nowak, R. 2006. Socio-spatial criteria are important for the establishment of maternal preference in lambs. Applied Animal Behaviour Science. 96:269-280.

Val-Laillet, D., Giraud, S., Tallet, C., Boivin, X., Nowak, R. 2006. Non nutritive sucking: One of the major determinants of filial love. Developmental Psychobiology. 48:220-232.

13

APPARENT DIGESTIBILITY OF RHODE ISLAND RED HENS' DIETS CONTAINING *Leucaena leucocephala* AND *Moringa oleifera* LEAF MEALS

F.M.K. Abou-Elezz[a,b], L. Sarmiento-Franco[a], R. Santos-Ricalde[a] and F. Solorio-Sanchez

[a] *Faculty of Veterinary Medicine and Animal Science, University of Yucatan (UADY), Mérida, Yucatán, Mexico.*
[b] *Faculty of Agriculture, Assiut University. Assiut (71526), Egypt.*
Email: abollez@aun.edu.eg
**Corresponding author*

SUMMARY

This study consisted of two experiments aimed to evaluate the dietary digestibility by Rhode Island Red (RIR) hens fed different levels of *Leucaena leucocephala* (LLM) or *Moringa oleifera* (MOLM). In each experiment, four groups each of nine birds were assigned to four dietary treatments containing 0, 5, 10 and 15 % of LLM (Exp 1) or MOLM (Exp 2). All groups received smashed diets, containing similar metabolizable energy and crude protein for six weeks and during the last four days, feed intake was recorded daily and excreta was totally collected twice a day and weighed individually. The dietary treatments had no significant effects on the intake of dry matter (DM), organic matter (OM), gross energy (GE), crude protein (CP) or neutral detergent fiber (NDF) in both experiments, while the acid detergent fiber (ADF) intake increased linearly with the increase of LLM and MOLM levels. The apparent digestibility of DM, CP, ADF, and OM decreased linearly with the increase of LLM and MOLM levels. The apparent digestibility of gross energy decreased linearly in LLM experiment, while it was not affected in MOLM experiment. LLM and MOLM could be recommended for RIR hens between 5 and 10% of the diet.

Keywords: Digestibility; nutrients intake; forages; hens; *Leucaena leucocephala; Mo*ringa oleifera; Rhode island hens.

INTRODUCTION

The current acute short supply of animal protein for humans in most developing countries along with the expensive prices of the conventional feedstuffs have necessitated several investigations on the potentials of some novel feed resources for livestock; such as leaf meals which are usually cheap and easily available (Sarmiento, 2001; Odunsi *et al.*, 2002; Ige *et al.*, 2006; Nworgu and Fasogbon, 2007; Atawodi *et al.*, 2008). The feed cost represents 60 to 80% of the total costs involved in poultry enterprises. Therefore, any reduction in feed cost will reduce remarkably the total production costs. In addition, using leaf meals could help in alleviating the competition between humans and animals for some conventional feedstuffs, such as soybean meal and maize (Ige *et al.*, 2006; Horsted, 2006).

Leucaena leucocephala and *Moringa oleifera* leaf meals (LLM and MOLM) are seemingly good feed alternatives to commercial livestock and poultry in the tropics (Makkar and Becker, 1997; Agbede, 2003). *Leucaena leucocephala* and *Moringa oleifera* leaves are available and cheap in vast areas worldwide (Brewbaker and Sorensson, 1990; Morton, 1991; Price, 2000). The LLM and MOLM contain considerable amounts of crude protein, essential amino acids, metabolizable energy, vitamins and minerals. Therefore, the LLM and MOLM are expected to be sustainable resources for commercial poultry production in the tropics (Makkar and Becker, 1997; Price, 2000; Elkhalifa *et al.*, 2007; Kakengi *et al.*, 2007; Atawodi *et al.*, 2008). In comparison with ruminants and other monogastric animals like rabbits and pigs, the digestive system of chickens is very simple and has limited capacity to digest high fibrous materials (Esonu *et al.*, 2006; Ige *et al.*, 2006). Poor laying and growing performances of chickens which fed diets with high inclusion levels of leaf meals were reported. Therefore, including the leaf meals in the chicken diets might go through good knowledge of their nutritional value and its impact on nutrients digestibility (Angkanaporn *et al.*, 1994; Esonu *et al.*, 2006; Kakengi *et al.,* 2007; Iheukwumere *et al.* 2008; Ayssiwede *et al.*, 2010).

Measuring digestibility is a way to evaluate the availability of nutrients. In addition, an accurate prediction of overall nutritive value can be made when digestibility is combined with nutrients intake data (Khan *et al.*, 2003). The objective of this study was to evaluate the effect of dietary inclusion of either *Leucaena leucocephala* or *Moringa oleifera* leaf meal on nutrients digestibility by Rhode Island Red hens, which is expected to present a recommendation about the possible inclusion level of LLM and MOLM in the laying hen diets.

MATERIALS AND METHODS

The experimental site

The experimental work was carried out using the poultry facilities of the Faculty of Veterinary Medicine and Animal Science (FMVZ), University of Yucatan (UADY), Yucatán, Mexico. The climate is sub-humid, with an average annual rainfall (highly variable) of 960 mm, and 6-7 months of dry period; the annual average temperature is 26 °C. The soils, are calcareous and mainly shallow (<10 cm in depth), are classified as Rendzinas and Lithosols and are of moderate fertility, with 1.0 to 1.5 % organic carbon content and a pH range of 7.5 to 7.8.

Birds, treatments and experimental design

The study consisted of two feeding experiments, which were carried out simultaneously following similar design. In each experiment, thirty six Rhode Island Red hens at 36 weeks of age with 60% egg laying rate on average and 2 kg body weight, were randomly divided into four groups each of nine birds which were allocated in individual cages (40 x 40 x 40 cm). The four groups were corresponded to four dietary treatments containing 0 (control), 5, 10 and 15 % of *LLM* (Exp1) or MOLM (Exp 2), respectively. All groups received diets in the form of meal, containing similar metabolizable energy and CP levels (16% and 2900 kcal ME/kg diet, as fed basis).

Preparing the leaf meals

L. leucocephala and *M. oleifera* fresh leaves were harvested from trees (two- year old, last harvest was 4 months before) growing at the FMVZ farm, under the tropical conditions of the Yucatán Peninsula, southern Mexico. As described by Morton (1991) and Shelton and Brewbaker (1994), the young branches were cut from trees, leaves were separated from branches, spread out and dried under shade for a period of 1 day. Thereafter, they were dried in ovens (60ºC) for two days. The dried leaves were grounded with a hammer mill (3.0 mm sieve) to make the LLM or MOLM, which were incorporated to the experimental diets.

Digestibility study

The hens were fed on the experimental diets for six weeks, and during the last four days (Ige *et al*., 2006), feed intake was recorded daily for the individual birds. Fresh droppings (feces + urine) were collected and weighed two times a day for the individual birds, kept individually, oven dried at 60°C and weighed. At the end, the collected excreta per bird were pooled, mixed, grounded and stored respectively for chemical analyses.

Laboratory analyses

Chemical composition of experimental diets (Table 1), leaf meals (Table 2) and excreta samples including; dry matter (DM), crude protein (CP), acid detergent fiber (ADF), neutral detergent fiber (NDF), gross energy (GE), ash, calcium and phosphorus were determined at FMVZ nutrition lab, following the standard methods of AOAC (1990). A graded volumetric flask was used to measure the volume of each diet, in which five replicates (each of 100 g) of each diet were included (Table 3).

Determination of apparent nutrients digestibility

The apparent coefficients of nutrients digestibility were determined according to the following formula:

Apparent nutrient digestibility= [(NI-NE) / NI] x100

Where, NI represented the nutrient intake and NE expressed the nutrient excreted (Ayssiwede *et al.*, 2010). In case of energy, this estimate was called apparent metabolizability of energy (AME), the average daily ME intake was calculated by multiplying the average GE intake by its AME coefficient (Kendeigh *et al.*, 1977; Karasov, 1990). Similarly, the ME for the experimental diets was calculated by multiplying its GE (Kcal/kg diet) content by its AME coefficient.

Statistical analyses

The complete randomized design was used and data in each experiment were analyzed by response surfaces using orthogonal contrasts to evaluate the linear, quadratic and cubic effects of the LLM or MOLM levels, using the general linear model (GLM) procedure of SAS (2002).

Table 1. Composition of experimental diets.

Items	Exp. 1 diets (LLM[1] treatments)				Exp. 2 diets (MOLM[2] treatments)			
	0% Control	5%	10%	15%	0% Control	5%	10%	15%
Ingredients (% diet)								
Leaf meal	0.00	5.00	10.00	15.00	0.00	5.00	10.00	15.00
Sorghum	63.94	60.32	56.22	52.13	61.23	59.34	57.43	55.52
Soya bean meal	18.32	16.22	14.21	12.21	19.30	15.96	12.63	9.30
$CaCO_3$ (38%)	9.10	8.75	8.48	8.21	9.00	8.67	8.34	8.00
Canola meal	5.00	5.00	5.00	5.00	5.00	5.00	5.00	5.00
Soya oil	1.00	1.92	3.13	4.33	2.83	3.23	3.62	4.02
Di-Cal-phosphate	1.53	1.56	1.59	1.62	1.52	1.50	1.50	1.50
Lysine	0.10	0.18	0.27	0.35	0.11	0.23	0.35	0.47
Methionine	0.17	0.21	0.26	0.31	0.17	0.23	0.29	0.35
NaCl	0.30	0.30	0.30	0.30	0.30	0.30	0.30	0.30
Flavomicine 4%	0.25	0.25	0.25	0.25	0.25	0.25	0.25	0.25
Mycosorb	0.10	0.10	0.10	0.10	0.10	0.10	0.10	0.10
Funginat 42	0.05	0.05	0.05	0.05	0.05	0.05	0.05	0.05
Choline chloride	0.05	0.05	0.05	0.05	0.05	0.05	0.05	0.05
Minerals premix [a]	0.05	0.05	0.05	0.05	0.05	0.05	0.05	0.05
Vitamins premix [b]	0.03	0.03	0.03	0.03	0.03	0.03	0.03	0.03
Anti-oxidant	0.01	0.01	0.01	0.01	0.01	0.01	0.01	0.01
Total	100.0	100.0	100.0	100.0	100.0	100.0	100.0	100.0
Chemical composition (on DM basis)								
DM (%)	89.55	89.82	90.68	90.68	89.96	89.8	89.9	90.24
GE (Kcal/kg)	3886.3	4010.9	3889.1	4010.8	3999.4	4086.1	4123.6	4128.2
CP (%)	18.06	18.24	18.58	18.83	18.39	17.66	17.88	17.45
ADF (%)	7.87	9.39	10.32	10.49	7.87	9.5	10.16	10.16
NDF (%)	54.35	58.19	59.14	66.85	52.34	59.74	61.04	60.29
Calcium (%)	3.47	3.69	3.42	3.64	3.1	3.45	3.38	3.3
Phosphorus (%)	0.64	0.65	0.62	0.68	0.64	0.66	0.67	0.51
Ash (%)	13.49	12.5	12.09	12.88	13.04	12.47	12.4	12.78

[a]: Content kg^{-1} of diet: Manganese, 65 mg; iodine, 1 mg; iron, 55 mg, copper, 6 mg; zinc, 55 mg; selenium, 0.3 mg; [b]: Content kg^{-1}: vitamin A, 8000 UI; vitamin D, 2500 UI; vitamin E, 8 UI; vitamin K, 2 mg; vitamin B12, 0.002 mg; riboflavin, 5.5 mg; pantothenate of calcium, 13 mg; niacin, 36 mg; choline, 500 mg; folic acid, 0.5 mg; thiamine, 1 mg; pyridoxine, 2.2 mg; biotin, 0.05 mg. [1] *Leucana leucocephala* leaf meal [2] *Moringa oleifera* leaf meal.

Table 2. Chemical analysis of *Leucaena leucocephala* (LLM) and *Moringa oleifera* (MOLM) leaf meals.

Leaf meal	DM %	Nutrients content (on dry matter basis)						
		CP%	ADF%	NDF%	Ca %	P %	Ash %	GE(Kcal/kg)
LLM	91.0	23.61	25.69	40.38	1.62	0.18	8.27	4731
MOLM	91.22	19.76	27.11	44.42	2.13	0.24	9.61	4456

Dry matter (DM), crude protein (CP), acid detergent fiber (ADF), neutral detergent fiber (NDF), gross energy (GE), calcium (Ca), available phosphorus (P)

RESULTS

The results showed considerable amounts of CP (23.61 %) and GE energy (4731 Kcal/kg DM) in LLM. The corresponding values in MOLM amounted 19.76% CP and 4456 Kcal/kg DM, respectively. In addition, they contain good levels of calcium and phosphorus. The volumes of dietary treatments (Table 3) increased linearly with the increase in LLM and MOLM levels in the diets.

The results of feed and nutrients intake of RIR hens fed on diets containing different levels of LLM and MOLM are showed in tables 4 and 5, respectively. The dietary treatments had no significant effect on the ingested feed, dry matter (DM), organic matter (OM), gross energy (GE), crude protein (CP) or neutral detergent fiber (NDF) in both experiments, while the acid detergent fiber (ADF) consumption increased linearly with the increase of either LLM or MOLM levels.

Table 3. The volume averages (of 100 g) of the experimental diets contained *Leucaena leucocephala* (LLM) or *Moringa oleifera* (MOLM) leaf meal.

Experiments	Dietary treatments				SEM	*P*- value		
	0%	5%	10%	15%		Linear	quadratic	Cubic
Exp 1. LLM diets volumes (ml)	142.85	151.42	157.53	163.68	1.20	0.0070	0.7830	0.8673
Exp 2. MOLM diets volumes(ml)	141.91	157.01	168.82	179.47	1.66	0.0050	0.8366	0.6751

SEM: Standard error of mean, P: Probability

Table 4. Feed and nutrients intake of Rhode Island Red Hens fed diets containted *Leucaena leucocephala* leaf meal (LLM).

Nutrients	LLM treatments				SEM	*P* –value		
	0%	5%	10%	15%		Linear	quadratic	Cubic
FI (g/d)	106.81	105.50	101.25	98.56	4.18	0.1307	0.8698	0.8117
DM (g/d)	95.65	94.76	91.81	89.37	3.76	0.2054	0.8384	0.8801
OM (g/d)	82.74	82.91	80.71	78.37	4.57	0.4597	0.7862	0.914
GE (Kcal/d)	371.73	380.07	357.08	359.87	20.76	0.5307	0.8939	0.5408
CP (g/d)	17.27	17.28	17.05	16.89	0.95	0.754	0.9282	0.9441
ADF (g/d)	7.52	8.89	9.47	9.41	0.50	0.0096	0.165	0.9448
NDF (g/d)	51.98	55.14	54.30	59.90	3.16	0.1146	0.7007	0.4652

SEM: standard error of mean, P: probability, FI: feed intake, DM: dry matter, OM: organic matter, GE: gross energy, CP: crude protein, ADF: acid detergent fiber, NDF: neutral detergent fiber

Table 5. Feed and nutrients intake of Rhode Island Red Hens fed diets contained *Moringa oleifera* leaf meal (MOLM).

Nutrients	MOLM treatments				SEM	*P* –value		
	0%	5%	10%	15%		Linear	quadratic	Cubic
FI (g/d)	113.61	4.68	111.22	107.43	4.68	0.3411	0.7409	0.9629
DM (g/d)	102.20	4.21	99.99	96.95	4.21	0.3687	0.79	0.9612
OM (g/d)	88.87	3.68	87.59	84.56	3.68	0.3973	0.6928	0.9646
GE (Kcal/d)	408.75	17.45	412.31	379.52	17.45	0.9859	0.8181	0.923
CP (g/d)	19.81	0.75	17.87	16.91	0.75	0.0142	0.5353	0.4136
ADF (g/d)	8.04	0.41	10.15	9.85	0.41	0.0028	0.027	0.8954
NDF (g/d)	53.49	2.51	60.28	59.17	2.51	0.1455	0.1118	0.5591

SEM: standard error of mean, P: probability, FI: feed intake, DM: dry matter, OM: organic matter, GE: gross energy, CP: crude protein, ADF: acid detergent fiber, NDF: neutral detergent fiber

The results of apparent nutrients digestibility (tables 6 and 7) revealed a linear decrease for DM, CP, ADF, and OM with the increase of LLM and MOLM levels. MOLM treatments did not show any significant effect on GED, while it decreased linearly in LLM trial. The dietary treatments had no effect on NDF digestibility in both trials.

The GED coefficients were used to calculate the metabolizable energy (ME) of the experimental diets, which amounted to 2933, 2831, 2585, and 2600 Kcal/kg diet at 0, 5, 10 and 15% LLM, respectively. The corresponding ME values at the same levels of MOLM treatments were 3104, 3073, 2873 and 3025 Kcal/kg diet, respectively. The average ME intake amounted, 313, 300, 262 and 260 (Kcal/hen/d) at 0, 5, 10 and 15% LLM, respectively. Meanwhile, the ME intake was 353, 347, 330 and (Kcal/hen/d), at 0, 5, 10 and 15% MOLM, respectively.

Table 6. Apparent digestibility of Rhode Island Red Hen diets contained *Leucaena leucocephala* leaf meal (LLM).

Nutrients	Nutrients digestibility (%) of LLM treatments				SEM	*P* –value		
	0%	5%	10%	15%		Linear	quadratic	Cubic
DM	75.90	69.46	65.34	62.34	2.51	0.0004	0.498	0.9165
OM	80.89	74.36	70.88	66.76	2.22	0.0001	0.5913	0.7129
GED	84.28	78.59	73.31	71.49	1.90	0.0001	0.3197	0.7244
CP	60.95	47.54	45.21	39.35	5.56	0.0075	0.4776	0.5393
ADF	45.59	39.94	35.48	25.63	4.31	0.0021	0.6317	0.7399
NDF	83.01	81.73	80.53	79.44	1.42	0.0703	0.9478	0.9957

SEM: standard error of mean, P: probability, DM: dry matter, OM: organic matter, GED: Gross energy digestibility, CP: crude protein, ADF: acid detergent fiber, NDF: neutral detergent fiber

Table 7. Apparent digestibility of Rhode Island Red Hen diets contained *Moringa oleifera* leaf meal (MOLM).

Nutrients	Nutrients digestibility (%) of MOLM treatments				SEM	*P* –value		
	0%	5%	10%	15%		Linear	quadratic	Cubic
DM	83.97	75.34	72.90	72.94	2.20	0.0011	0.0587	0.7101
OM	84.04	80.50	77.86	78.03	2.63	0.0189	0.4872	0.8738
GED	86.29	83.77	80.15	81.22	2.19	0.0741	0.4195	0.5603
CP	70.30	63.96	62.57	56.70	4.96	0.0128	0.1411	0.6149
ADF	64.94	49.74	41.56	40.69	4.92	0.0009	0.1559	0.989
NDF	86.26	84.56	84.25	82.76	2.26	0.2937	0.9635	0.8016

SEM: standard error of mean, P: probability, DM: dry matter, OM: organic matter, GED: Gross energy digestibility, CP: crude protein, ADF: acid detergent fiber, NDF: neutral detergent fiber

DISCUSSION

The feed and nutrients intake of chickens is affected by the increased diet bulkiness (D'Mello, 1991). However high bulkiness results were observed in the diets of high levels of LLM and MOLM, no significant effects were detected on FI or nutrients intake (except for ADF) in both trials. The FI results were in agreement with FI results reported by Ayssiwede *et al.* (2010), who reported similar FI values of the indigenous Senegal hens fed diets with 0, 7, 14 or 21% LLM. When chickens receive diets with high bulkiness, which is commitment to high fiber content, they may increase their FI in order to meet their nutrients requirements as reported by Kakengi *et al.* (2007). Indeed, the hens are likely try to eat until they meet their energy demand (Smith, 1999). Low energy utilization in CF component of MOLM or LLM was reported (D'mello and Acamovic, 1989; Kakengi *et al.*, 2007). As a solution, Kakengi *et al.* (2007) recommended the use of high energy materials when the MOLM is added in high levels in the laying hen's diets. In the current study, the oil was used to formulate Iso-caloric dietary treatments.

The obtained digestibility coefficients revealed a linear reduction in CP, DM, OM and ADF by the increase of MOLM and LLM levels. In addition, GED decreased linearly by the increase of LLM levels. In contrast of our results, Ayssiwede *et al.* (2010) reported feeding hens at 0, 7, 14 and 21% LLM and had no significant effect on utilization values of DM, OM, CP and energy. However, they concluded that the controversy of their results with others is due to the fact that they have treated the

LLM with additive ferric-sulphate to form a complex with mimosine, which led to eliminate its toxicity. Indeed, Mimosine is considered toxic for poultry and constitutes 7% of the total protein existed in *LLM* (Meulen *et al.*, 1979; D'Mello and Acamovic, 1982). Detoxification of LLM could be possible by using ferric sulfate, but the low AME remains the major challenging problem with LLM (D'Mello and Acamovic, 1989), which could be associated with the decreased AME values obtained in LLM treatments in the current study. Bhatnagar *et al.* (1996) and Kakengi *et al.* (2007) showed that including high levels (10 and 15%) of LLM and MOLM in the diets of laying hens has contributed to low available energy and CP. Similar results with broilers are reported by Iheukwumere *et al.* (2008), where the utilization values of DM, CP and CF were lower in 10 and 15% Cassava leaf meal than its control. The obtained CP and DM digestibility values were comparable to those found by Ige *et al.* (2006), where the hens fed on 0, 5, 10, and 15% of gliricidia leaf meal, respectively.

The obtained ME (Kcal/kg diet) as well as the daily ME intake at all MOLM treatments were in the recommended ranges by NRC (1994), which reported to be amounted to 2900 Kcal/Kg diet; and 292 Kcal/hen/day. The corresponding values which were obtained at 10 and 15% LLM were lower than the recommended range, while the achieved values at 0 and 5% LLM levels were within the recommended range. That could indicate that the hens fed on 10% and 15% LLM of their diet could not meet their ME demands. The same conclusion was reported by D'Mello and Acamovic (1989).

The low digestibility of crude protein observed in high fiber diets is attributed to the indigestible cell wall which reduces the digestion and absorption of protein overall (Nyman *et al.*, 1990). In addition, Bolton and Blair (1974) reported that up to 20 percent of the nitrogen in dried forages maybe present as non-protein compounds. Poor nutrient utilization observed could be in general due to the increased fiber content of the diets with higher LLM and MOLM levels, also attributed to the low capacity of the chicken digestive system in utilizing the high fibrous feeds. The caeca, which is the unique place for digesting plant fiber and where the major concentration of intestinal micro-organisms is found in mature birds, is very short. In addition, previous reports revealed no more than 25% of the total dry matter passing through the intestines enters the caeca (Son *et al.,* 2002; Odunsi *et al.*, 2002 and Ige *et al.*, 2006). Unlike ruminants, and even some species like pigs and rabbits, chickens lack the anatomic adaptations and the necessary digestive microorganisms to support processes of fermentation, which is necessary for efficient utilization of fibrous feeds in order to cover part of their nutrient needs (Esonu *et al.*, 2006; Ige *et al.*, 2006).

CONCLUSION

Including LLM or MOLM upto 15% of the diets of RIR hens had no effect on the ingested feed, DM, GE, CP, or NDF. According to the obtained digestibility coefficients of dietary nutrients, the incorporation of LLM or MOLM in the diets of Rhode Island Red hens could be recommended between 5 and 10% of the diet.

AKNOWLEDGEMENT

This research work was partially funded by PROMEP/SEP through research project (Tree fodder mixture for improving production and nutrient cycling in agro-forestry systems) and by means the scholarship awarded to the first author by the government of Mexico through the Ministry of Foreign Affairs (SRE).

REFERENCES

Agbede, J.O. 2003. Equi-protein replacement of fishmeal with Leucaena leaf protein concentrate: An Assessment of Performance Characteristics and Muscle Development in the Chicken. International Journal of Poultry Science. 2: 421-429.

Angkanaporn, K., Choct, M., Bryden, W.L., Annison, E.F. 1994. Effects of wheat pentosans on endogenous amino acid losses in chickens. Journal of the Science of Food and Agriculture. 66: 399 – 404.

A.O.A.C. 1990. Official methods of analysis. Association of Official Analytical Chemists (15th ed.), Arlington, VA.

Atawodi, S. E., Mari, D., Atawodi, J. C., Yahaya, Y. 2008. Assessment of *Leucaena leucocephala* leaves as feed supplement in laying hens. African Journal of Biotechnology. 7: 317-321.

Ayssiwede, S. B., Dieng, A., Chrysostome, C., Ossebi, W., Hornick, J.L., Missohou, A. 2010. Digestibility and metabolic utilization and nutritional value of *Leucaena leucocephala* (Lam.) leaves meal incorporated in the diets of indigenous seneghal chickens. International Journal of Poultry Science. 9: 767-776.

Bhatnagar, R., Kataria M., Verna, S.V.S. 1996. Effect of dietary *Leucaena leucocephala* leaf meal on the performance and egg charateristics in white leghorn hens. International Journal of Animal Science. 66:1291-1294.

Bolton, W., Blair, R. 1974. Poultry nutrition. Book, fourth edition. Ministry of Agriculture, fisheries and food. London: her majesty's stationery office.

Brewbaker, J.L., Sorensson, C.T. 1990. New tree crops from interspecific Leucaena hybrids. In: Janick, J. and Simon, J.E. (eds), Advances in New Crops. Timber Press, Portland, pp. 283-289. Online in URL address: [http://www.hort.purdue.edu/newcrop/proceedings1990/V1-283.html] Consulted, 24 May 2011.

D'Mello, J.P. 1991. Nutritional potentialities of fodder trees and fodder shrubs as protein sources in monogastric nutrition. Proceedings of the FAO Expert Consultation held at the Malaysian Agricultural Research and Development Institute (MARDI) in Kuala Lumpur, Malaysia. Online in URL address: [http://www.fao.org/docrep/003/t0632e/T0632E08.htm] Consulted, 24 May 2011.

D'Mello, J.P.F., Acamovic, T. 1989. *Leucaena leucocephala* in poultry nutrition. A review. Animal Feed Science and Technology. 26: 1-28

D'Mello, J.P.F., Acamovic, T. 1982. Growth performance and mimosine excretion by young chicks fed on *Leucaena leucocepha*. Animal Feed Science and Technology. 26: 1-28

Elkhalifa, A.E.O., Ahmed, S.A., Sara, A. 2007. Nutritional evaluation of *Moringa Oleifera* leaves and extract. Report. Ahfad Journal, 01-DEC-07.

Esonu, B.O., Opara, M.N., Okoli, I.C., Obikaonu, H.O., Udedibie, C., Iheshiulor, O.O.M. 2006. Physiological response of laying birds to Neem (Azadirachta Indica) leaf meal-based diets: Body weight organ characteristics and haematology. Online Journal of Health and Apllied Sciences. 2:4. Online in URL address: [http://www.ojhas.org/issue18/2006-2-4.htm] Consulted: 24th May 2011

Horsted, K. 2006. Increased foraging in organic layers. PhD thesis, University of Aarhus, Faculty of Agricultural Sciences, Department of Agroecology. Online in URL address: [http://orgprints.org/10463/01/10463.pdf] Consulted: 24th May 2011

Ige, A.O., Odunsi, A.A., Akinlade, J.A., Ojedapo, L.O., Ameen, S.A., Aderinola, O.A., Rafiu, T.A., 2006. Gliricidia leaf meal in layer's diet Effecast on performance, nutrient digestibility and economy of production. Journal of Animal and Veterinary Advances. 5: 483-486.

Iheukwumere, F.C., Ndubuisi, E.C., Mazi, E.A., Onyekwere, M.U., 2008. Performance, nutrients utilization and organ characteristics of broilers fed Cassava leaf meal (Manihot esculenta Crantz). Pakistan Journal of Nutrition. 7: 13- 16.

Kakengi A.M.V., Kaijage, J.T., Sarwatt, S.V., Mutayoba, S.K., Shem, M.N. Fujihara, T. 2007. Effect of *Moringa oleifera* leaf meal as a substitute for sunflower seed meal on performance of laying hens in Tanzania. Livestock Research for Rural Development. 19(8). Online in URL address: [http://www.lrrd.org/lrrd19/8/kake19120.htm] Consulted: 18th December, 2011.

Karasov, W.H. 1990. Energetics of foraging. Digestion in Birds: Chemical and Physiological determinants and ecological implications. Studies in Avian Biology. 13: 391-415

Kendeigh, S.C., Dol'nik, V.R., Gavirilov, V.M. 1977. Avian energetics, p. 127-204. In J. Pinowski and S.C. Kendeigh (eds.), Granivorous birds in ecosystems. Cambridge Uni. Press, New York.

Khan, M.A., Nisa, M.U., Sarwar, M. 2003. Techniques measuring digestibilities for the nutritional evaluation of feeds. International Journal of Agriculture and Biology. 1: 91-94

Makkar, H.P.S., Becker, K. 1997. Nutrient and anti quality factors on different morphological parts of the Moringa tree. Journal of Agricultural Science. 128: 311-322

Meulen, U.T., Struck, S., Schulke, E., El-Harith, E.A. 1979. A review on the nutritive value and toxic aspects of *leucaena leucocephala*. Tropical Animal Production. 4: 113: 126

Morton, J.F. 1991. The Horseradish Tree, *Moringa pterygosperma* (Moringaceae) - A Boon to Arid Lands. Economic Botany. 45:318-333

NRC, 1994. Nutrient requirements of poultry. Ninth revised edition. National Academic Press, Washington D.C., USA.155.

Nworgu, F.C., Fasogbon, F.O. 2007. Centrosema (Centrosema pubescens) Leaf Meal as Protein Supplement for Pullet Chicks and Growing Pullets. International Journal of Poultry Science. 6: 255-260

Nyman, M., Schweizer, T.F., Tyren, S., Reimann, S., Asp, N.G. 1990. Fermentation of vegetable fiber in the intestinal tract of rats and effects on fecal bulking and bile acid excretion. Journal of Nutrition. 120: 459-466

Odunsi, A.A., Ogunleke, M.O., Alagbe, O.S., Ajani, T.O. 2002. Effect of feeding Glicridia sepium leaf meal on the performance and egg quality of layers. International Journal of Poultry Science. 1: 26-28

Price, L.M. 2000. The moringa tree. ECHO, 17391 Durrance RD., Ft. Myers FL 33917, USA. echo@echonet.org; Online in URL address: [http://www.echonet.org] Consulted: 18th December, 2011.

Sarmiento, L.F. 2001. Evaluation of Chaya (Cnidoscolous aconitifolius) leaf meal as an ingredient in poultry diets: digestibility and performance of chickens. Ph.D. Thesis, University of Edinburgh, 218 pp

SAS Institute, 2002. SAS User's Guide Statistics Ver. 6.12 Edition SAS Institute. Inc., Cary, Nc.

Shelton, H.M., Brewbaker, J.L. 1994. *Leucaena leucocephala* - the most widely used forage tree legume. In Gutteridge, R.C. & Shelton, H.M. (Eds) Forage trees as legumes in tropical agriculture, pp. 15-29. CAB International, Wallingford.

Smith, A.J. 1999. Poultry. The tropical agriculturist First edition. Macmillan Publishers. Wageningen. 218 pp.

Son, J.H., Ragland, D., Adeola, O. 2002. Quantification of digesta flow into de caeca. British Poultry Science. 43: 322-324.

14

EVALUATING THE GROWTH PERFORMANCE OF LOCAL KEI CHICKENS AND THEIR F_1-CROSSES WITH RHODE ISLAND RED AND FAYOUMI BREEDS IN WATERSHED AREAS OF *GURAGHE* ADMINISTRATIVE ZONE, SOUTHERN ETHIOPIA

Misba Alewi[1] and Aberra Melesse[2*]

[1]*Debrezeit Agricultural Research Center, P.O. Box 32, Debre Zeit, Ethiopia*
[2]*Department of Animal and Range Sciences, Hawassa University,*
P.O. Box 5, Hawassa, Ethiopia
[*] *Corresponding author*

SUMMARY

This study describes the on-farm performance of local Kei chicken and its F_1 crosses with Fayoumi and Rhode Island Red (RIR) breeds in *Beresa* watershed of *Guraghe* administrative zone, southern Ethiopia. The Kei paternal line was mated with maternal lines of Fayoumi and RIR chickens to produce F_1-crosses. Body weight developments and feed intake were determined on weekly basis. Then body weight gain and feed efficiency ratios (FER, weight gain: feed) were calculated. Hatchability was 59.0% and 73.7%, 54.7% for Kei, Fayoumi-crosses and RIR-crosses, respectively. In male chicks, the mean day-old weight ranged from 28.8 to 40.1 g and in female chicks from 26.2 to 35.8 g. The average body weight of matured male and female RIR-crosses was 1682 and 1227 g, respectively. The corresponding values for Fayoumi-crosses were 1310 and 1054 g and that of local Kei 1273 and 987 g. During the brooding period, the average daily feed intake was 25.9, 27.0 and 24.4 g for Kei, Fayoumi-crosses and RIR-crosses, respectively. The mean FER was 0.197, 0.213 and 0.243 g for Kei, Fayoumi-crosses and RIR-crosses, respectively. The average age at sexual maturity for Kei, Fayoumi-crosses and RIR-crosses was 183, 154 and 162 days, respectively. The results of the present study suggest that both F_1-crosses showed a significant improvement in body weight, weight gain and FER under farmers' management condition of the watershed areas suggesting the suitability of both genetic combinations with local Kei chickens for on-farm chicken productivity.

Key words: Farmers' management; Growth performance; F_1-crosses; Rhode Island Red chicken; Fayoumi chicken; Local Kei chicken

INTRODUCTION

In Ethiopia, there are about 49.3 million chickens in the country of which 96.6% are local chickens (CSA, 2011), indicating the significance of indigenous chicken ecotypes as principal potential farm animal genetic resources of the country. These chickens have been reported to adapt very well to the traditional small-scale production system of the rural community (Hassen et al. 2007; Moges et al. 2010; Melesse and Negesse, 2011). Comparatively little research and development work has been carried out on village chickens, despite the fact that they are more adapted to various production constraints and numerously available than commercial chickens.

Improving the poultry productivity would improve protein nutrition and could increase the income levels of the rural population. In addition, consumers prefer meat from indigenous chickens, because of its leanness. They also like the multi-colored plumage of these birds, particularly the Kei ecotype due to its reddish feather color. Evaluation of the egg production performance of crossbreeds between local and exotic birds was conducted by different research and development organizations. Most results showed that the overall performance of the crosses was better than either the native or the exotic parents under the existing management condition (Haile-Mariam, 1998; Melesse *et al.,* 2005). However, limited information is available on the comparative growth performance traits of local chickens such as Kei and their F_1 crosses with exotic chicken breeds under farmers' management conditions. Therefore, the study was designed to evaluate the genetic potential of local Kei chicken and their F_1 crosses with RIR and Fayoumi breeds under farmers' management condition of *Guraghe* watershed areas.

MATERIALS AND METHODS

Description of the study area

The present study was conducted in *Beresa* watershed areas of *Guraghe* administrative zone of the southern Regional State of Ethiopia. The watershed comprises two Peasant Associations (PAs), namely *Beresa* and *Dubo-Tuto* and covers an area of 1000 ha. The watershed areas are characterized as food insecure; highly degraded land and poor soil fertility. Agro-ecologically, the watershed is classified under the category of mid-altitude. The average annual rainfall is 1308 mm, while the mean annual temperature is 18.5 °C (BoARD, 2010).

On-station breeding plan

Acquisition of female and male parental lines.

The project started by purchasing 800 fertile eggs of RIR and Fayoumi breeds. The fertile eggs of RIR breed were obtained from Hawassa's Ministry of Agriculture Poultry Farm while that of Fayoumi from Debre Zeit Agricultural Research Centre (DZARC). At the similar time, 400 fertile eggs of local Kei chicken were purchased from farmers of the study area. Immediately after hatching, chicks were vaccinated against New Castle Disease, Infectious Bursal Disease (IBD) and Marek's disease. During the brooding and rearing periods, all birds were fed *ad libitum* with standard commercial starter and grower rations.

Mating plan to produce experimental birds.

Mating was started at 20 weeks of age using the two exotic breeds (RIR and Fayoumi) as the maternal-lines and local Kei as the paternal-line. Fifty cocks were randomly picked from the local Kei chicken population. The cocks were then divided randomly into two groups of which the first group of 25 cocks was crossed with 100 RIR pullets and the second group of 25 cocks with 100 Fayoumi pullets with the ratio of 1 male to 4 females to ensure better fertilization. The local Kei male and female chickens were mated at similar ratio to produce the Kei chickens as control group. Mating was carried out in separate blocks of houses with a deep-litter housing system, each block had five pens. Each pen was equipped with group laying nests, perches, feeder and drinker. During mating, birds were offered with standard layer ration and water was offered ad libtum.Eggs were collected on daily basis for a period of 10 consecutive days and stored broad-end up-position in a cool dry place to ensure better air exchange until they were set in the incubator. Two-thousand eggs were set from all three genotypes (Local Kei, RIR-crosses and Fayoumi-crosses) in an incubator. Candling was performed on the 10th day of incubation and repeated on the 18th day. Newcastle disease vaccination was given at hatch (ocular administration) and repeated on the 7th and 21st day of age. Chickens were regularly vaccinated every three months thereafter.

On-farm management of chickens

The experiment was conducted on 720 unsexed day-old leg-tagged chicks consisting of 240 chicks from each genotype (Local Kei and its F_1 crosses with RIR and Fayoumi breeds). Seventy-two hay-box brooders fitted with 72 chick-runs of 10 chicks' capacity were constructed adopting the dimensions recommended by Demeke (1999). The hay-box brooders and chick-runs had dimensions of 30 x 26 x 26 cm and 30 x 56 x 56 cm, respectively. Hay was stuffed very loosely between the sides of the boxes and the top were covered with sacks filled with hay. Twenty-four randomly selected households from watershed area were involved and each received 30 chicks (10 chicks from each genotype). For each household, 3 hay-box brooders fitted with 3 chick-runs were provided to keep the three breeds separate during the brooding period. The hay-box brooders were used during the night whereas chick-runs were used to hold and feed the chicks during the day.

During the brooding period (up to 8 weeks old), chicks were fed on commercial starter ration. A measured quantity of feed was provided to chicks using plastic cups that had been marked and distributed to the participating farmers. Feed refusal from each genotype was collected separately on daily basis using marked plastic bags. Water was provided to chickens at *ad libitum*. After 8 weeks of age, feeding with commercial starter ration was discontinued and chicks were left to scavenge and farmers provided chickens with some whole maize and/or kitchen leftovers occasionally. Female chickens were separated from males by the size of their combs, presence or absence of wattle and tail feathers. Male birds were kept to evaluate body weight development and female birds to evaluate the egg production performance until 52 weeks of age.

Data collection

The fertility and hatchability of eggs were calculated for each genotype. Hatchability was calculated for total eggs set and fertile eggs. Individual body weights were taken immediately after hatch. During the brooding period, individual body weight of chicks was recorded on weekly basis and every two weeks after 8th week until they were 20 weeks old. After 20th week, body weight was recorded bimonthly. Body weight gain was calculated as the difference between weights measured in consecutive measurements. Mortality of chickens was recorded as it occurred during the entire experimental period.

Statistical analysis

Data were subjected to analysis of variance (ANOVA) using the General Linear Models (GLM) Procedure of Statistical Analysis System (SAS, 2002). Single factor ANOVA model was used to assess the effects of genotypes on different response variables. Identification number of each chicken and the households in which groups of chickens were kept has been considered as random effects when there were repeated observations per animal. Moreover, observations on egg fertility, hatchability and mortality of the investigated birds were analyzed using the frequency procedure of Chi-square test. Differences between group means were evaluated by Duncan's Multiple Range Test.

The effects of genotype and sex of chicks on body weight and body weight gain was assessed using two between subject and one within subject (week) repeated measure ANOVA design. Chicks were the subjects and the weeks of body weight measurements repeated measure factors The specific model statement has been described as follows:

$$Y_{ijkl} = \mu + A_i + B_j + + AB_{ij} + e_{ijkl}$$

Where:

Y_{ijkl} output (body weight and body weight gain)
μ overall mean
A_i fixed effect of genotype. i = 3 (Fayoumi-crosses, RIR-crosses and local Kei)
B_j fixed effect of sex. j = 2 (male and female)
AB_{ij} interaction of fixed effects of genotype and sex
e_{ijkl} random error of households

To analyze feed intake and feed efficiency ratio (FER), one between subject factor (genotype) and one within subject factor (experimental week) repeated measure ANOVA design was used. The model statement is expressed as follows:

$$Y_{ijk} = \mu + A_i + B_j + AB_{ij} + e_{ijk}$$

Where:

Y_{ijk} output (Feed intake and FER)
μ overall mean
A_i fixed effect of genotype. i = 3 (Fayoumi-crosses, RIR-crosses and local Kei)
B_j fixed effect of age (experimental weeks). j = 8 (i.e., 1 – 8 weeks of age)
AB_{ij} interaction of fixed effects of genotype and age (experimental weeks)
e_{ijk} random error of households

RESULTS

Fertility and hatchability of eggs

As presented in Table 1, egg fertility on 18th day of candling was 85.8%, 80.5% and 73.4% for Fayoumi-crosses, RIR-crosses and Kei, respectively and did not differ significantly between genotypes ($\chi^2 = 0.153$, $p > 0.05$). The highest fertility was observed in Fayoumi-crosses whereas the lowest in Kei chickens. Hatchability of total eggs set for Fayoumi-crosses, RIR-crosses and Kei was 73.7 %, 54.7 % and 59.0 %, respectively while hatchability of fertile eggs was 85.8 %, 80.4 % and 67.9 % for Fayoumi-crosses, Kei and RIR-crosses, respectively (Table 1). Although hatchability of total eggs set and that of fertile eggs was higher for Fayoumi-crosses and Kei than for RIR-crosses, the differences were insignificant ($\chi^2 = 0.102$, $p > 0.05$).

Day old and post-hatch body weights and weight gains

The body weight of day old chicks was significantly ($p<0.05$) higher for RIR-crosses than Fayoumi-crosses and Kei in both sexes (Table 2).

During brooding period, genotype and sex had a significant ($p<0.05$) effect on the overall mean body weight (Table 2). Accordingly, the RIR-crosses had the highest body weight (201.8 g) followed by Fayoumi-crosses (186.9 g) and Kei (168.1 g). The overall mean body weight of male chicks (204.0 g) was significantly ($p<0.05$) higher than female chicks (167.2 g).

Table 1. Hatchability of total eggs set and fertile eggs of local Kei and its crosses with Rhode Island Red and Fayoumi chicken breeds.

Breed combinations	Egg number			Hatchability (%)	
	Set	Fertile	Hatched	of eggs set	of fertile eggs
Fayoumi (♀) x Kei (♂)	600	515	442	73.7	85.8
RIR (♀) x Kei (♂)	600	483	328	54.7	67.9
Local Kei	800	587	472	59.0	80.4

Table 2. Least square means of body weight development of birds from hatch, brooding, post-brooding and maturity phases.

Growth phases	Growth traits	Sex	Kei	Kei x Fayoumi	Kei x RIR	Pooled S.E.M
At hatching	Day old weight (g)	Male	28.8^c	29.7^b	40.1^a	0.22
		Female	26.2^c	27.0^b	35.8^a	0.20
Brooding (Hatch-8 wks)	Body weight (g)	Male	183^c	206^b	222^a	3.89
		Female	153^c	167^b	179^a	3.61
	Weight gain (g)	Male	5.69^c	6.33^b	6.76^a	0.08
		Female	4.73^c	5.06^b	5.39^a	0.08
Post-brooding (9-20 wks)	Body weight (g)	Male	762^c	852^b	968^a	12.0
		Female	641^c	704^b	778^a	9.46
	Weight gain (g)	Male	8.20^c	9.12^b	10.8^a	0.07
		Female	6.75^c	7.67^b	8.52^a	0.06
Matured birds (21-52 wks)	Body weight (g)	Male	1273^b	1310^b	1682^a	13.9
		Female	987^c	1054^b	1227^a	9.60
	Weight gain (g)	Male	2.06^b	1.86^b	2.76^a	0.31
		Female	1.52^a	1.42^a	1.76^a	0.17

[a,b,c] Means within genotypes and sex in the same row with different superscript letters are significantly ($p<0.05$) different; RIR = Rhode Island Red; S.E.M = Standard error of the mean

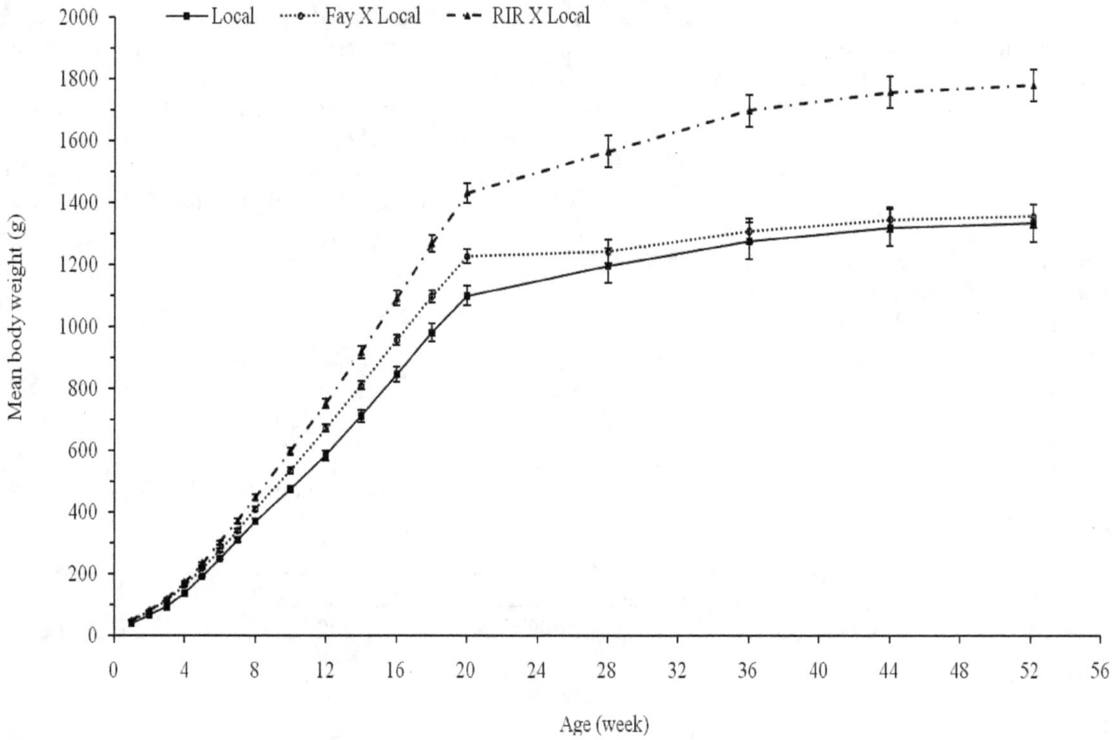

Figure 1. Body weight development of male birds from hatch to 52 weeks of age. Error bars are 95% CI values. Note that non-overlapping error bars indicate significantly different mean values

During post-brooding period, the overall mean body weight of RIR-crosses was significantly ($p<0.05$) higher than those of Fayoumi-crosses and Kei for both sex groups. A significant ($p<0.05$) increase in body weight was observed for both sexes as birds grew older. After 20 weeks of age onwards, the RIR-crosses had still significantly ($p<0.05$) higher body weight than Fayoumi-crosses and local Kei for both sexes. However, no significant ($p>0.05$) difference was observed between Fayoumi-crosses and Kei male genotypes. The mean body weight of male birds was significantly ($p<0.05$) higher than that of females.

As presented in Figures 1 and 2, both male and female birds showed a significant ($p<0.05$) increase in body weight from hatch to 20 weeks of age during which the body weight was increased at a higher rate. However, the body weight development became slower after 20 weeks of age onwards in which the males showed better body weight development as compared to females.

Weight gain

During the brooding period, the Fayoumi-crosses of male and female chicks had significantly ($p<0.05$) higher mean daily weight gain than those of RIR-crosses and Kei (Table 2). However, the difference in daily gain was not significant ($p>0.05$) between F_1 crosses. Male chicks generally gain more weight than female chicks.

During the post-brooding phase, the overall daily weight gain of male and female RIR-crosses was significantly ($p<0.05$) higher than those of Fayoumi-crosses and Kei chickens. The male and female Fayoumi-crosses had significantly ($p<0.05$) higher daily weight gain than local Kei. During this period, the male chickens had significantly ($p<0.05$) higher daily weight gain than females.

During the adult phase, the mean daily weight gain of male RIR-crosses was significantly ($p<0.05$) higher than those of Fayoumi-crosses and local Kei. However, no significance differences were observed amongst the three genotypes in weight gain of female birds. The female birds showed a similar trend in weight gain performance as that of males. During this period, the rate of gain declined consistently.

Figures 3 and 4 show the trend of daily weight gain of male and female birds, respectively during the entire growth period. During the brooding period, a sharp increase in mean daily weight gain was observed among the three genotypes. A low body weight gain was observed during week 8 through10 due to an outbreak of coccidiosis during this period of time.

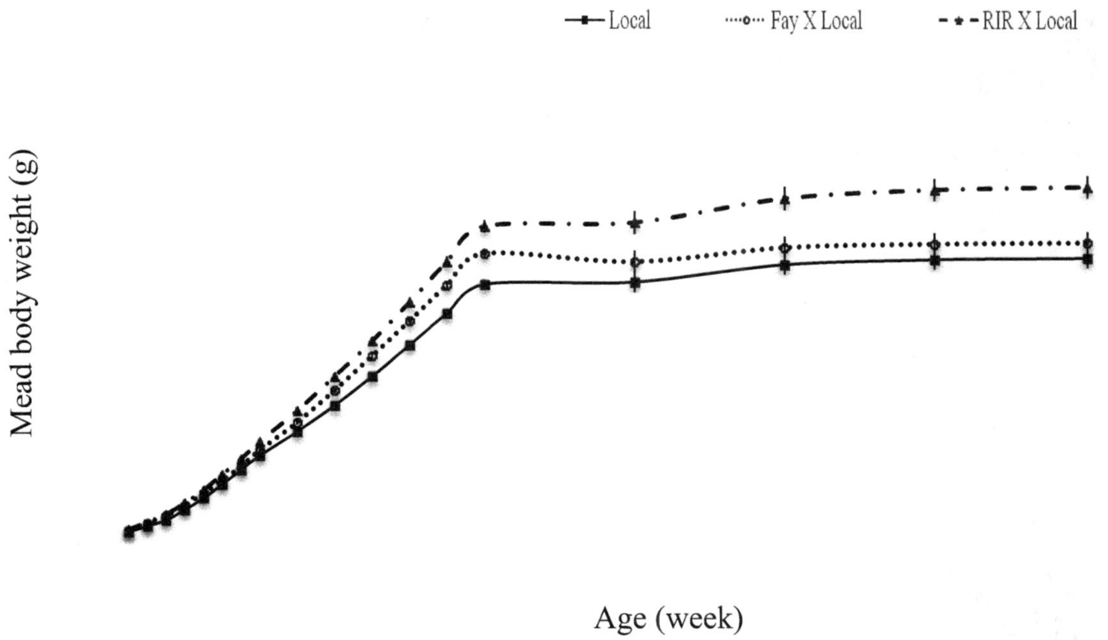

Figure 2. Body weight development of female birds from hatch to 52 weeks of age. Error bars are 95% CI values. Note that non overlapping error bars indicate significantly different mean values

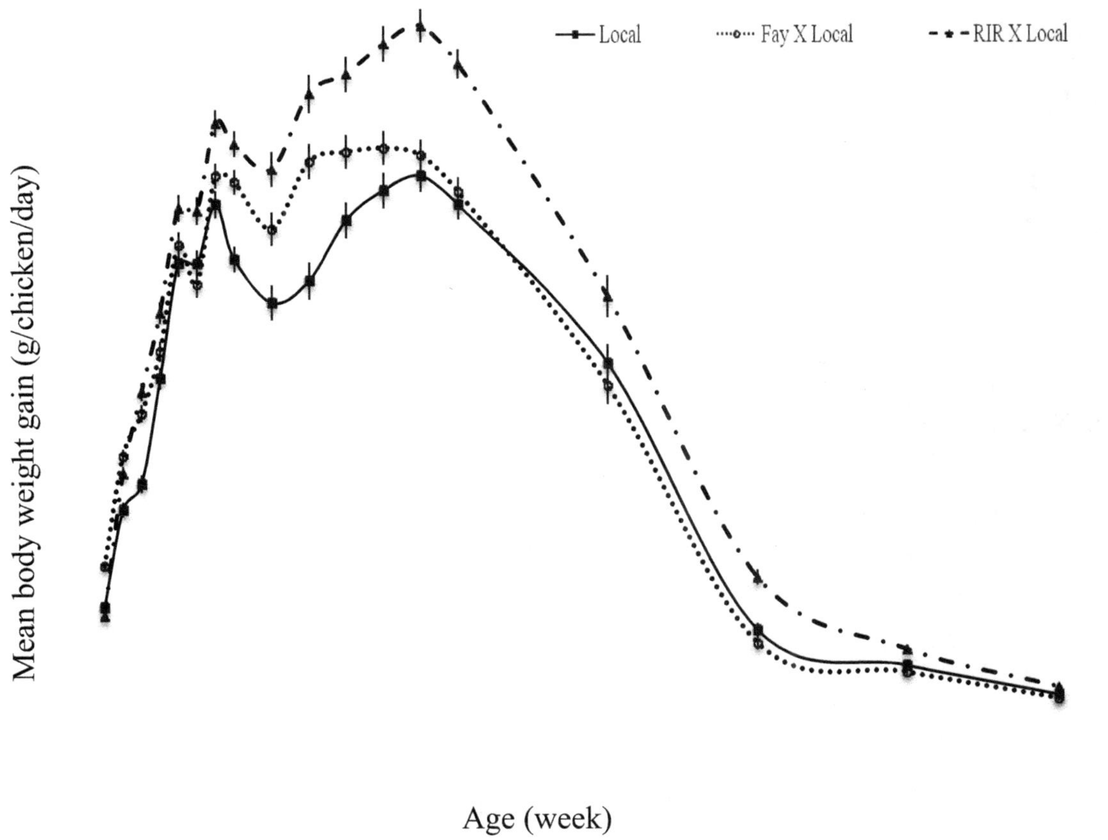

Figure 3. Pattern of body weight gain (g/chicken/day) of male birds from hatch to 52 weeks of age. Error bars are 95% CI values. Note that non-overlapping error bars indicate significantly different mean values

The rate of weight gain in male and female RIR-crosses was comparably higher than both Fayoumi-crosses and Kei chickens. However, the male and female Fayoumi-crosses and Kei chickens showed a comparable daily weight gain performance after 18 weeks of age.

Feed intake

As shown in Table 3, a significant ($p<0.05$) difference in feed intake was observed between the three genotypes at different age points. The Fayoumi-crosses had the highest feed intake followed by Kei and RIR-crosses.

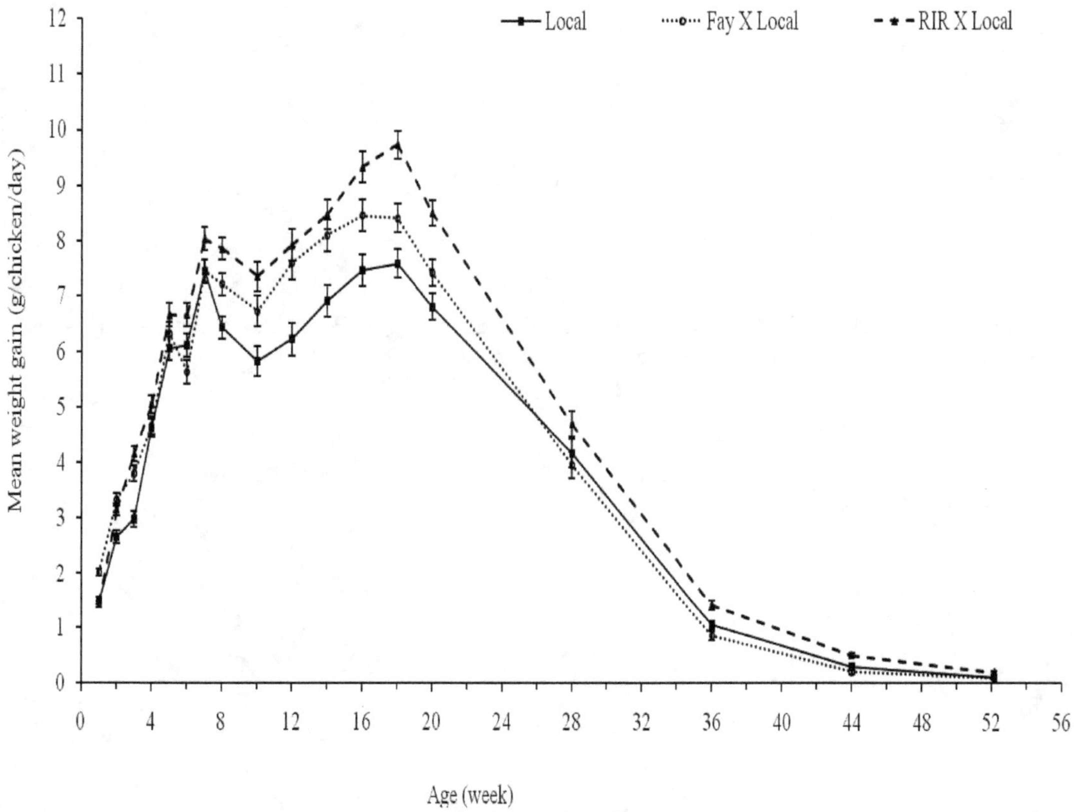

Figure 4. Pattern of body weight gain (g/chicken/day) of female birds from hatch to 52 weeks of age. Error bars are 95% CI values. Note that non-overlapping error bars indicate significantly different mean values

Table 3. Feed intake (g/chicken/day) of the three genotypes at different developmental ages of birds during the brooding period.

Age (wks)	Kei	Kei x Fayoumi	Kei x RIR	Pooled S.E.M
0-1	7.4^b	8.2^a	6.6^c	0.16
1-2	13.3^b	14.3^a	12.3^c	0.21
2-3	21.0^b	22.8^a	20.8^b	0.32
3-4	26.0^b	27.4^a	24.5^c	0.30
4-5	29.0^a	30.2^a	26.8^b	0.32
5-6	32.2^a	32.3^a	29.6^b	0.34
6-7	36.2^a	36.8^a	34.0^b	0.31
7-8	42.3^a	43.6^a	40.5^b	0.42
Overall mean	25.9^a	27.0^a	24.4^b	0.30

^{a,b,c} Means between genotypes in the same row with different superscript letters are significantly ($p<0.05$) different; RIR = Rhode Island Red; S.E.M= Standard error of mean

The Fayoumi-crosses significantly ($p<0.05$) consumed more feed from hatch to 4 weeks of age than both RIR-crosses and local Kei chickens (Table 3). During the same period, the feed intake of Kei was significantly ($p<0.05$) higher than those of RIR-crosses except between weeks 2 and 3. After week 4 to the end of the brooding period, the feed intake of both Fayoumi and Kei chickens was significantly ($p<0.05$) higher than RIR-crosses. However, no significant ($p>0.05$) difference was observed in feed consumption between Kei and Fayoumi-crosses.

Feed efficiency ratio (FER)

A significant ($p<0.05$) difference in FER was observed between genotypes in which the RIR-crosses had the highest value followed by Fayoumi-crosses and local Kei (Table 4). At week 1, the FER value in Fayoumi-crosses was significantly ($p<0.05$) higher than both RIR-crosses and Kei genotypes. At week 2, there was no significant ($p>0.05$) difference in FER between F_1 crosses. However, the FER in both F_1 crosses was significantly ($p<0.05$) higher than that of Kei chickens. After week 2 to the end of brooding, RIR-crosses had the highest FER than other two genotypes.

At week 3, Fayoumi-crosses had significantly ($p<0.05$) higher FER than local Kei. After week 3 through 7 weeks of age, no significant ($p>0.05$) difference was observed between Kei and Fayoumi genotypes. However, at the end of brooding, the FER in Fayoumi-crosses was significantly ($p<0.05$) higer than local Kei.

Mortality

As shown in Table 5, no mortality was recorded during the brooding period due to predation. However, mortality rates of 6.7%, 2.5% and 3.8% were observed for RIR-crosses, Fayoumi-crosses and local Kei chickens respectively during the same growth period. Similarly, a higher mortality rate was recorded from 8 to 20 weeks of age, which was mainly caused by diseases. The mortality rate for RIR-crosses (25.9%) was significantly higher (χ^2 = 9.731, $p < 0.01$) than Fayoumi-crosses (15.8%) and local Kei chickens (17.8%). However, the mortality rate due to predation was very low during this period in the three genotypes (Table 5).

Table 4. Feed efficiency ratio (g weight gain/g feed) of the three genotypes at different developmental ages of birds during the brooding period.

Age (wks)	Local Kei	Kei x Fayoumi	Kei x RIR	S.E.M
0-1	0.213^b	0.270^a	0.222^b	0.031
1-2	0.222^b	0.263^a	0.282^a	0.030
2-3	0.159^c	0.189^b	0.223^a	0.032
3-4	0.192^b	0.192^b	0.234^a	0.024
4-5	0.227^b	0.233^b	0.275^a	0.026
5-6	0.208^b	0.196^b	0.258^a	0.033
6-7	0.217^b	0.222^b	0.260^a	0.023
7-8	0.164^c	0.182^b	0.212^a	0.024
Overall mean	0.201^b	0.218^b	0.246^a	0.023

^{a,b,c} Means between genotypes in the same row with different superscript letters are significantly ($p<0.05$) different; RIR = Rhode Island Red; S.E.M= Standard error of mean

Table 5. The overall mortality (%) of local Kei, Kei x Fayoumi and Kei x Rhode Island Red chickens at different stages of development.

Genotypes	Brooding		Post-brooding		Adult stage	
	Disease	Predator	Disease	Predator	Disease	Predator
Local Kei	3.75^b	0	17.8^b	2.60	4.17^a	8.33^b
Kei x Fayoumi	2.50^b	0	15.8^b	2.14	2.08^b	4.17^c
Kei x RIR	6.67^a	0	25.9^a	2.23	4.17^a	14.6^a
Overall mean	4.30	0	19.8	2.32	3.47	9.03

$^{a,b,c}\chi^2<0.05$ values differ significantly between genotypes within a column
RIR = Rhode Island Red chicken breed; 0 = Not observed

DISCUSSION

Fertility and hatchability

The hatchability rates of total eggs set and fertile eggs obtained from RIR-crosses were comparable to those reported by Hassen (2007) for RIR chicken breed raised under intensive management. The percentage hatchability from fertile eggs reported by Wondmeneh *et al.* (2011) for Fayoumi breed under intensive management was in good agreement with the current results obtained from Fayoumi-crosses. However, the average percentages fertility and hatchability from the total eggs set reported by the same authors was higher than obtained from Fayoumi-crosses in the current study. On the other hand, values for fertility and hatchability of eggs set and fertile eggs for local Kei chickens in present study were higher than those reported for local chicken ecotypes reared under farmers' management condition (Hassen, 2007). These variations in fertility and hatchability rates of eggs can be explained by the age of the birds, nutrition, disease and environmental factors (North and Bell, 1990).

Body weight and weight gain

The highest body weight observed in F_1-crosses at hatch suggests that crossing of local Kei chickens with RIR and Fayoumi breeds has significantly improved the body weight at hatch. Similarly, an improvement in body weight at hatch of indigenous chickens crossed with RIR and Fayoumi was reported by Bekele *et al.* (2010). The body weight of local chickens at hatch in Northwest Ethiopia reported by Hassen (2007) was 27.2 g for Tilili, 27.9 g for Gelila, 27.1 g for D/Elias and 27.9 g for Mecha under intensive management, which was comparable with the current findings. The same author reported that the body weight of RIR at hatch to be 35.2 g, which was lower than that of RIR-crosses (38.0) in the present study.

The high body weight observed in RIR-crosses compared with Fayoumi-crosses and Kei chickens might be attributed to the genetic superiority of the RIR in body weight that is a highly heritable trait, and known for its non-additive genetic response to crossbreeding. Moreover, RIR is a dual purpose chicken breed recognized for better weight gain compared with egg layer type breeds. Similarly, Zaman *et al.* (2004) reported higher body weight for RIR-crosses than Fayoumi-crosses under semi-scavenging condition in Bangladesh. Furthermore, Forsido (1986), Shanawany (1987) and Bekele *et al.* (2010) reported that heavier eggs produce heavier chicks at hatch.

The body weight of local female chickens reported by Tadesse (2007) between 10 and 20 weeks age was lower than that of the chickens in the current study. Similarly, the pure RIR breed showed relatively lower body weight than the RIR-crosses under the same age category. However, the Fayoumi breed had relatively higher body weight than the Fayoumi-crosses obtained from the current study.

Melesse *et al.* (2005) reported 952 g body weight of Ethiopian female naked-neck chicken under intensive management system, which is comparable to the current results for local female chickens (924 g) of similar age. The same authors reported an average body weight of 1169 g at 20 weeks of age for F_1 crosses of Ethiopian naked-neck and White Lohmann, which is comparable to the body weight of RIR-crosses (1132 g) at similar age.

The average body weight of female local Kei chickens obtained at 52 weeks of age in the present study was 1020 g, which is much lower than reported by Melesse *et al.* (2005) for Ethiopian female naked-neck chickens at similar age. Under scavenging conditions, Melesse and Negesse (2011) reported 1510 and 1204 g of body weight for male and female local Kei chickens, respectively, which is higher than that obtained from the current study. The reported differences in body weight can be explained by the variations in availability of scavenging materials, management and environmental factors. Under the scavenging system, the non-genetic factors in particular poor nutrition and health care have much larger effects on production parameters than the genetic characteristics of the birds (Sazzad *et al.*, 1988).

The daily weight gain of Tilili, Gelila, D/Elias, Melo-Hamusit, Gassay/Farta, Guangua and Mecha chickens between day old and 4 weeks of age reported by Hassen (2007) under intensive management was lower than the average of male and female Kei chickens of the current study at similar age. Similarly, the mean daily weight gain observed by Hassen (2007) from the RIR was relatively lower than the average body weight of male and female RIR-crosses of similar age. The same author reported an average daily body weight gain of 9.8 g from 5 to 8 weeks of age for RIR chickens under intensive management which was comparable to the average body weight of RIR-crosses at similar age in the current study. This indicates that under farmers' management conditions, the RIR-crosses can perform better than pure RIR breeds kept under intensive type of production system.

Under intensive management condition, Dessie *et al.* (2003) reported an average daily weight gain of 7.2 g for local chickens between 6 and 12 weeks of age, which was comparable with that of local chickens in the current study.

Feed intake and utilization

The highest feed intake observed in Fayoumi-crosses could be attributed to their active physical activity, which might have required additional feed consumption to meet the maintenance requirement as reported by Melesse (2008). In agreement with the present result, Akhtar *et al.* (2007) reported higher feed intake for Fayoumi breeds than for Lyallpur Silver Black and RIR breeds. The daily feed intake reported for Tilili, Melo-Hamusit, Guangua and Mecha chickens from day old to 4 weeks age by Hassen (2007) under intensive management were comparable with that of local Kei chickens in the current study at similar age. The daily feed intake of RIR chicks reported by Melesse *et al.* (2011) was comparable with that of RIR-crosses at similar age category. The daily feed intake of Tilili, D/Elias, Melo-Hamusit and Guangua chickens reported by Hassen (2007) under intensive management were comparable with that of local Kei chickens in the present study at similar age. However, the daily feed intake of RIR reported by the same author was lower than that of RIR-crosses at similar age.

The better efficiency in feed utilization (higher FER) of RIR-crosses might be due to their improved body weight gain performance along with low feed intake. Akhtar *et al.* (2007) reported better efficiency in feed utilization of RIR breeds as compared with Lyallpu Silver Black and Fayoumi breeds. The FER reported by Hassen (2007) for Tilili, Melo-Hamusit, Guangua and Mecha local chickens from day old to 4 weeks age under intensive management was lower than that of the current study observed in local Kei chickens at similar age. This could be explained by the fact that chickens kept under intensive management condition might have better chance to express their genetic potential (Sazzad, 1992; Melesse *et al.,* 2005).

Mortality

The low mortality observed during the brooding period might be attributed to the use of brooding box fitted with chicken-run feeding, which could have protected chickens from both aerial and ground predators which have been reported as the main causes of mortality in southern regional state of Ethiopia (Melesse and Negesse, 2011). The higher rate of mortality due to disease observed from 8 to 20 weeks of age was caused by the outbreak of coccidiosis which occurred during this period. The highest mortality rate observed in RIR-crosses due to predation could be explained by the exposure of chickens to the rainy season (June to August) during which time maize and sorghum crops were used as hideout for ground predators such as wild cats and foxes. Moreover, searching for feed resources around the homestead might have exposed birds to be easily spotted and attacked by predators. Beside this, RIR-crosses were relatively less active to escape from both aerial and ground predators compared with Fayoumi-crosses and local chickens.

CONCLUSIONS

The overall results of the present study revealed that both F_1-crosses demonstrated a significant improvement in body weight, weight gain and feed efficiency ratio under farmers' management condition as compared with local Kei chickens. It can be concluded that the exotic blood of Rhode Island Red and Fayoumi chicken breeds had played a significant role in upgrading the productive potentials of the local Kei chicken. Thus, using both exotic chicken breeds as a strategy for upgrading the poor growth performance of indigenous chickens might be an alternative option in *Guraghe* and other watershed areas along with improved feeding, housing and health care schemes. It should be however noted that the Rhode Island Red chicken breeds should be used with cautious as they are highly susceptible to local predators and diseases.

Acknowledgments

The authors would like to thank Operational Research Project, funded by the Irish Government through Irish Aid Programme for the funding support. The authors also express their sincere admiration to all farmers in *Beresa* watershed district of *Guraghe* administrative zone, who actively participated on this study and highly acknowledge their great collaboration, hospitality and willingness to share their experiences. Dr Yosef T/Giorgis is highly acknowledged for conducting part of the statistical analysis.

REFERENCES

Akhtar, N., Mahmood, S., Hassan, M., Yasmeen, F. 2007. Comparative study of production potential and egg characteristics of Lyallpur Silver Black, Fayoumi and Rhode Island Red breeds of poultry. Pakistan Veterinary Journal. 27(4): 184-188.

Bekele, F., Ådnøy, T., Gjøen, H.M., Kathle, J., Abebe, G. 2010. Production Performance of Dual Purpose Crosses of Two Indigenous with Two Exotic Chicken Breeds in Subtropical Environment. International Journal of Poultry Science. 9 (7): 702-710.

Bureau of Agriculture and Rural Development (MoARD), 2010. Woreda general information. Bureau of Agriculture and Rural Development. Meskan Woreda, SNNPRS, Ethiopia.

Central Statistics Authority (CSA). 2011. Agricultural sample survey 2010-2011. Central Statistics Authority. Report on livestock and livestock characteristics, Vol. II. Statistical Bulletin No. 505. Addis Ababa, Ethiopia.

Demeke, S. 1999. Suitability of homemade hay-box chick brooder to the Ethiopian household poultry production system. In: Proceedings of the 7th Annual Conference of Ethiopian Society of Animal Production (ESAP), 26-27 May 1999, Addis Ababa, Ethiopia. Pp. 115-124.

Dessie, T., Kijora, C., J. Peters, K. 2003. Indigenous chicken ecotypes in Ethiopia, Growth and feed utilization potential. International Journal of Poultry Science. 2(2): 144-152.

Forsido, T. 1986. Studies on the meat production potential of some local strains of chicken in Ethiopia. Ph.D Thesis, J. L. Giessen University. 210 pp.

Haile-Mariam, M. 1998. Egg laying performance of White leghorn and their crosses with local birds at Debre Zeit, Ethiopia. In: Proceedings of 6th Conference of Ethiopian Society of Animal Production (ESAP), 14-15 May 1998, Addis Ababa, Ethiopia, pp.141-150.

Hassen, H. 2007. Phenotypic and genetic characterization of indigenous chicken populations in North-West Ethiopia. Ph.D Thesis. Submitted to the faculty of natural and agricultural sciences department of animal, wildlife and grassland Sciences. University of the Free State, Bloemfontein, South Africa.

Hassen, H., Neser, F.W.C., VanMarle-Koster, E., DeKock, A. 2007. Village-based indigenous chicken production system in north-west Ethiopia. Tropical Animal Health and Production. 39: 189-197.

Melesse, A., Maak, S., von Lengerken, G. 2005. The Performance of Naked Neck and their F1 crosses with Lohmann White and New Hampshire chicken breeds under long-term heat stress conditions. Ethiopia Journal of Animal Production. 5(1): 91-106.

Melesse, A. 2008. Physiology of Animal Adaptations in Stressful Environments. Teaching Book. Hawassa University, Hawassa, Ethiopia, 391 pp.

Melesse, A., Tiruneh, W., Negesse, T. 2011. Effects of feeding *Moringa stenopetala* leaf meal on nutrient intake and growth performance of Rhode Island Red chicks under tropical climate. Tropical and Subtropical Agroecosystems. 14: 485-492.

Melesse, A., Negesse, T. 2011. Phenotypic and morphological characterization of indigenous chicken populations in southern region of Ethiopia. Animal Genetic Resources Information Journal. 49: 19-31. .

Moges, F., Melesse A., Dessie, T. 2010. Assessment of village chicken production system and evaluation of the productive and reproductive performance of local chicken ecotype in Bure district, Northwest Ethiopia. African Journal of Agricultural Research. 5 (13): 1739-1748.

North, M.O., Bell, D.D. 1990. Commercial chicken production manual, 4th edition, Published by Van Nostrand Reinhold, New York, USA.

Statistical Analysis System (SAS). 2002. SAS/STAT® Guide version 9.0. SAS, Institute Inc., Cary, NC.

Sazzad, M.H. 1992. Comparative study on egg production and feed efficiency of different breeds of poultry under intensive and rural conditions in Bangladesh. Livestock Research for Rural Development. (4) 3.

Sazzad, M.H., Momtazul, S.M.H., Asceduzaman, A.U. 1988. Growing Pullets of 1,968 Deshi and Kakki Cambell ducks under rural condition. Indian Journal of Poultry Science. 23:165-166.

Shanawany, H.S. 1987. Hatching weight in relation to egg weight in domestic birds. World's Poultry Science Journal. 43:107-115.

Tadesse, A. 2007. Poultry management practices and on farm performance evaluation of Rhode Island Red, Fayoumi and local chicken in Umbulo Wachu watershed in Sidama Zone, unpublished MSc Thesis, Hawassa University, Hawassa, Ethiopia. 126 pp.

Wondmeneh, E., Dawud I., Adey, M. 2011. Comparative evaluation of fertility and hatchability of Horro, Fayoumi, Lohmann Silver and Potchefstroom Koekoek breeds of chicken. Asian Journal of Poultry Science. 5 (3): 124-129.

Zaman, M.A., Sørensen, P., Howlider, M.A.R. 2004. Egg production performance of a breed and three crossbreeds under semi-scavenging system of management. Livestock Research for Rural Development. 16 (8). *http://www.cipav.org.co/lrrd/lrrd16/8/zama16060.htm.*

15

PRODUCTIVE AND QUALITATIVE PERFORMANCE OF NATURALIZED AND NATIVE FORAGE LEGUMES IN THE TEMPERATE ZONE OF PUEBLA STATE, MEXICO

Juan de Dios Guerrero-Rodríguez[1*], Rocío Margarita Flores-Galván[2], Numa Pompilio Castro-González[2], Ángel Bustamante-González[1] and Ramiro Escobar-Hernández[2]

[1] Colegio de Postgraduados, Campus Puebla. Km. 125.5 Carretera Federal México-Puebla, San Pedro Cholula, Puebla, México. CP 72760 Tel. (222) 2 85 14 42 Ext. 2208. Email: grjuan2000mx@yahoo.com; rjuan@colpos.mx

[2] Benemérita Universidad Autónoma de Puebla. Unidad Académica de Ingeniería Agrohidráulica-Plantel de Ingeniería Agronómica y Zootecnia. Avenida Reforma No. 165 CP 73900, Tlatlauquitepec, Puebla. México. Tel (233) 31800 07.

[*] Corresponding author

SUMMARY

The aim of this study was to evaluate seven species of native and naturalized legumes in terms of forage production and nutritive quality. The control species, *Vicia sativa* was represented by two varieties, which maintained a high dry matter production at two locations, in one of them, matched by *Melilotus albus*. The latter species also had high yields of dry matter in two locations, but in one of them the varieties of *V. sativa* were not successful. Less yielding species were those that had lower fiber concentration, a situation that was in part due to a higher leaf:stem ratio. *Medicago polymorpha* had the lowest digestibility, which coincided with higher concentrations of neutral and acid detergent fiber. The crude protein concentration was different among species ($P<0.0001$), where *M. polymorpha* consistently had low ($P<0.05$) concentration (16.8%) as well as *M. albus* (17%). Among the species tested in this study, several of them have potential yield and quality to improve the diet of ruminants in the highland region of Puebla State and can replace the vetches. Additionally, even when the climate may be the same, the soil conditions also determine which species can thrive in a region.

Key words: herbaceous legumes; nutritive value; ruminants; temperate climate.

INTRODUCTION

In the temperate region of Puebla, mainly in the valleys of Libres, Puebla and Serdán, there is a ruminant population of around 513,000 heads, where sheep represents 65% (INEGI, 2008). Feeding these animals relies on the consumption of native pastures and crop residues, which usually have low concentrations of protein, leading to poor liveweight gain and in some species like goats and cattle, milk production has not yet reached potential (Church *et al.*, 2004). To cover the shortfall on requirements, sometimes commercial concentrates are supplied. This means increased production costs, which affects the profitability and stability of the production system. An alternative to the deficit of protein in the diet, is the inclusion of legumes, which generally have, compared to grasses, the highest concentration of crude protein (which in several cases exceeds 20%), and higher concentrations of calcium than grasses. This makes them important elements in improving the diet and reducing production costs by way of diminishing the purchase of concentrates.

Within the main legume group grown for forage in the humid temperate region of Puebla state, there are some perennials such as *Medicago sativa* (alfalfa), *Trifolium repens* (white clover) and *Trifolium pratense* (red clover), and annuals such as, *Vicia villosa* (winter vetch) and *Vicia sativa* (common vetch). However, these species have special requirements and do not suit all environments. There are some other naturalized and native legumes, which can be options for ruminant feeding, but their productive potential and quality is unknown. Among them it can be mentioned some naturalized such as *Melilotus albus*, *Medicago polymorpha*, *Medicago lupulina* and *Vicia sativa sp nigra*, some other native as *Crotalaria pumila* and *Cologania tenuis*. Data reported for some of these species are across a wide range, but all studies have been carried out in other countries. For example, crude protein (CP) in *M. polymorpha* has a range from 11 to 25%, neutral detergent fiber (NDF) varies between 27 and 70% and acid detergent fiber (ADF) between 22 and 58%; for *M. lupulina* CP reported ranges are from 14 to 27%, NDF from 22 to 54% and the FDA between 16 and 39% (Zhu *et al.*, 1996; Sherestha *et al.*, 1998; Alford *et al.*, 2003; Muir *et al.*, 2003). In *M. albus* ranges reported for CP are from 17 to 24.7%, NDF from 32 to 34.7% and ADF from 19.8 to 24.4% (Guerrero-Rodríguez, 2006).

In the highlands valleys of Puebla, the prevailing climate is temperate in combinations of semi-dry, sub-humid and humid. Variation in soil pH and texture is broad, from saline to acid soils and from clay to sandy. It is important to know the performance of the species under different conditions to be able to give informed recommendations. Therefore, this study aimed to evaluate the performance in yield and quality of seven species of forage legumes (biennial and annual types) in three temperate regions of the state of Puebla, Mexico.

MATERIALS AND METHODS

Location

This study was conducted in the municipalities of San Salvador El Verde, Tlatlauquitepec, and Libres, Puebla, Mexico. The experiments were located at: San Salvador el Verde at 19° 15.374' N and 98° 31.115' W at an altitude of 2434 meters above the sea level, in a sandy loam soil of pH 6.3; in Tlatlauquitepec at 19° 50.411' N and 97° 29.839' W at an altitude of 1959 masl and a sandy loam soil with pH of 6.7; en Libres at 19° 20.479' N and 97° 32.577' W at an altitude of 2374 masl, in a sandy soil pH 7.2. Planting dates were 12, 14 and June 24, 2008, for Libres, San Salvador El Verde and Tlaltlauquitepec, respectively.

Establishing the experiments

Seven herbaceous forage legume species were planted. Five of them are found naturally in the grasslands and roads in the temperate region of Puebla. These were: *Melilotus albus*, seeds collected from the mountains of Tentzo in the Municipality of Tecali de Herrera; *Medicago lupulina*, seeds collected from the grazing areas of San Pedro Cholula and San Salvador El Verde; *Crotalaria pumila*, *Medicago polymorpha*, *Vicia sativa sp nigra* and *Cologania tenuis*, seeds collected from pastures at San Salvador El Verde. The control species was vetch (*Vicia sativa sp sativa*) with two varieties (white flower –WF- and purple flower –PF-), seeds for both purchased from a commercial source.

Experimental design

At each site, the species were distributed in a randomized complete block design with four replicates. The experimental unit consisted of a plot of 2x3 m. The statistical model was: $Yij = \mu + \beta_j + \tau_i + \varepsilon_{ij}$ where μ = the population mean, τ_i = species effect i, β_j = effect of block j, ε_{ij} = experimental error. When considering locations, a combined analysis was performed where the model included the effect of Locality, Blocks nested into Locality and the interaction Species*Locality. Comparisons among averages were carried out using the Tukey's test with α=0.05. All data analyses were performed using the GLM procedure of SAS programme version 9.2 (SAS Institute, 2002).

Crop management

The experiments were kept free of weeds, which were removed manually for all cases. There was only one sampling, in all the replicates. Quadrants of 0.5 x 0.5 m were used for harvest choosing the most representative part of the plot, taking the cut to 3 cm from the ground. It is noteworthy that with the exception of the species *Vicia sativa sp nigra* all the others were in bloom.

Variables evaluated

To determine the yield of dry matter (DM) per hectare, the fresh herbage weight was recorded and placed in paper bags, and then the bags were dried in a forced air oven at a temperature of 60 °C for at least 72 hours until constant weight was reached.

The determination of crude protein was done following the procedure of AOAC (1975) by the Kjeldahl method. The neutral detergent fiber (NDF) and the acid detergent fiber (ADF) were conducted under the protocols of Ankom (Ankom Technology, 2006) sequentially.

The *in vitro* digestibility (IVDDM) was carried out by the enzymatic method, according to the two-stage pepsin-cellulase technique (Jones and Hayward, 1975) following the recommendations by Clarke *et al.* (1982); the enzymes pepsin and cellulase (Onozuka RS) were purchased from SIGMA-ALDRICH.

RESULTS AND DISCUSION

The species differed (P <0.0001) in dry matter production. *M. albus* and vetches, on average, were the most productive (P<0.05) compared to other species (Figure 1). There was an interaction Species*Locality (P <0.0001). *M. albus* showed a marked superiority at the locality of Miravalles (Libres) where it was the best producer in the presence of salinity in the soil, but vetches did not prosper. This situation was not replicated in the locality of San Salvador El Verde where *M. albus* productive performance was poor compared to that of vetches. *M. polymorpha* followed a similar trend to *M. albus* across locations. *M. lupulina* also had a better production in the locality of Miravalles where its performance was 51% higher than in the locality of Tlatlauquitepec. The nigra vetch and *C. tenuis* did not survive in Miravalles, most likely by maladjustment to the alkaline soil pH. These two species together with *M. lupulina* had the lowest average yield. Although these species are well adapted to temperate climate and can be found in the wild in some of these sites, their adaptation may be limited by the absence of specific symbionts that are not found in all soils.

In relation to fiber contents, there were marked differences among species (P<0.0001) (Figure 2). Less yielding species had lower fiber content, a situation that was in part due to the higher leaf: stem ratio.

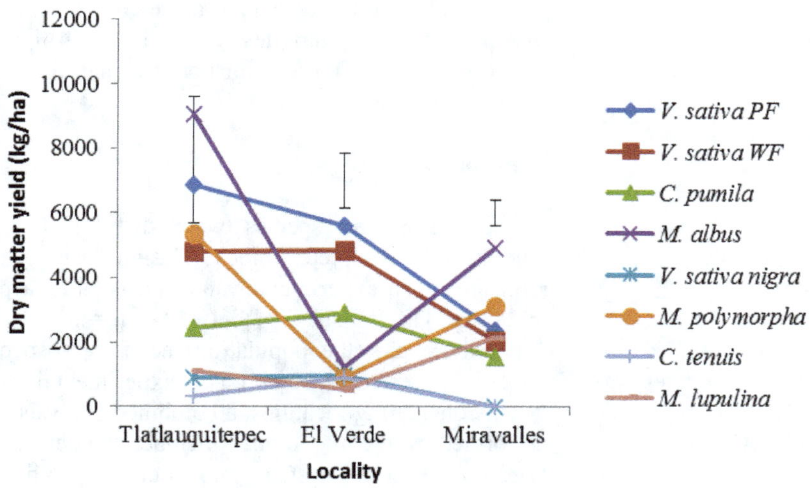

Figure 1. Dry matter production of biennial and annuals legumes in three different localities of the High Plateau of Puebla State. Bars represent the least significant difference of Tukey ($\alpha = 0.05$).

Species interactions with locality were observed in *M. albus* which had low fiber concentration in the localities El Verde and Miravalles. This impacted on its IVDDM values (Figure 3) where the highest were recorded in the locality of San Salvador El Verde. On average, *M. polymorpha* had the lowest IVDDM, which coincided with higher concentrations of both fractions of fiber. Species differ (P<0.0001) in their crude protein concentration (Figure 4), an aspect which was expressed mostly in the locality San Salvador El Verde, followed by Miravalles. *M. polymorpha* consistently had lower (P <0.05) protein content (16.8%) as well as *M. albus* (17%).

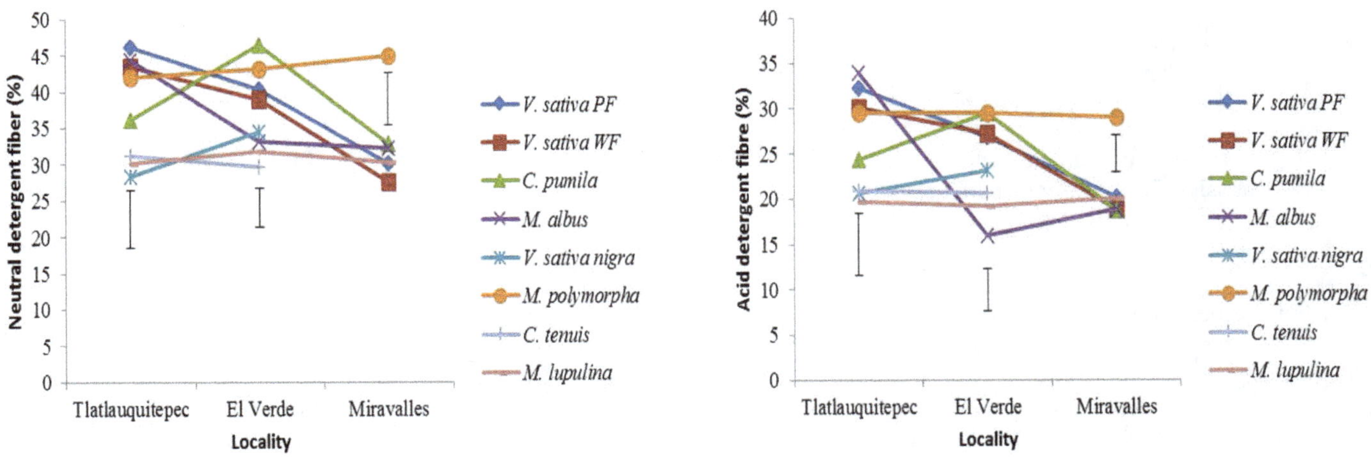

Figure 2. Concentrations of neutral detergent fiber and acid detergent fiber in annual and biennial legume species on the High Plateau of Puebla. Bars represent the least significant difference of Tukey (α = 0.05).

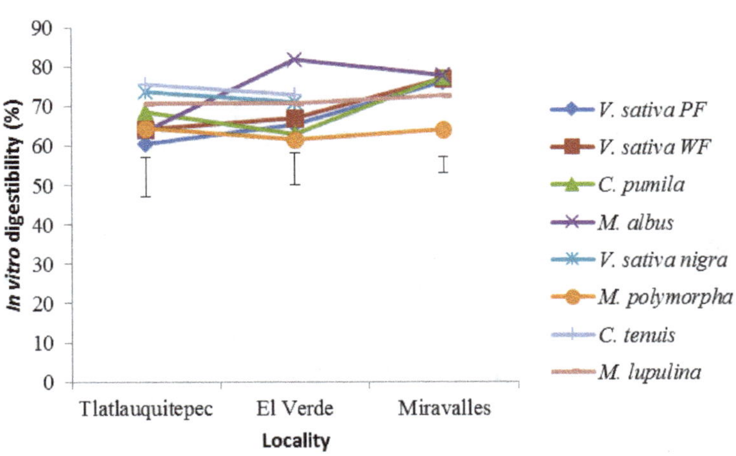

Figure 3. *In vitro* digestibility of the dry matter of annual and biennial legume species in three highland regions of Puebla. Bars represent the least significant difference of Tukey (α = 0.05).

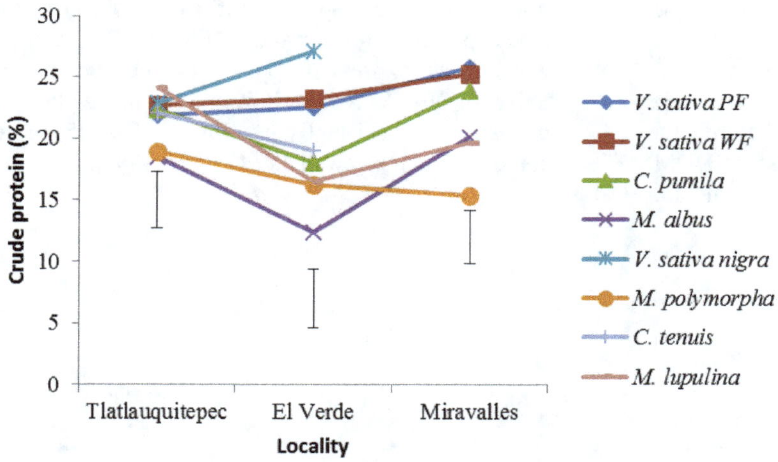

Figure 4. Crude protein content in annual and biennial forage legumes on the High Plateau of Puebla. Bars represent the least significant difference of Tukey ($\alpha = 0.05$).

For the best-known naturalized species (*M. lupulina*, *M. polymorpha* and *M. albus*) the results for crude protein and fibers coincide with the data reported worldwide. There are no values reported for *Crotalaria pumila* and *Cologania tenuis*, but their values fall within the acceptable range for legumes, which is high crude protein content and lower fiber content than grasses. All these species have somewhat contrasting growth habits, having potential for a mixed of legume species pasture. For example, *M. albus* is a biennial erect growth species with high dry matter production and acceptable digestibility. This species compared with vetches which are annual, can be extended to grazing areas in rangelands and croplands, as it has good adaptation to poor soils and drought conditions found in the temperate region. Likewise, the *M. polymorpha*, although it had a somewhat lower nutritional quality than the other species, its forage yield was the same as vetches in two of the environments tested; raising the possibility of improvement trough management. Knowledge in relation to the rhizobia associated with all these species is lacking, an important issue that needs to be resolved to improve the productive performance of these species by establishing effective inoculations and thereby improving the supply of nitrogen to the plant.

CONCLUSION

Several of the species tested in this research have the qualitative and productive potential to improve the diet of ruminants in the highland region of the state of Puebla. The *Melilotus albus* and *Medicago polymorpha* surpassed or equaled the vetches which were the control species in this study. Similarly, when forage selection by the ruminants is considered, diet quality is improved; thus, by ensuring high dry matter production and proper management of the species listed, animal production may improve. More studies are needed on the species that in the wild show abundant growth but in this study performed poorly in order to determine the limiting factors affecting their development.

Acknowledgements

The authors are grateful to the Joint Fund National Council for Science and Technology - Government of Puebla State for the financial resources granted for this research as part of the project 77239 "Strategy for boosting competitiveness of the chain-value sheep-meat through infrastructure and technological innovation in the State of Puebla, Mexico".

REFERENCES

Alford, C.M., Krall, J.M., Miller, S.D. 2003. Intercropping irrigated corn with annual legumes for fall forage in the high plains. Agronomy Journal. 95: 520-525.

Ankom Technology. 2006. Operator´s Manual ANKOM$^{200/220}$ Fiber Analyzer Ankom Technology procedures. Macedon, NY, USA. 27p.

A.O.A.C. 1975. Official Methods of Analysis of the Association off Official Agricultural Chemists. 12th Edition. Washington, D. C.

Clarke, T., Flinn, P.C., McGowan, A.A. 1982. Low-cost pepsin-cellulase assays for prediction of digestibility of herbage. Grass and Forage Science. 37:147-150.

Church, D.C., Pond, W.G., Pond, K.R. 2004. Fundamentos de nutrición y alimentación de animales. Segunda Edición Limusa Wiley. México, D. F., 438 pp.

Guerrero-Rodríguez, J. de D. 2006. Growth and nutritive value of lucerne (Medicago sativa L.) and Melilotus (Melilotus albus Medik.) under saline conditions. Thesis submitted for the degree of Doctor of Philosophy, The University of Adelaide, Australia. 151 p.

INEGI. 2008. Anuario Estadístico del Estado de Puebla. Tomo II, Edición 2008. México. 981 p.

Jones, D.I.H, Hayward, M.V. 1975. The effect of pepsin pretreatment of herbage on the prediction of dry matter digestibility from solubility in fungal cellulose solutions. Journal of the Science of Food and Agriculture. 26: 711-718.

Muir, J.P., Ocumpaugh, W.R., Read, J.C. 2003. Spring forage yield and nutritive value of Texas black medic accessions. Agronomy Journal. 95: 908-912.

SAS Institute. 2002. SAS User's Guide: Statistics. Version 9.0. Statistical Analysis System Institute. Cary, North Carolina, USA.

Shrestha, A., Hesterman, O.B., Squire, J.M., Fisk, J.W., Sheaffer, C.C. 1998. Annual medics and berseem clover as emergency forages. Agronomy Journal. 90: 197-201.

Zhu Y., Sheaffer, C.C., Barnes, D.K. 1996. Forage yield and quality of six annual Medicago species in the North-Central USA. Agronomy Journal. 88: 955-960.

16

EVALUATION OF OXIDATIVE STRESS IN DAIRY GOATS LACTATING NATURALLY OR BY HORMONAL INDUCTION

Rocío Teresa García-Balcázar [a], Miriam Plata-Rodríguez [a], Antonio Díaz-Cruz [b], Karla Rodríguez-Hernández [d], Joel Hernández-Cerón [c] and Alejandro Villa-Godoy [a*]

[a] Universidad Nacional Autónoma de México, Facultad de Medicina Veterinaria y Zootecnia (UNAM-FMVZ), Departamento de Fisiología y Farmacología. Edif. 2, 2313, Ciudad Universitaria, Ciudad de México, CP 04510, México.

E. mail aavillagodoy@gmail.com
[b] UNAM-FMVZ, Departamento de Nutrición Animal y Bioquímica. Ciudad Universitaria, Ciudad de México, 04510, México.
[c] UNAM-FMVZ, Departamento de Reproducción Animal. Ciudad Universitaria, Ciudad de México, 04510, México.
[d] Instituto Nacional de Investigaciones Forestales, Agrícolas y Pecuarias, CIR-Norte Centro. Matamoros, Coahuila, 27440, México.
**Corresponding author*

SUMMARY

The objective was to evaluate the oxidative stress (OS) and its association with performance and serum cortisol in dairy goats under induced (IL) and natural lactation (NL). Six goats programmed to be eliminated due to reproductive problems were IL as follows: a) days 1-7, estradiol plus progesterone; b) days 8-14, estradiol; c) days 18-20, flumetasone; d) every seven days bovine somatotropin; f) on day 21 milking began. Cortisol, antioxidant capacity and Glutathione Peroxidase activity (GHS-Px) were determined in serum. During the first two weeks in milk, NL goats had higher cortisol ($P<0.05$) and lower ($P<0.0001$) antioxidant capacity than NL goats (week1: NL= 3.5±1.2, IL= 14.4±1.1nM/ml; week2: NL= 9.7±0.8, IL= 34.4±0.7nM/ml) but groups did not differ in GHS-Px activity ($P>0.05$). Milk yield/day between groups (NL=2.54±0.2; IL= 2.12±0.2kg) and proportion of pregnant goats (NL=100%; IL=83.3%) were similar ($P>0.05$). The treatment was effective as inductor of lactation; besides most of the IL animals became pregnant despite their previous reproductive problems, prolonging their productive life with a natural lactation plus the additional benefit of at least one more kid. Because the antioxidant capacity increased and cortisol decreased in IL goats, apparently their welfare was more adequate than in NL animals.

Key words: Oxidative stress, serum cortisol, induced lactation, dairy goats.

INTRODUCTION

Oxidative stress (OS) affects negatively the productive performance and health of ruminants (Celi, 2010; Lykkesfeldt and Svendsen, 2007). In dairy cows, OS is associated to the transition period (Spears and Wise, 2008; Sordillo and Aitken, 2009), when most cows experience a series of metabolic adjustments that result in a state of negative energy balance (Ingvartsen and Andersen, 2000). In goats, studies relative to OS during the transition interval and lactation are scarce but overall they show that under conventional diets, dairy goats experience negative energy balance in early lactation (Donnem et al., 2011), phenomenon that could explain the OS observed during their periparturient period (Celi et al., 2008). On the other hand, serum cortisol has been implicated as a mechanism that mediates effects of stress inducing factors in the low reproductive performance of dairy cows and some of the metabolic diseases observed in those animals (Stoebel and Moberg, 1982), thus cortisol has been proposed as an indicator of nonspecific stress. Elevated serum cortisol concur with OS in cows at the end of pregnancy and during parturition (Ingvartsen and Andersen, 2000), and an association between OS and circulating cortisol may exist as it was suggested by Dimri et al. (2010). An issue of concern for some sectors of society is the potential negative effect that the veterinary and production related procedures may have on farm animals welfare (Mormède et al., 2007), particularly under intensive production systems. Goats milk farming under intensive management is becoming more frequent (Celi et al., 2008), and reproductive problems is a major drawback in those systems as indicated by an annual culling rate of 14.5% due to infertility in dairy goats (Malher et al., 2001). Similarly, production of proteins of interest for humans is generated in transgenic goats more frequently than in other animal models (Houdebine, 2009), and dairy goats are usually recognized as a preferable animal model to cattle for the production of transgenic proteins in milk, due to their shorter gestation, earlier sexual maturation, and smaller size (Zhang et al., 2010). However, relative to other systems devised to produce recombinant pharmaceutical proteins, biopharming with transgenic goats offer some disadvantages (Houdebine, 2009), including a relatively long time for collection of products and elevated infertility rate. The induction of lactation by simulating the maternal hormonal changes near parturition was originally designed to reduce economic losses due to the high culling rate due to reproductive problems of heifers and cows in intensive bovine dairy systems (Centeno et al., 2012). Therefore, hormonal induction of lactation may reduce the culling rate, and make possible to obtain milk from goats with a high milk yield potential but with problems to become pregnant, prolonging their productive life. In a previous study (Mellado et al., 1996), goats induced to lactate hormonally had similar total solids and protein yield than goats under natural lactation. Based on these results and that protocols that induce lactations do not modify importantly the composition of milk from transgenic goats (Zhou et al., 2005), we tested in dairy goats programmed for leaving the farm due to reproductive problems, the effects of a protocol designed to induce lactation, with the objective of determining whether or not the oxidative stress and serum cortisol levels were different than in goats under natural lactation.

MATERIALS AND METHODS

Animals and general management

This experiment was approved by the Institutional Subcommittee for Caring of Experimental Animals (Subcomité Institucional para el Cuidado de Animales en Experimentación, FMVZ-UNAM). The study was performed from January to November, 2008 in an experimental farm located in the central plateau of Mexico, at 20° 36′ North and 99° 56′ West. The location altitude is 1,920 meters over the sea level and the mean annual values for temperature and rainfall are 17.5°C and 388 mm, respectively. Multiparous dairy goats (n=11) were housed in pens designed for optimal comfort; each treatment group was in a separate pen. Housing consists in a circular unit and pens are projected in a radial form from the center. Pens are trapezoidal and the external side (15 m long) gives access to a feeder bunk for 34 animals/pen; the inner side of pens (3 m long) gives access to the management area and lateral sides are 23 m in length. The shade covers the central management unit and extends for 7 m into each pen inner area which is equipped with automatic drinker; floor in the shaded area and near the feeder bunk is cemented and the rest is dust. Animals were milked twice a day (0700 and 1700 h) in a tandem parlor equipped with 4 milking units. Feed was offered two h prior each milking time. Diet consisted in a completely mixed ration based on ingredients from the same lot to minimize variation throughout the experiment. Ingredients were alfalfa hay, corn silage, ground corn, commercial concentrate and a minerals-vitamins mix. Proximate analysis and atomic spectroscopy of mixed ration samples indicated that the chemical composition of 1 kg of diet (dry matter) was CP, 19.8%; NEl, 2.97 Mcal; ADF, 24.4%; NDF, 33.7%; Ca, 10.95 g; P, 3.26 g; Vit. A, 10.50 IU; Vit. E, 0.04 IU; Vit. D, 105 IU; Se, 0.38 mg; Mn, 1063.7 mg; Zn, 1331 mg; Fe, 1967.02 mg; Cu, 8.96 mg; Co, 6.2 mg; and I, 87.33 mg. Based on feed left over from each pen, we estimated an average individual intake of 3kg/animal/day. During the reproductive season (Figure 7), goats estrus was synchronized by installing an intravaginal device (CIDR: Progesterone

0.3 g; Pfizer, New Zealand) which remained *in situ* for 11 days; additionally an intramuscular injection was applied containing eCG (400 IU/animal; Foligon, Intervet, Mexico). Goats were separated by breed and were exposed for five consecutive days, along with the rest of the farm herd, to bucks previously evaluated for adequate libido and fecundity and equipped with marker aprons. Mounting activity was observed every 12 h and crayon marks were recorded. Seven days later, bucks were penned again with does and remained with them for seven days. Pregnancy was determined by ultrasonography (ultrasound Aloka 550, 7.5 Mhz probe, Aloka Co., Walinford, CT, USA) between 45 and 60 days after the beginning of the mating period.

Treatments

Six multiparous, non-pregnant goats (3 Alpine and 3 Toggenburg; 2.8 parturitions) that had been programmed for culling due to reproductive problems, received a hormonal treatment for induction of lactation (IL). Additionally, 5 does (3 Alpine and 2 Toggenburg; 2.6 parturitions) of natural lactation (NL) that kidded during the application period of the IL treatment, were considered as control animals. Hormonal treatment given to IL animals consisted of: a) from days 1 to 7 a subcutaneous injection (SC) of estradiol cipionate (0.05 mg/kg; ECP, Pfizer, Mexico) plus SC progesterone (0.6 mg/kg; Progesterona, Fort Dodge, Mexico); b) from days 8 to14 SC estradiol cipionate (0.05 mg/kg); c) from day 18 to 20 an intramuscular injection of flumetasone (0.25 mg/day; Fluvet, Fort Dodge, Mexico); d) on days 1, 7, 14 and 21 a SC administration of bovine recombinant somatotropin (125 mg/goat; Lactotropina, Elanco, Mexico).

Measurements and samples

Individual body weight (BW) and body condition (BC; 1 to 5; 1=extremely thin, 5=obese) were recorded weekly from day 1 of hormonal treatment to day 140 of lactation. Individual milk yield was measured weekly during the whole lactation. Serum was obtained from blood samples taken from a jugular vein during early (weeks 1 to 4) and mid lactation of goats (weeks 11 and 15).

Laboratory analysis

Serum cortisol was determined by a solid phase radioimmunoassay (Coat-A-Count, Siemens Health Diagnostics Inc., Los Angeles, CA, USA); based on one assay, the C. V. was 5.6% (control serum mean$\pm$SD= 34$\pm$1.9 ng/ml). Ferric Reducing Antioxidant Power (FRAP spectrophotometry assay) was used to quantify the non-specific antioxidant capacity (Benzie et al., 1996). To determine the specific Glutathione Peroxidase (GHS-Px) enzyme activity in serum, we used a modified coupled assay described by Lawrence and Burk (1976).

Design and statistical analysis

Data from continuous variables were examined by analysis of variance for fixed effects models either for one way analysis (duration of lactation, milk yield/lactation, milk yield in the lactation previous to the experiment) or for repeated measures over time (milk yield/week, changes in BW and BC transformed to Arc-sine, cortisol, FRAP and GHS-Px). In the repeated measures analysis, the model included effects of treatment, sample and their interaction; both, one way ANOVA and repeated measures analysis were performed by the PROC GLM (SAS, 2002). The correlation of changes in serum cortisol levels and variations of FRAP, GHS-Px, body measurements and milk yield related variables was determined by the PROC CORR (SAS, 2002). Homogeneity of groups (breed and kidding number) was determined by the Fisher´s Exact Test (Agresti, 1992). During the lactation that preceded the experiment, goats assigned to IL had a lower milk yield than NL goats. Similarly, IL goats had lower BW and BC at the beginning of the experimental lactation than NL animals. However, when covariance analysis were performed, only milk yield in previous lactation was significant as a covariant ($P<0.01$), thus this variable but not BW or BC was used as a covariance to analyze all data relative to milk production collected during the experimental lactation.

RESULTS

As it was indicated above, IL goats had lower ($P<0.05$) BC and BW on the first day of milking than NL goats; in addition IL animals produced less milk ($P<0.01$) during the lactation previous to the experiment than NL goats (Table 1).

Animals from both treatments had similar pregnancy rate and milk production/day of lactation. However NL goats had a longer lactation and higher total milk yield/lactation (Table 2). Nevertheless, the analysis of repeat measures over time of data relative to milk yield (Figure 1) did not identify an interaction between treatments and samples ($P>0.05$).

From the beginning of the study, IL goats were lighter and had lower BC than the NL goats, however as the lactation progressed, both BW and BC of IL animals increased (Figures 2 and 3), and were similar to those of NL goats from week 5 through 30 of lactation. In both body measurements, no effects of treatment were detected ($P>0.05$) but they were influenced by day of measurement (P<0.05) and by the interaction between treatments and day of measurement (P<0.01).

Table 1.- Milk yield (kg) during the previous lactation, as well as body condition and body weight (kg) on the day of first milking in goats that were in natural (NL) or induced lactation (IL). Data are presented as mean ± standard error.

Treatment	Milk lactation	Body condition	Body weight
NL	531.4±32.5^{a}	3.2±0.25^{c}	62.7±2.0^{a}
IL	337±35.6^{b}	2.3±0.23^{d}	47.9±1.9^{b}

$^{a, b}$ Within column, distinct letters indicate difference between means (P<0.01).
$^{c, d}$ Within column, distinct letters indicate difference between means (P<0.05).

No effect of sample (P>0.05) was detected in serum cortisol, however an effect of treatment and the interaction of treatment by sample was found ((P<0.05); thus concentrations of serum cortisol were higher in the NL animals than in IL goats during the first two weeks of lactation; afterward levels of cortisol were similar between groups (Figure 4). Variations in serum concentrations of cortisol were not correlated (P>0.05) with changes in FRAP and GHS-Px values or with variations in milk yield or body measurements related data; however when the analysis was performed within treatment, only in the NL goats a negative correlation was found of cortisol with milk yield per lactation (r = -0.40; P<0.05) and lactation length (r = -0.38; P<0.05). Additionally in the same group of goats, a positive correlation (r = 0.54; P<0.01) was observed between GHS-Px and lactation length.

Data from FRAP analysis, indicating antioxidant capacity, were not affected (P>0.05) by treatment but they were significantly altered (P<0.001) by sample and the interaction between treatments and samples. Accordingly, the general antioxidant capacity (Figure 5) was considerably lower (P<0.01) in NL goats during weeks 1, 2 and 4 of lactation but in week 3 they had a higher value than IL goats (P<0.01); later FRAP values were parallel but again in week 11 they were greater in NL goats (P<0.01). As in FRAP, data of GHS-Px activity were not affected by treatment (P>0.05) but they were significantly affected by sample (P<0.06) and the interaction between treatments and samples (P<0.01). Consequently GHS-Px values in serum (Figure 6) were similar in both groups during the first three weeks of lactation but the NL animals had higher GHS-Px values on week 4 (P<0.01), whereas the IL goats had higher values on week 15 than NL goats (P<0.05).

DISCUSSION

It was expected that IL and NL animals would differ in body measurements at the beginning of milking and in milk yield during the lactation previous to the experiment as randomization procedure was not used to assign goats to lactation treatments, rather IL animals were those that had been programmed for elimination of the herd. Similarly the NL goats were animals that kidded during the time when IL treatments were applied. Nevertheless the use of milk production in the previous lactation allowed us to adjust for those differences and to examine the effects of lactation type without bias. Under these conditions, IL and NL goats were similar in reproductive performance and milk production per day of lactation. However, milk yield during lactation as well as lactation length were greater in NL than in IL goats. Our results are consistent with a report (Mellado et al., 1996) in which goats hormonally induced to lactate equaled the NL animals in at least one of the variables associated with productive performance.

Table 2.- Productive and reproductive performance of goats during a natural (NL) or induced lactation (IL); data related to milk yield are mean ± standard error.

Treatment (kg)	Pregnant (%)	Milk/day (kg)	Lactation length (day)	Milk/lactation (kg)
NL	100.0 (5/5)*	2.54±0.2	248.8±3.8^{a}	638.4±36.8^{a}
IL	83.3 (5/6)*	2.12±0.2	232.6±4.4^{b}	502.3±42.1^{b}

$^{a, b}$ Indicates difference between means (P<0.05). * Inside the parenthesis: pregnant goats/exposed goats

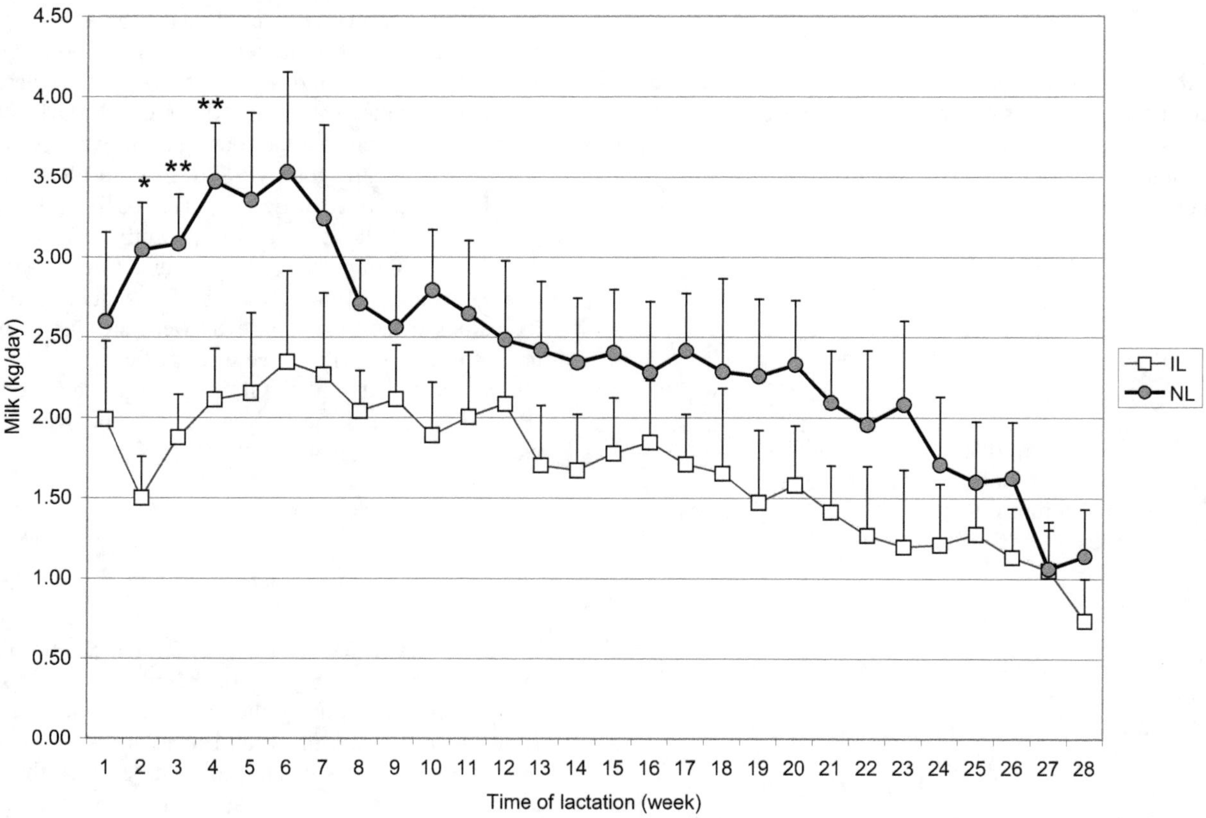

Figure 1.- Milk yield in goats during a natural (NL) or induced lactation (IL); each point is the weekly mean ± standard error. * and ** indicate difference between contemporary means ($P<0.01$ and $P<0.05$, respectively). Treatment ($P>0.05$); Time ($P<0.01$); Treatment x Time ($P<0.01$).

Despite that NL goats had higher BW and BC during early lactation, serum cortisol levels remained elevated during the first two weeks of milking when compared with IL animals. The increased serum cortisol in NL animals may be attributed to the homeorhetic mechanisms that occur during the final stage of gestation and beginning of lactation (Bauman and Currie, 1980), when accelerated growth of fetuses takes place along with stress induced by parturition, phenomena that are not experienced by IL goats. Those effects may be combined with the rapid increment in milk production in NL animals versus the slow increments in milk yield observed in IL goats. It has been reported that high serum concentrations of cortisol are one of the mechanisms inherent to lactation and indicate the metabolic response induced for the increased demands of nutrients imposed by the milk synthesis process (Bauman and Currie, 1980; Tucker, 2000).

The results from the FRAP assay in serum indicate that the NL goats were exposed to an increased risk of OS during the first two weeks postpartum, and the profile describing the variation in the antioxidant capacity follows a similar pattern to that recorded in dairy goats fed with energy deficient diets (Celi et al., 2010).

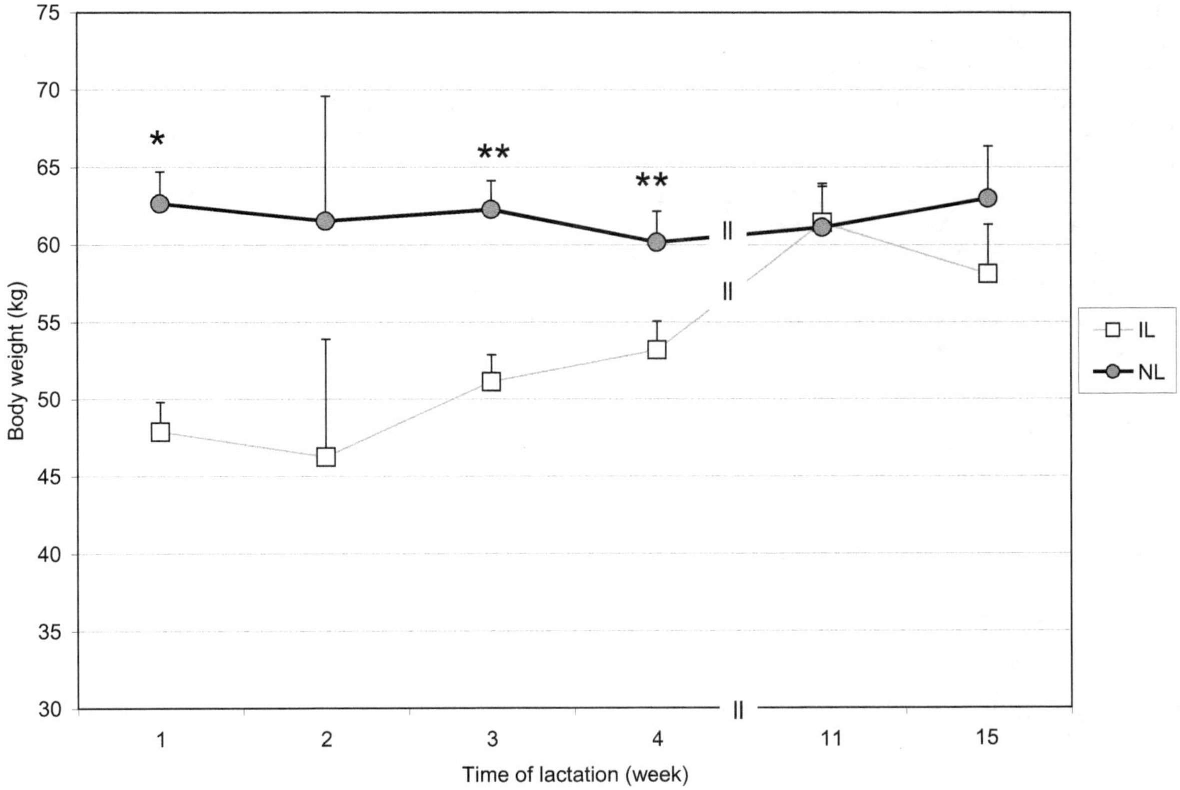

Figure 2.- Changes in body weight of goats during a natural (NL) or induced lactation (IL); each point is the weekly mean ± standard error at early (weeks 1 to 4) and mid lactation (weeks 11 and 15). * and ** indicate difference between contemporary means (P<0.01 and P<0.05, respectively). Treatment (P>0.05); Time (P<0.05); Treatment x Time (P<0.01).

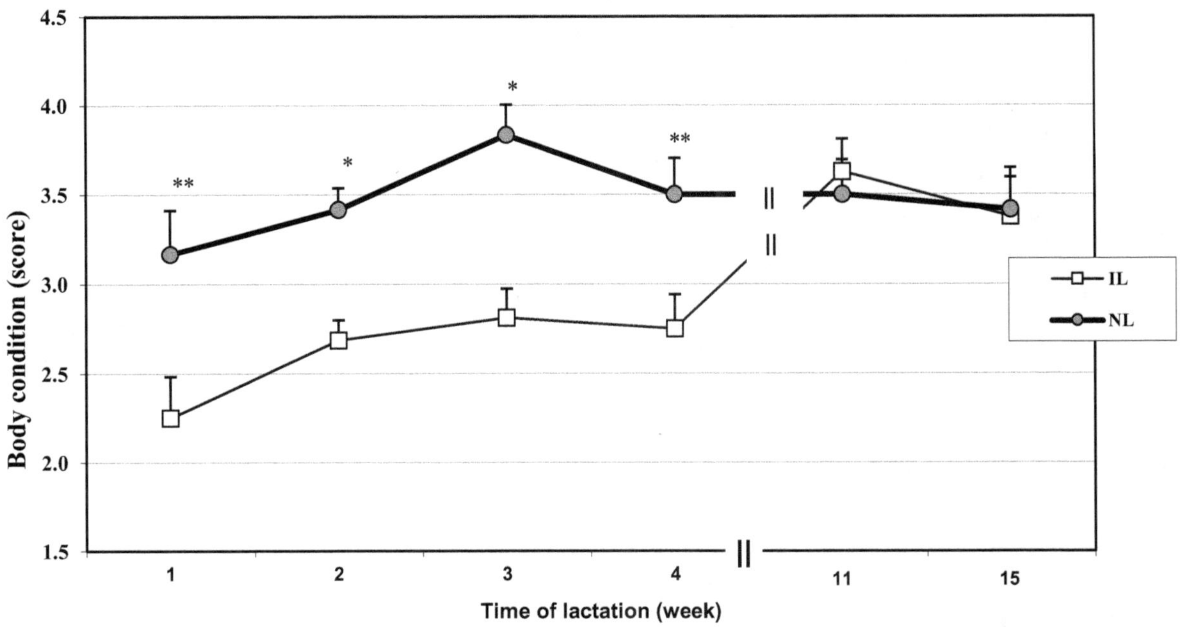

Figure 3.- Changes in body condition of goats during natural (NL) or induced lactation (IL); each point is the weekly mean ± standard error at early (weeks 1 to 4) and mid lactation (weeks 11 and 15). *, ** indicate difference between contemporary means (P<0.01 and P<0.05, respectively). Treatment (P>0.05); Time (P<0.05); Treatment x Time (P<0.01).

Relative to NL animals, IL goats showed a higher general antioxidant capacity during the two first weeks of lactation, an observation that according to our best knowledge is informed for the first time. The high levels of serum cortisol observed here may indicate an association with the decreased antioxidant capacity; this is possible because corticosterone fed to rats daily during four weeks, decreased the repair capacity of the hepatic mitochondrial antioxidant system (Caro et al., 2007). It is also possible that the NL goats in this study had some dietary antioxidant deficiencies because they were fed with forages preserved as hay or silage, and there are reports that at least β-carotene is low in this type of diets (Spears and Weiss, 2008). Goats fed high-energy diets (Celi et al., 2010) had an elevated antioxidant capacity in comparison with the insufficiently fed goats, and was similar to that recorded in IL animals in the present study; thus it is feasible that the relationship between ingestion and nutrients demand in these goats, due to the slower increase in milk production, was more balanced than in NL animals. Perhaps for the same reason, cortisol was lower and the general antioxidant capacity was higher in IL than in NL goats.

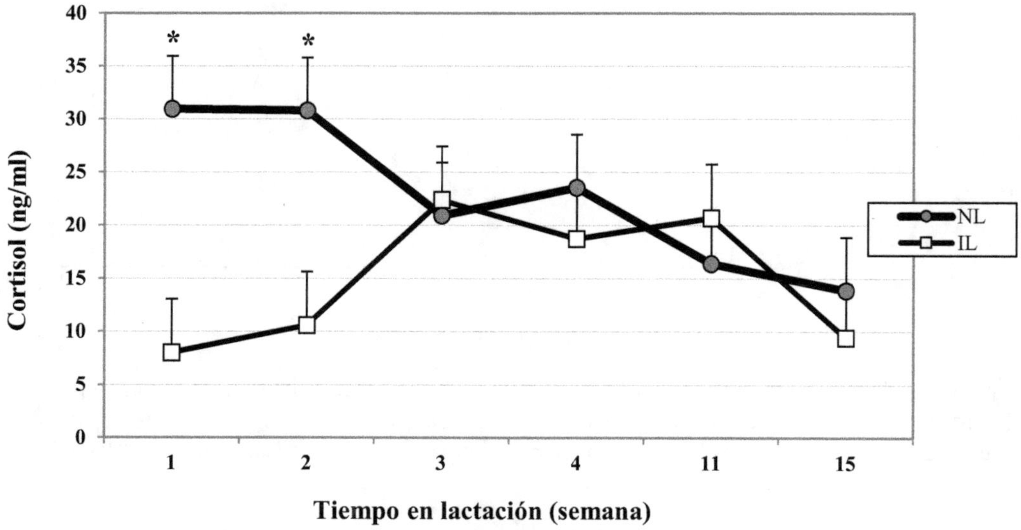

Figure 4.- Serum concentrations of cortisol in goats during natural (NL) or induced lactation (IL); each point is the weekly mean ± standard error during early (weeks 1 to 4) and mid lactation (weeks 11 and 15). * indicates difference between contemporary means (P<0.05). Treatment (P<0.05); Time (P>0.05); Treatment x Time (P<0.05).

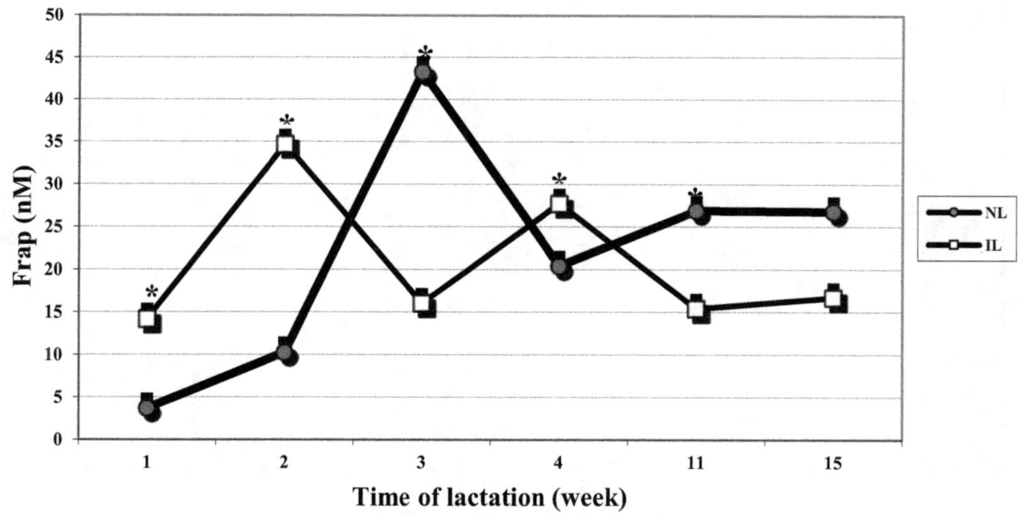

Figure 5.- Global antioxidant activity in goats during natural (NL) or induced lactation (IL); each point is the weekly mean ± standard error during early (weeks 1 to 4) and mid lactation (weeks 11 and 15). * indicates difference between contemporary means (P<0.01). Treatment (P>0.05); Time (P<0.001); Treatment x Time (P<0.001).

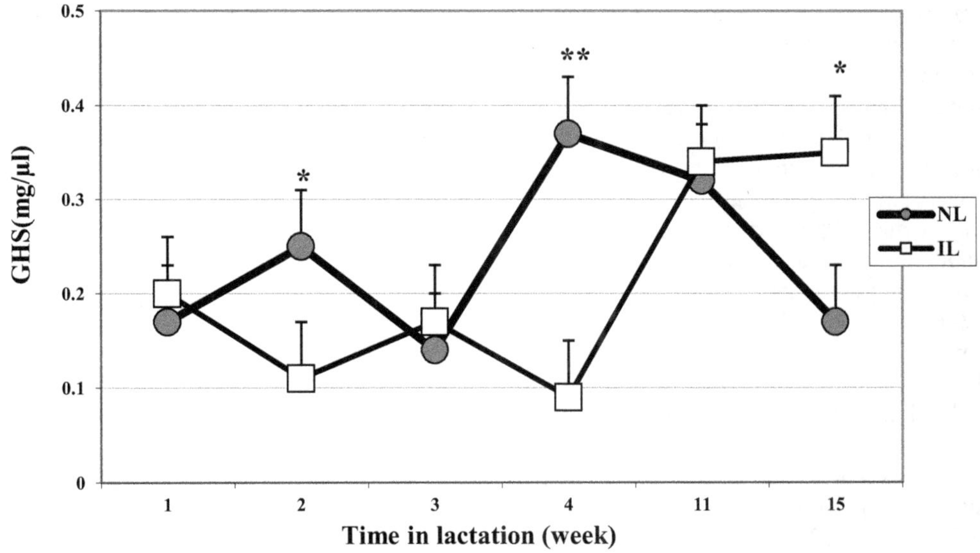

Figure 6.- Glutathione Peroxidase activity in goats during natural (NL) or induced lactation (IL); each point is the weekly mean ± standard error during early (weeks 1 to 4) and mid lactation (weeks 11 and 15). *, ** indicate difference between contemporary means (P<0.05 and P<0.01, respectively). Treatment (P>0.05); Time (P<0.06); Treatment x Time (P<0.001).

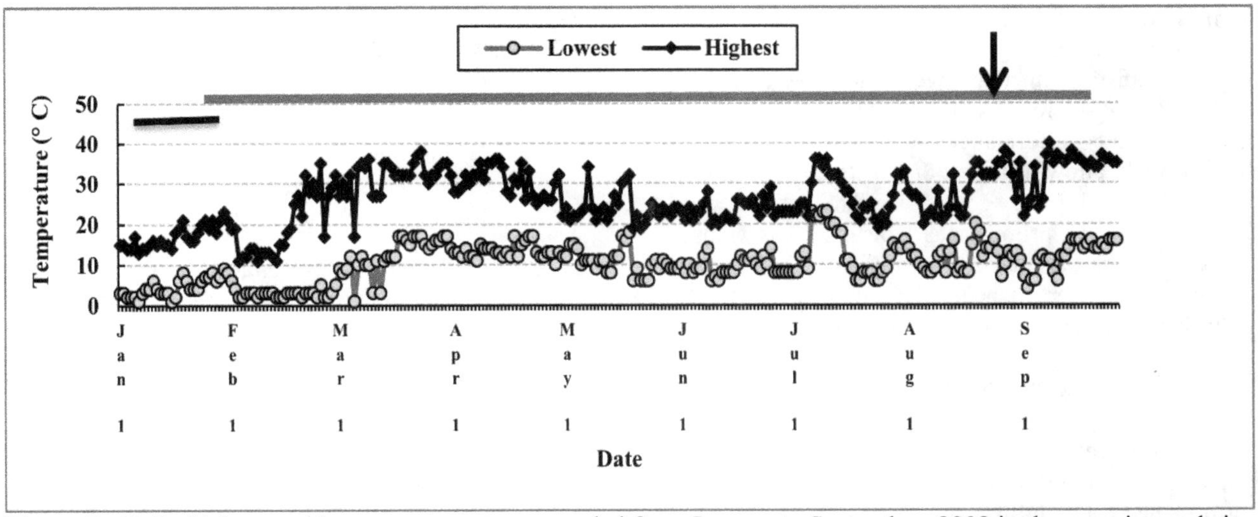

Figure 7.- Daily highest and lowest temperature recorded from January to September, 2008 in the experimental site. The black bar indicates intervals of treatment application for goats induced to lactate and parturitions of natural lactation animals. The gray bar shows the milking interval and the black arrow indicates the beginning of breeding.

In contrast with the general antioxidant capacity, the specific antioxidant function of GSH-Px was similar between IL and NL animals during the first three weeks of lactation. However, GSH-Px was higher in week four of lactation in NL than in IL goats but became higher again in IL goats in samples collected on week 15. Perhaps the proximity of the lactation peak in NL animals evoked a higher production of reactive oxygen metabolites and the consequence was the increment of GSH-Px activity in week four. However, data from this study do not provide evidence to explain the differences between groups detected at mid lactation. It may be speculated that an interaction between physiological state and environmental temperature (Figure 7) occurred since samples from weeks 11 (33°C and 17°C; highest and lowest temperature, respectively) and 15 (32°C and 18°C; highest and lowest temperature, respectively) were collected in April and May, when maximal mean temperatures during the day were frequently above 30°C and minimal temperatures were the highest during the experimental interval. This is possible because in studies carried out in a Mediterranean, semiarid region, lactating goats presented a moderate OS during the hottest season relative to values recorded during cooler seasons (Di Trana et al., 2006). It should also be noticed that NL and IL animals differed from the phenotypic standpoint since in the lactation previous to this experiment, milk yield was significantly higher in the former; thus metabolic capacity could be different between groups as it has been suggested for high producing dairy cows in comparison with mediocre yielders (Fenwick, et al., 2008; Bauman and Currie, 1980); consequently the differences relative to the antioxidant capacity recorded here may be associated, at least partially, with the productive potential of goats. To clarify these questions, it is necessary to study the variations of antioxidant capacity in IL and NL goats under different scenarios of environmental conditions and genetic background.

Despite the differences found relative to cortisol, GHS-Rx activity and the antioxidant capacity, goats from both groups did not show evidence of reproductive or health problems; so the OS recorded, even in animals with the lowest antioxidant values, could be considered as mild. All physiological, metabolic and performance data generated in the present work, allow us to say that IL treatment do not affect negatively the goat's welfare, as it was documented in IL primiparous Holstein cows whose cortisol concentrations in serum and hair were significantly lower than in the contemporary NL controls (González-de-la-Vara et al., 2008).

CONCLUSIONS

In summary, relative to NL animals, IL goats had similar milk yield/day of lactation and reproductive performance but during the first two weeks of lactation had lower circulating levels of cortisol and higher antioxidant capacity, thus our conclusion is that the protocol for induction of lactation is not damaging for goats health and that apparently the IL goats welfare was more adequate than in NL goats.

ACKNOWLEDGMENTS

This experiment was supported partially by the grant SDEI-UAIFE-PTEI-UNAM 07.03.04. The first author was under the program Early Initiation in Research-FMVZ-UNAM. The authors would like to thank Dr. Clara Murcia Mejía for running the cortisol RIA, and Dr. Cuauhtémoc Nava Cuellar for teaching the FRAP and GHS-Px tests to the first author. This paper was edited by Lorena Villa-Parkman (Northwestern University, Evanston, Ill, USA), a native English speaker with six year experience in writing and editing.

REFERENCES

Agresti, A. 1992. A survey of exact inference for contingency tables. Statistical Sciences. 7:131–153.

Bauman, D.E., Currie, W.B. 1980. Partitioning of nutrients during pregnancy and lactation: A review of mechanisms involving homeostasis and homeorhesis. Journal of Dairy Science. 63:1514-1529.

Benzie, I.F.F., Strain, J.J. 1996. The ferric reducing ability of plasma (FRAP) as a measure of "antioxidant power": The FRAP assay. Analytical Biochemistry. 239:70-76.

Caro, P., Gómez, J., Sanz, A., Portero-Otín, M., Pamplona, R., Barja, G. 2007. Effect of graded corticosterone treatment on aging-related markers of oxidative stress in rat liver mitochondria. Biogerontology. 8:1-11.

Celi, P. 2010. The role of oxidative stress in small ruminants` health and production. Revista Brasileira de Zootecnia. 39:348-363.

Celi, P., Di Trana, A., Claps, S. 2010. Effects of plane of nutrition on oxidative stress in goats during the peripartum period. The Veterinary Journal. 184:95-99.

Celi, P., Di Trana, A., Quaranta, A. 2008. Metabolic profile and oxidative status in goats during the peripartum period. Australian Journal of Experimental Agriculture. 48:1004-1008.

Centeno, B.S., Luna, P.C., Aguilar, C.J.A., Pérez-Mora, A., Rodríguez, H.K., Villa-Godoy, A., Trueta, S.R. 2012. Economic evaluation of two variations of a hormonal induction treatment in replacement heifers in a tropical dual purpose system. Veterinaria México. 43:295-316.

Dimri, U., Ranjan, R., Sharma, M.C., Varshney, V.P. 2010. Effect of vitamin E and selenium supplementation and oxidative stress indices and cortisol level in blood in water buffaloes during pregnancy and early postpartum period. Tropical Animal Health and Production. 42:405-410.

Di Trana, A., Celi, P., Claps, S., Fedele, V., Rubino, R. 2006. The effect of hot season and nutrition on the oxidative status and metabolic profile in dairy goats during mid lactation. Animal Science. 82:717-722.

Donnem, I., Eknaes, M., Randby, A.T. 2011. Energy status measured by computer tomography (CT)-scanning, and milk quality of dairy goats fed rations with various energy concentrations. Livestock Science. 142:235-244.

Fenwick, F.A., Llewellin, S., Fitzpatrick, R., Kenny, D.A., Murphy, J.J., Patton, J., Wathes, D.C. 2008. Negative energy balance in dairy cows is associated with specific changes in IGF-binding protein expression in the oviduct. Reproduction. 135:63-75.

González-de-la-Vara, M., De Anda, F., Vázquez, C.J.C., Romero, A., Romano M.C., Villa-Godoy, A. 2008. Grouping primiparous dairy cows: Production performance and cortisol in serum and hair. 42nd Congress of the International Society of Applied Ethology: Applied Ethology, Addressing Future Challenges in Animal Agriculture. Dublin, Ireland. Pp 82.

Houdebine, M.L. 2009. Production of pharmaceutical proteins by transgenic animals. Comparative Immunology, Microbiology and Infectious Diseases. 32:117-121.

Ingvartsen, K.L., Andersen, J.B. 2000. Integration of metabolism and intake regulation: A review focusing on periparturient animals. Journal of Dairy Science. 83:1573-1597.

Lawrence, R.A., Burk, R.F. 1976. Glutathione peroxidase activity in selenium-deficient rat liver. Biochemical and Biophysical Research Communication. 71, 952-958.

Lorhke, B., Viergutz, T., Kanitz, W., Losand, B., Weiss, D.G., Simko, M. 2005. Short communication: Hydroperoxides in circulating lipids in dairy cows, implications for bioactivity of endogenous oxidized lipids. Journal of Dairy Science. 88:1708-1710.

Lykkesfeldt, J., Svendsen, O. 2007. Oxidants and antioxidants in disease: Oxidative stress in farm animals. The Veterinary Journal. 173:502-511.

Malher, X., Seegers, H., Beaudeau, F. 2001. Culling and mortality in large dairy goat herds managed under intensive conditions in western France. Livestock Production Science. 71:75-86.

Mellado M, Bernal A; Mendoza R, Carrillo E. 1996. Hormonal induction of lactation in prepuberal and multiparous crossbred goats under extensive conditions. Small Rum Res 19;143-147.

Mormède, P., Andanson, S., Aupérin, B., Beerda, B., Guémené, D., Malmkvist, J., Manteca, X., Manteuffel, G., Prunet, P., van Reenen, C.G., Richard, S., Veissier, I. 2007. Exploration of the hypothalamic-pituitary-adrenal function as a tool to evaluate animal welfare. Physiology and Behavior. 92:317-339.

SAS. 2002. Statistical Analysis System Institute Inc., Cary, NC, USA.

Sordillo, L.M., Aitken S.L., 2009. Impact of oxidative stress on the health and immune function of dairy cattle. Veterinary Immunology and Immunopathology. 128:104-109.

Spears, J.W., Weiss, W.P. 2008. Role of antioxidants and trace elements in health and immunity of transition dairy cows. The Veterinary Journal. 176:70-76.

Stoebel, D.P., Moberg, G.P. 1982. Effect of adrenocorticotropin and cortisol on luteinizing hormone surge and estrous behavior of cows. Journal of Dairy Science. 65:1016-1024.

Tucker, H.A. 2000. Symposium: Hormonal regulation of milk synthesis. Hormones, mammary growth and lactation: A 41 years perspective. Journal of Dairy Science. 83:874-884.

Zhang, Y.L., Wan, Y.J., Wang, Z.Y., Xu, D., Pang, X.S., Meng, L., Wang, L.H., Zhong, B.S., Wang, F. 2010. Production of dairy goats embryos, by nuclear transfer, transgenic for human acid β–glucosidase. Theriogenology. 73:681-690.

Zhou, Q., Kyazike, J., Echelard, Y., Meade, H.M., Higgins, E., Cole, E.S., Edmunds, T. 2005. Effect of genetic background on glycosilation heterogeneity in human antithrombine produced in the mammary gland of transgenic goats. Journal of Biotechnology. 117:57-72.

THE USE OF AGRO INDUSTRIAL BY-PRODUCTS IN THE DIET OF GROWER TURKEYS

O.J. Makinde [1*] and M. Inuwa[2]

[1]Department of Animal Science, Federal University, Gashua, Nigeria.
Email: johyinmak@yahoo.com
[2]Department of Animal Production Technology, Federal College of Wildlife Management, PMB 268, New Bussa, Niger State, Nigeria.
**Corresponding author*

SUMMARY

An experiment was conducted to evaluate the response of grower turkeys to diets containing different agro industrial by-products. A total of 120grower turkeys were used for this study. Four experimental diets were formulated and designated as T1, T2, T3, and T4. T1 (control diet) contained none of the agro industrial by products used while T2, T3 and T4 contained15% each of Wheat offal, Maize offal, Rice offal respectively. The turkeys were randomly allocated to four experimental treatments of thirty birds per treatment while each treatment was replicated three times (3) with ten (10) birds per pen in a completely randomized design. The result of growth performance revealed that average feed intake, average weight gain and feed conversion ratio of turkeys fed the control and 15% Wheat Offal diets were not significantly different (P>0.05). The result of carcass characteristics and organ weights followed similar trend. Turkeys fed 15% Maize Offal and 15% Rice Offal diets recorded the lowest values (P<0.05). There were no significant differences (P>0.05) in all the haematological parameters and serum biochemical indices measured except PCV which decreased across the treatment groups. It was concluded that 15% wheat offal can be included in the diet of grower turkeys without adverse effect on growth performance, carcass yield and health status of birds.

Key words: grower turkeys; wheat offal; maize offal; rice offal; growth performance.

INTRODUCTION

Turkey (*Meleagris gallopavo*) is an integral part of poultry. It is a very important bird usually raised for economic benefit (Adene and Oguntade, 2006). The population of turkey in Nigeria had grown from 1.5 to 2.0 million (Morgan, 1991). In poultry farming, feeding accounts for 65 – 80 % of the production cost and the poultry industry has suffered more than any other livestock industry as a result of the problem arising from inadequate supply of feed (Lepaideur, 2004). Energy and protein feedstuff, which constitute about 80% of poultry feedstuff have been the major hindrances to effective poultry production in Nigeria (Uchegbu *et al.*, 2004). Cereal grains constitute the major source of energy in poultry diet in the tropics (Oluyemi and Roberts, 2000). Maize has remained the chief source of energy in compounded diet and it constitutes about 50% of poultry ration (Ajaja *et al.*, 2002). Pressure on maize and recently cassava has been on the increase worldwide with emphasis being placed on export in Nigeria, since recent trends have seen massive exploitation of corn in ethanol production as an alternative source of fuel (Doki, 2007). This trend has necessitated the use of agro industrial byproducts such as wheat offal, maize offal, rice offal, palm kernel meal, brewers dried grains etc. in formulating feed for the livestock. Maize offal is a by-product of maize milling processes, second to wheat offal as the most preferred and utilized conventionally in livestock feeds in Nigeria (Babatunde and Oluyemi, 2002). Maize offal contains about 110 to 120 g/kg crude protein and 80 to 90 g/kg crude fibre (Onifade and Babatunde, 1998). The relatively low crude fibre content compared to other agro industrial by-products could be an advantage in fibre nutrition whereas the low protein content appears to be a limitation. Rice offal has high lysine and methionine content and it is available in large quantities all year round in many towns in the rice growing areas of Nigeria (Makinde *et al.*, 2014). However, its phytate content, enzyme inhibitor, high fibre content and oxidative rancidity may have deleterious effects on poultry. Rice offal may become rancid and decrease poultry growth performance and lipid stability of meat (Chae *et al.*, 2002). It is suggested to include it at relatively low levels (up to 15% only) in poultry diets (Makinde *et al.*, 2014) as higher levels may result in low calcification, decreased feed intake and even mortality above 80% inclusion (ArunaTomar and Singh, 1999). More recently, Aduku (1993) in Nigeria reported that wheat offal contains 1256 and 2320 Kcal of metabolizable energy per kg for poultry and swine respectively, 15.6% crude protein and mineral elements such as calcium and phosphorus. These reports have portrayed wheat offal somewhat, as a product with a fair potential for use in the diet, even if it is to replace a portion of maize, which is used conventionally in the diet of chicken. Nuhu *et al.* (2008) reported that the FCR of birds fed on wheat offal diets were efficiently more utilized than those birds fed on maize offal-based diets. These workers concluded that 20% wheat offal inclusion in broiler diets is recommended to farmers. Despite the availability of various nutrients in these agro industrial byproducts, information on their utilization in the turkey diets is scanty, therefore, this study aimed at assessing the use of agro industrial byproducts in the diet of grower turkeys.

MATERIALS AND METHOD

Experimental site

This study was conducted at the Poultry Unit of the Teaching and Research Farm, Federal College of Wildlife Management (FCWM), New Bussa, Niger State, Nigeria. The poultry building is an open sided type that permits adequate ventilation in the house, with a concrete floor and zinc-roofing sheet. New Bussa sits at 9°53′N 4°31′E, and the original town of Bussa is located about 40 km North of New Bussa at 10°13′51″N 4°28′31″E (altitude 170 m.a.s.l.). The climate of the area is tropical with monthly average temperature of 34°C and mean annual relative humidity of 60%.

Sources of Experimental Materials

The agro industrial by-products used in this study were wheat offal, rice offal and maize offal which were purchased around Neighbouring market within New- Bussa and its catchment areas.

Experimental Design and Management of Birds

One hundred and twenty day old unsexed turkeys (Narraganset and bronze breed) were purchased from a reputable hatchery in Ibadan, Nigeria. The birds were raised on deep litter pen with wood shaving as litter material. They were fed on a common diet for the first four (4) weeks of life. Feed and water was supplied *ad-libitum*. Management practices and vaccination programme was followed strictly. At the end of the four (4) weeks, the grower turkeys (initial average weight = 330g) were randomly allocated to four experimental treatments of thirty birds per treatment, while each treatment was replicated three times (10birds per replicate) in a completely randomized design (CRD).

Experimental diets

Four experimental diets were formulated and designated as T1, T2, T3, and T4. T1(control diet) contained none of the agro industrial by products used while T2, T3 and T4 contained 15% each of Wheat offal (WO), Maize offal (MO), Rice offal (RO) respectively. The gross composition and proximate composition of the experimental diets arepresented on Table 1 and 2 respectively.

Table1. Gross Composition of Experimental Diets

Ingredients	T1 (control)	T2 (15% W.O)	T3 (15% M.O)	T4 (15% R.O)
Maize	52.00	46.00	46.00	46.00
Wheat offal	-	15.00	-	-
Maize offal	-	-	15.00	-
Rice offal	-	-	-	15.00
GNC	24.00	10.00	10.00	10.00
Fish meal	13.45	18.45	18.45	18.45
Palm oil	2.00	2.00	2.00	2.00
*Premix	0.40	0.40	0.40	0.40
Limestone	3.00	3.00	3.00	3.00
Bone meal	4.50	4.50	4.50	4.50
Salt	0.25	0.25	0.25	0.25
Methionine	0.30	0.30	0.30	0.30
Lysine	0.10	0.10	0.10	0.10
TOTAL	**100**	**100**	**100**	**100**
Calculated Nutrients				
Metabolizable energy (Kcal/Kg DM)	2999	2876	2784	2691
Crude Protein (%)	25.34	25.01	24.42	24.09
Crude fibre (%)	2.50	3.89	4.08	4.59
Ether Extract (%)	3.80	3.80	3.80	3.80
Calcium (%)	1.20	1.20	1.20	1.20
Methionine (%)	0.80	0.80	0.80	0.80
Lysine (%)	1.60	1.60	1.60	1.60

Wheat offal (WO), Maize offal (MO), Rice offal (RO)
*Composition of vitamin premix per kg is as follows: Vitamin A, 8000 iu; Vitamin D_3, 1600 iu; Vitamin E 5 iu, Vitamin K 0.200 mg; Vitamins B, Thiamine B_1 0.5mg; Riboflavin B_2 4mg; Pyridoxine B_6 0.015 mg; Niacin 0.015mg; B_{12} 0.01mg; Pantothenic acid 0.5mg; folic acid 0.5mg and Biotin 0.020 mg; Chlorine chloride 0.02 mg; Anti-oxidant 0.125g and Minerals (Mn, Zn, Fe, Cu, I, Si, Co) 0.156g.

Table 2. Proximate Composition of Experimental Diets

Nutrients %	Control	15% W.O	15% M.O	15% R.O
Dry Matter	92.93	93.48	94.51	93.68
Crude Protein	25.25	24.97	23.90	23.08
Crude fibre	3.78	4.01	4.56	5.02
Ash	5.68	6.71	6.98	7.28
Ether Extract	6.21	6.52	6.75	6.98
N F E	61.17	61.93	59.42	60.45

NFE-Nitrogen free extract

Performance data

The amount of feed given and left over was recorded on daily basis and it was used to calculate the feed intake. Before the commencement of the experiment, the initial weight of the birds were taken and the birds were weighed weekly thereafter to obtain weekly weight gain.Feed intake and weight recorded were used to calculate feed conversion ratio (FCR) using the formula below.

Feed conversion ratio (FCR) =feed intake/weight gain

Blood collection

At the end of the study period, 5ml of blood was collected from three birds per treatment through the wing vein and put into bottles containing Ethylene Diaminetetra- acetic Acid (EDTA) to determine the packed cell volume (PVC), red blood cell (RBC), haemoglobin (Hb), and white blood cell (WBC). Blood sample meant for serum biochemical studies were collected into plane bottles (without Anti-coagulant) to enhance serum separation. The blood serum obtained was used to determine total protein (TP), Albumin, Globulin,Glucose and Urea. All the analysis was done at the College Research Laboratory according to the methods described by Kohn and Allen (1995); Schalm *et al.* (1975) and Peters *et al.* (1982).

Carcass and Organs Weight determination

At the end of the study, two turkeys per replicate were selected at random and starved for about 12h to empty thecrops. They were then slaughtered, scalded, plucked and eviscerated. The carcass and internal organs (liver, heart, kidney, gizzard and intestines) were removed, weighed and expressed as a percentage of live weight.

Chemical analysis

Proximate composition of experimental diets was analysed using the methods described by AOAC (2000).

Statistical analysis

Data collected were subjected to analysis of Variance using SAS software (SAS 2008) while significant means were separated with Duncan multiple range test at 5% level of significance.

RESULTS

Table 3 shows the result of the growth performance of grower turkeys fed diets containing different agro industrial by-products. There were significant ($P<0.05$) differences in the daily weight gain, daily feed intake and feed conversion ratio of turkeys. Daily weight gain of turkeys fed 15% W.O was not different ($P>0.05$) from those fed the control diet. Turkeys fed 15% M.O and 15% R.O recorded the lowest daily weight gain ($P<0.05$). Average daily feed intake of turkeys fed control and 15% W.O diets were similar ($P>0.05$) and this was followed by birds fed 15% M.O diet.Birds fed 15% R.O diet recorded the lowest daily feed intake. Feed conversion ratio of turkeys fed control, 15% W.O and 15% M.O diets were not different ($P>0.05$). Turkeys fed 15% R.O recorded the highest value for feed conversion ratio. There were no differences ($P>0.05$) in feed cost per kg and feed cost per kg gain.

Table 4 shows the result of haematological parameters and serum biochemical indices of grower turkeys fed diets containing different agro industrial by-products. There were no differences ($P>0.05$) in the values obtained for haemoglobin, white blood cell and red blood cell across the treatment groups. Packed cell volume was however different ($P<0.05$). Birds fed control and 15% W.O diets recorded higher values for Packed Cell Volume than those fed 15% M.O and 15% R.O based diets. The results of serum biochemical indices show that there were no differences in all the parameters measured ($P>0.05$). The values obtained for Total Protein, Glucose, Albumin, Globulin and urea were not affected ($P>0.05$) by the dietary treatments.

Table 5 shows the results of carcass characteristics of grower turkeys fed diets containing different agro industrial by-products. There were differences ($P<0.05$) in all the parameters measured across the treatment groups. Turkeys fed 15% W.O diet were comparable with the turkeys fed the control diet in all the carcass parameters measured. Turkeys fed 15% M.O and 15% R.O diets performed poorly and were similar ($P>0.05$).

DISCUSSION

There have been wide variations in responses of Chickens to the use of Agro industrial byproducts in poultry diets. These were attributed to differences in quality, varieties, storage periods and climatic conditions to mention but a few. However, there are several literature reports on the inclusion levels of these unconventional, agro by-products in broiler diets without adverse effect on performance in Nigeria (Maikano, 2005; Duru, 2010).

Table 3. Growth Performance of grower Turkeys fed diets containing different agro industrial by-products

Parameters	Control	15% W.O	15% M.O	15% R.O	SEM
Initial weight, g/b	330.90	330.55	330.70	331.04	0.16
Final weight, g/b	1186.67[a]	1100.00[a]	786.67[b]	583.33[c]	22.29
Total weight gain, g/b	855.77[a]	769.45[a]	455.96[b]	252.30[c]	22.33
Average daily weight gain, g/b/d	20.38[a]	18.32[a]	10.86[b]	6.00[c]	4.79
Total feed intake, g/b	2990.00[a]	2750.70[a]	1555.50[b]	1084.0[c]	54.89
Average daily feed intake, g/b/d	71.19[a]	65.49[a]	37.04[b]	25.81[c]	10.13
Feed conversion ratio	3.53[a]	3.60[a]	3.49[a]	4.35[b]	0.27
Cost per kg feed, ₦/ kg	98.85	90.45	89.17	85.91	4.31
Cost per kg gain	178.52	182.41	185.34	188.12	3.20

SEM= Standard Error of Mean, abc means on the same row having different superscripts are significantly different ($p > 0.05$). WO=Wheat offal, MO=Maize offal, RO=Rice offal

Table 4. Haematological parameters and Serum biochemical indices grower Turkeys fed diets containing different agro industrial by-products

Parameters	Control	15% W.O	15% M.O	15% R.O	SEM
Packed cell volume, %	30.21[a]	31.66[a]	23.42[b]	24.30[b]	2.75
Haemoglobin, g/dl	8.97	10.55	8.81	8.77	0.93
White blood cell, $X\ 10^6$	44.60	48.25	49.24	50.68	3.86
Red blood cell, $X\ 10^{12}$	4.16	5.60	5.37	5.89	0.43
Glucose, g/dl	129.50	105.50	126.27	114.00	14.67
Albumin, g/dl	3.06	3.13	2.87	2.39	0.25
Globulin, g/dl	3.59	3.03	3.97	3.46	0.31
Urea, mg/dl	52.16	55.97	53.91	65.78	6.95
Total Protein, g/dl	6.64	6.33	6.84	6.85	0.12

abc = mean with different superscripts within the same row are significantly ($P<0.05$) different. SEM=standard error of mean. WO=Wheat offal, MO=Maize offal, RO=Rice offal

Table 5: Carcass Characteristics of grower Turkeys fed diets containing different agro industrial by-products

Parameters	Control	15% W.O	15% M.O	15% R.O	SEM
Liveweight, g	1376.67[a]	1450.00[a]	850.00[b]	826.67[b]	137.51
Dressedweight, g	1300.00[a]	1326.67[a]	750.00[b]	726.67[b]	150.56
Carcass weight, g	1167.47[a]	1116.02[a]	500.00[b]	470.47[b]	174.25
Dressingpercentage, %	69.10[a]	69.31[a]	51.05[b]	49.62[b]	4.87
Backweight, %	14.57[a]	14.12[a]	13.89[b]	11.88[b]	0.90
Thigh, %	12.17[a]	11.76[a]	9.22[b]	8.68[b]	1.16
Wing, %	13.05[a]	12.55[a]	9.04[b]	8.42[b]	1.15
Drumstick, %	10.44[a]	10.20[a]	8.29[b]	8.01[b]	0.61
Breast, %	18.20[a]	18.03[a]	16.64[b]	15.39[b]	0.94
Neck, %	9.46[a]	9.23[a]	7.06[b]	5.21[b]	1.06
Liver, %	27.15[a]	26.70[a]	18.67[b]	17.26[b]	2.47
Heart, %	7.05[a]	7.45[a]	4.55[b]	4.00[b]	0.86
Spleen, %	0.90[a]	1.10[a]	0.75[b]	0.67[b]	0.70
Gizzard, %	56.70[a]	54.66[a]	47.55[b]	34.40[b]	5.58
Intestine weight, %	11.10[a]	12.10[a]	5.88[b]	4.91[b]	1.55

abc = mean with different superscripts within the same row are significantly ($P<0.05$) different. SEM=standard error of mean. WO=Wheat offal, MO=Maize offal, RO=Rice offal

In this present study, the average daily weight gain, average daily feed intake and feed conversion ratio of turkeys fed the control and 15% W.O diets were similar indicating that grower turkeys at 5-12 weeks of age can utilize wheat offal up to 15% in their diets. Turkeys fed control and 15% W.O diets had higher feed intake than those fed 15% M.O and 15% R.O based diets. The improved weight gain and feed intake observed in 15% W.O based diet may be attributed to the low fiber content and other available nutrients in wheat offal compared with maize offal and rice offal based diets. Morrison (1975) had earlier reported that the protein content of wheat offal is of higher quality than that of maize and it has about 66.9% total digestible nutrients (TDN). The author reported that Wheat offal has 16.4% crude protein and 4.5% fat and usually does not contain more than 10% fibre (Morrison, 1975). Reduction in body weight of grower turkeys fed 15%M.O and 15% R.O diets may be attributed to depressed feed intake that led to low availability of nutrients needed for tissue, organ and body development (Ani and Okorie 2008). The depressed intake may be because of anti-nutrients present in maize offal and rice offal diets which cause irritation and burning of the throat (Sakai, 1979), thereby lowering the intake (Okon *et al*., 2007). Anti-nutritional factors such as trypsin inhibitors and tannins can cause decrease in voluntary feed intake, digestibility of protein and impair the absorption of nutrients in the ingesta thereby resulting in depressed weight gain (Esminger *et al*., 1996). Anti-nutrients interfere with nutrient utilization by forming complexes with the substrate at the site of digestion (Abeke *et al*., 2008). The high content of fibre present in the 15%M.O and 15% R.O diets could also be responsible for the poor performance of the turkeys fedthesediets. Kocher (2001) reported that rice offal is known to contain high level of fibre and low protein and energy and high dietary fibre depresses apparent digestibility of dry matter and nitrogen, decreases daily body weight and increase feed to gain ratio (Longe and Adekoya, 1988). The depressing effect on apparent digestibility has been found to be due to greater rate of passage. Abdelsamie *et al.*(1983) observed that growth and feed utilization decreased with increase in fibre contents in chickens' diet. The authors observed that feed utilization decreased with increasing dietary fibre content of the diet. The result of this study however contradicts the reports of Maikano (2005) who reported that broiler chickens can tolerate up to 20% dietarylevels of rice offal with no adverse effect on growthperformance.

The observation on haematological parameters and serum biochemical indices revealed that there were no differences (P>0.05) in the parameters measured except for PCV. The increase in the values of WBC across the treatment groups is an indication of health hazard. This could be due to infections, inflammatory conditions, stress condition or conditions induced by A.N.Fs content of the feed (Brij*et al*., 1977). Higher levels of White blood cell have been produced in farm animals in a bid to fight against the foreign bodies. Oduguwa (2006) reported that an elevated White blood cell showed that the birds were reacting to one or more dietary factors which in this study could be tannin, NSP and HCN residues present in the fibre sources used. White blood cell played protective roles and associated with the production of antibodies and recognition of foreign substance such as bacteria and viruses.

Turkeys fed 15% M.O and 15% R.O diets recorded lower values for PCV and Haemoglobin. Report had suggested that PCV is beneficial in assessing the protein status and possibly forecasting the degree of protein supplementation at different physiological states (Daramola *et al*., 2005). Keir *et al* (1982) reported that the reduced value of PCV is an indication of poor nutritional status and pointer to iron deficiency or nutritional anaemia. The poor nutritional value of maize offal and rice offal as compared to the control diet could be linked with the low PCV values obtained among turkeys fed 15% M.O and 15% R.O diets in this study. Also, reduced Haemoglobin values obtained among birds fed15% M.O and 15% R.O diets is a confirmation of the poor dietary protein quality. Udo (1987) reported that nutrition was the most important factor affecting haemoglobin levels of the blood. The range of 44.60 – 50.68 X 10^6 observed for WBC, 4.16 – 5.89 X 10^{12} for RBC, 23.42 – 31.66% for PCV and 8.07 – 10.55 g/dl for haemoglobin were within the ranges reported by Obadire *et al.* (2015) for grower turkeys fed diets containing malted sorghum sprouts supplemented with enzyme.All the parameters measured for serum biochemical indices were within the normal range reported by Elizabeth *et al.* (2010) and Obadire *et al.* (2015) for healthy domestic turkeys.

The similarity in the carcass parameters measured between the control and 15% W.O diets is a true reflection of the body weight performance indices and that of tissue development.This indicates that the nutrients supplied by these diets were adequate and produced comparable carcass parameters that were similar. The cut-up parts such as drumstick, thigh and wing weight (expressed as percentage of live weight) of the grower turkeys fed 15% W.O diet followed the performance pattern of the body weight development.The weight of different organs measured were (P<0.05) different across the treatment groups. The values of these organs indicate that there were metabolic differences in turkeys fed 15% M.O and 15% R.O diets. The lower weight of organs observed among turkeys fed15% M.O and

15% R.O diets could also be attributed to the pathological effect of residual of ANFs (tannin, phytin e.t.c) present in the maize offal and rice offal based diets. The lack of effect (P>0.05) in the values of spleen and other organs of grower turkeys fed the control and 15% W.O diets implies that wheat offal can be included in the diet of grower turkeys and the immunity of the birds against disease would not be compromised since spleen is the major source of lymphocytes and the storage site for white and red blood cells (Hetland and Svihus, 2001).

CONCLUSION

Based on the result of this study, it can be concluded that 15% wheat offal can be included in the diet of grower turkeys without adverse effect on growth performance, carcass yield and health status of birds.

Acknowledgement

The authors gratefully acknowledge Ogunwale Segun, Abdulahi Jimoh and Omotayo Oluwadamilola for their assistance in the collection of data pertaining to this study.

REFERENCES

A.O.A.C. 1990. Official Methods of Analysis.13th Ed. Association of Analytical Chemists, Washington DC.

Abeke, F.O., Ogundipe, S.O., Sekoni, A.A., Dafwang, I.I. and Adeyinka, I.A. 2008.Effect of dietary levels of cooked *Lablab purpureus* beans on the performance of broiler chickens. American Journal of Food Technology 3(1):42-49.

Abdelsamie, R.E., Ranawere, K.N.P. and Nano, W.E. 1983. The influence of fibre content and physical texture of the diet on the performance of broilers in the tropics. British Poultry Science 24:383-39.

Adene, D.F. and Oguntade, A.E. 2006. The Structure and Importance of the Commercial and Village based Poultry industry in Nigeria. FAO (Rome) Italy, pp 22 – 28.

Aduku, A.O. 1993. Tropical Feedstuffs Analysis Table. Department of Animal Science. Faculty of Agriculture, Ahmadu Bello University, Zaria. Pp. 196.

Ajaja, K., Agbede, J.O. and Aletor, V.A. 2002. Replacement Value of Sorghum Dust for Maize in Diets for Broiler Chicks.Proc., 27th Annual Conference of the Nigeria Society for Animal Production. March 17 – 21, Federal University of Technology, Akure, pp: 109 – 112.

Ani, A.O. and Okorie, A.U. 2008. Carcass and organ characteristics of Broiler finisher fed diets containing graded levels of processed castor oil bean (*Ricinus communis L*) meal and supplementary DL-methionine. Proceedings 33rd Annual Conference Nigerian Society Animal Production. Ayetoro, March pp:378-381.

ArunaTomar, S. and Singh, K.S. 1999. Femur bone mineralization of chicken fed rice offal based diets. Indian Journal Poultry Science, 34 (3): 332- 336.

Babatunde, B.B. and Oluyemi, A. 2002. Comparative digestibility of three commonly used in fibrous ingredient in maize - soya bean meal – fish meal diet by broiler chicks. Tropical Journal Animal Science, 3 (1): 33- 43.

Brij, M.A., Howard, M.R. and Bhuran, V.V. 1977. Clinical, biochemical and haematological values in normal experimental animals. Massan publishing co. U.S.A. Fr Gef. lugelkunde 54:34-37.

Chae, B.J., Lee, K.H., Lee, S.K. 2002. Effect of feeding rancid rice offal on growth performance and chicken meat quality on broiler chicks. Asian-Australasian Journal Animal Science, 15 (2): 266 – 273.

Daramola, J.O., Adeloye, A.A., Fatoba, T.A. and Soladoye, A.O. 2005. Haematological and biochemical parameters of West African Dwarf goats. Livestock Research for Rural Development, *17*(8): 95. Retrieved January 22, 2014, from http://www.lrrd.org/lrrd17/8/dara17095.htm.

Doki, T. 2007. Benue Gives Land to NNPC. In: Saturday Independent, January 27, Tunde Abatan (ed). 1(86): 44.

Duru. S. 2010. Evaluation of three types of rice offal in broiler diets Ph.D Thesis. Department of Animal Science, Ahmadu Bello University, Zaria. Nigeria.

Elizabeth, M.S., Antonio, C.P., Ivan, M.L., Gislaine, R.V.M., Leonildo, N.J. and Alan, P.T. 2010. Serum biochemical parameters of female bronze turkeys (*Meleagris gallopavo*) during Egg-laying season. Int'l Journal of Poultry Science 9(2): 177-179.

Ensminger, M.E., Oldfield, J.E. and Heinemam, W.N. 1996. Feed and Nutrition. The Ensminger publishing Coy, Clovis, USA.

Hetland, H. and Svihus, B. 2001. Effect of oat hulls on the performance, gut capacity and feed passage time in broiler chickens. British Poultry Science Journal 60: 63-69.

Keir,G.R., Lar, M.K., Horium, R.J. and Ranola, E.E. 1982. Relationship between dietary and biochemical measures of nutritional status. American Journal of Clinical Nutrition. 35: 294-308.

Kohn, R. A. and Allen, M. S. 1995. Enrichment of proteolytic activity relative to nitrogen in preparations from the rumen for in vitro studies. Animal Feed Science Technology, 52:1-14.

Kocher, A. 2001. Enzymatic degradation of non-starch polysaccharides in vegetable proteins in poultry diets. Recent Advances in Animal Nutrition in Australia, 13:163-168.

Leplaideur, M. 2004. Poultry farming – A disease called competition. In: SPORE Magazine, 114: 4 – 5.

Maikano, A. 2005. Effects of graded levels of rice offal in broiler diets M.Sc. Thesis, Department Animal Science, Ahmadu Bello University, Zaria. Nigeria.

Makinde, O.J., Enyigwe, P.C., Babajide, S.E., Atsumbe, J.A., Ibe, E.A. and Samuel, I. 2014. Growth performance and carcass characteristics of finisher broilers fed rice offal based diets supplemented with exogenous enzyme. Greener Journal of Agricultural Sciences 4 (4): 144-149.

Morgan, D.1991. African Farming and Food Processing, March/April, 1991.

Morrison, F.B. 1975. Feeds and feeding.9th (ed.) Morrison publishing company.Vail-Ballou press.Inc. Binghamton, New York, USA. Pp: 290-293.

Nuhu, B., Abba, A. and Amina, L. 2008. The profitability of broiler chickens raised on graded levels of maize offal and wheat offal-based diets. Proceedings of the 13th Annual Conference of the Annual Science Association of Nigeria (ASAN), 15th-19th Sept., 2008, ABU, Zaria, Nigeria Pp: 298-302.

Obadire, F.O., Oso, A.O., Osofowora, A.O., Oduguwa, O.O., Adeyemi, A.O., Jegede, A.V. and Olurunsola, R.A. 2015. Blood chemistry and carcass characteristics of growing turkeys fed diets containing malted gorghum sprouts supplemented enzyme and yeast. Nigeria Poultry Science Journal. 11:88-98.

Oduguwa, O.O. 2006. Utilization of whole pods of *Albiziasaman* in diets of growing rabbits. Nigerian Journal of Animal Production. 33(2): 197-202.

Okon, B.I., Obi, M.B. and Ayuk, A.A. 2007. Performance of quails (*Coturnix coturnix japonica*) fed graded levels of broiled sun-dried taro cocoyam (*Colocasia esculenta*) as a replacement to maize. Agricultural Journal 2(6) 654-657.

Oluyemi, J.A. and Roberts. R.A. 2000. Poultry Production in Warm Wet Climates. Macmillian Press, London. 2nd Edition, p. 692.

Onifade, A.A. and Babatunde, G.M. 1998. Comparison of the utilization of palm kernel meal, brewers dried grains and maize offal by broiler chicks. British Poultry Science. 39:245-250.

Peters, T., Biomont, C. T. and Doumas, B. T. 1982. Protein in serum, urine and cerebrospinal fluid, albumin in serum: In selected methods of clinical chemistry, W.R. Faulkner and S. Meites (eds.) Washington D.C. American Association of Clinical Chemist, Vol 9.

SAS.2008. User's guide: Statistics. Version 12.0. SAS Institute, Inc. Cary. NC, USA. 2008.

Sakai, W.S. 1979. Arid Root Crops, Acridity and Raphids. In: Inglett, G.E. and charalampous, G. (Eds) Trop. Foods chain and Nutrition, New York, Academic Press 1:265.278.

Schalm, O.W., Jain, N.C. and Carrol, E.J. 1975. Veterinary haematology 3rd Edn. Lea and Fabinger, Philadelphia.

Uchegbu, M.C., Etuk, E.B., Okpala, C.P., Okoli, I.C. and Opara, M.N. 2004. Effect of Replacing Maize with Cassava Root Meal and Maize/Sorghum Brewers Dried Grains on the Performance of Starter Broilers. Proceedings of the 9th Annual Conference of Animal Science Association of Nigeria. Sept. 13–16, 2004, Ebonyi State University, Abakiliki, pp. 49 – 51.

Udo, H. 1987. Effect of Cassava peel supplementation in swine ration on performance, haematology and serum metabolites in growing pigs. M.Sc. Thesis submitted to the Department of Animal Sci., University of Ibadan. Pp.112.

18

PERFORMANCE OF MASHONA COWS REARED ON NATURAL RANGELANDS WITH NON-CONVENTIONAL PROTEIN SUPPLEMENTATION IN THE DRY SEASON, ZIMBABWE

[1*]J. Gusha, [1]S. Katsande, [1]P.I. Zvinorova and [2]T. Chiuta

[1]Department of Para clinical Veterinary studies, Faculty of Veterinary Science, University of Zimbabwe, MP167, Mount Pleasant, Harare.
Email: jtgusha@gmail.com
[2]Makoholi Research Institute, P. Bag 9182, Masvingo, Zimbabwe.
**Corresponding author*

SUMMARY

Performance of animals in smallholder farming systems is hindered by feed availability and feed costs that make up 60-70% of total variable cost in a farming system. The use of non-conventional feedstuffs is an alternative which can be adopted to minimize feeding costs. The study was conducted to evaluate performance of animals fed with supplements formulated on-farm in comparison with a commercial prepared supplement. Thirty Mashona cows were grouped according to age and subjected to five treatments, given twice a week from the 1st of September until 31st December for three subsequent years. The five treatment diets, beef survival meal (BSM), urea treated maize stover (UTS), *Leucaena leucocephala* meal (LLM), mixed forage meal (MFM) and natural pastures (NP) were randomly assigned to cows in a complete randomised design (CRD). Average weight gains were determined. Performance in the first year was not significantly different across treatments. In the subsequent years, performance remained low in the NP. Average weight gains increased in the second and third years and was significantly different across treatments ($P<0.05$). Performance of animals supplemented with non-conventional feed was comparable to those offered commercial BFM; hence LLM and UTS can be used as alternative protein supplements especially in resource-constrained farming systems.

Key words: Resource-constrained farming systems; protein supplements; non-conventional protein sources.

INTRODUCTION

In Zimbabwe, communal livestock is a source of livelihoods for rural population (Sibanda and Khombe, 2006). However, livestock performance is constrained by inadequate nutrition (Gusha *et al.,* 2014; Zvinorova *et al.,* 2013). There is perennial feed deficit, both in quantity and quality during the April to November dry period in Zimbabwe. During this dry period most animals become thin and unable to conceive. More frequently, some die in large numbers during a cold spell that is experienced with some October rain showers. Due to the fact that the animals will be in poor body condition, any risk factors predispose to the animals may negatively affect them because of a weakened immunity system. Farmers experience very high animal mortality ranging 10% to 50% annually (Mavedzenge *et al.,* 2005; Gusha *et al.,* 2013a). Most communal farmers cannot afford high quality conventional supplements hence very few farmers supplement their cattle during the dry season (Ngongoni *et al.,* 2006). Animals that are reared in such management system fail to conceive annually and might have three or more years calving interval and calves born in those conditions take longer to reach puberty due to very slow growth rates. Animals reach slaughter weights after four to six years. The capacity of farmers to generate income from livestock production is compromised due low productivity of animals.

It is therefore imperative that research should evaluate any potential affordable supplements in order to drive farmers out of this predicament. Browse and forage legumes have been recommended from *in vitro* and *in vivo* studies potential supplements (Matenga *et al.,* 2003; Baloyi *et al.,* 2009; Gusha *et al.,* 2013b). Farmers use different form of supplements available at farm level and the optimum feeding levels that are sustainable according to level of production are yet to be evaluated or need further evaluation. This study was designed to evaluate performance of animals fed on supplements formulated on-farm. The study also aimed at comparing a commercial supplement to cheap farm-based resources.

MATERIAL AND METHODS

Study site

The experiment was conducted at Makoholi Research Institute located 32 km north of Masvingo town on 19°50' S, 30°47'E and altitude of 1 200 m. The area is characterised by erratic rainfall both within and between seasons; annual mean rainfall is 565 mm with a range of 133 to 1 155 mm. The granite-derived soils consist of 96% sand, 2% silt and 2% clay are inherently infertile and plant growth is severely limited by the unavailability of nitrogen and phosphorus.

Experimental diets

Leucaena leucocephala herbage was harvested from the trees that were grown around the farm. The herbage was harvested by cutting the trees at 30cm above the ground and then removed the twigs and leaves in October, February and in April. The harvested material was air dried under shade for at least a week. *M. atropurpureum* herbage was harvested at flowering stage and air-dried under shade after which it was milled prior to feeding. Maize stover was collected after harvesting and milled before treatment with dissolved urea fertiliser containing 46% nitrogen. Fifty kilograms of urea fertiliser was dissolved in 200 litres of water to produce 25%w/v. The solution was sprinkled over 1000 kg milled maize stover and stored in a pit silo covered with plastic and soil for six weeks to create anaerobic conditions after which the treated stover was sun-dried for three days prior to feeding. Mixed forage meal (MFM) was produced mixing a third of each of *L. Leucocephala, M. atropurpureum* and urea-treated maize stover. The experimental diets fed to experimental animals were as follows; beef survival meal (BSM), Urea treated maize stover (UTS), *L. Leucocephala* meal (LLM), Mixed forage meal (MFM) and those on natural pastures (NP) without feed supplement.

Animals feeding and experimental design

Thirty calf nursing cows were grouped according to age and randomly allocated to five treatments in a complete randomised design (CRD). Each animal was an experimental unit and was individually fed the experimental diets on Monday and Friday every week from the 1st of September until 31st December for three years. All experimental animals were dosed with a commercial dosing drug at the beginning of feeding experiment, in February and May each year. The feed was mixed with 100g of molasses to improve palatability. Table 1 shows the feeding procedure that was followed during the first year while Table 2 shows the procedure for year 2 and year 3. Initial weights of the animals were taken at the beginning of the trial and every fortnight for 4 months.

Table 1. Quantities of experimental diets offered to experimental animal in year 1

Treatment	Quantity offered		
	Monday	Wednesday	Friday
Beef survival meal (BSM)	3.5kg + 100g molasses	100g salt	3.5kg + 100g molasses
Urea-treated Stover (UTS)	3.5kg + 100g molasses	100g salt	3.5kg + 100g molasses
L. Leucocephala meal (LLM)	3.5kg + 100g molasses	100g salt	3.5kg + 100g molasses
Mixed forage meal (MFM)	3.5kg + 100g molasses	100g salt	3.5kg + 100g molasses
Natural pasture (NP)	100 grams molasses	100g salt	100g molasses

Table 2. Quantities of experimental diets offered during year 2 and year 3

Treatment	Quantity offered		
	Monday	Wednesday	Friday
Beef survival meal (BSM)	3.5kg + 100g molasses	100g salt	3.5kg + 100g molasses
Urea-treated Stover (UTS)	3.5kg + 100g molasses	100g salt	3.5kg + 100g molasses
L. Leucocephala meal (LLM)	3.5kg + 100g molasses	100g salt	3.5kg + 100g molasses
Mixed forage meal (MFM)	3.5kg + 100g molasses	100g salt	3.5kg + 100g molasses
Natural pasture (NP)			

Chemical analysis

Feeds were dried in a forced air oven at 60°C for 48 hours to determine DM. Nitrogen content in the feed and in the residues was analysed by Kjeldahl method (A.O.A.C, 1990). Neutral Detergent Fibre (NDF) and Acid Detergent Fibre (ADF) were analysed according to Goering and van Soest (1970). Total ash was obtained by igniting a dried sample in a muffle furnace at 600°C for 24 hours, calcium and phosphorous were determined by the EDTA (Kaur, 2007) and spectrophotometer (Danovaro, 2009) methods, respectively.

Statistical analysis

The data was analysed using the general linear models (GLM) procedure for repeated measures of SAS (SAS, 2010). Means were separated with the adjusted Tukey methods. The following model was used:

$$\mathbf{Y_{jk} = \mu + T_k + \varepsilon_{jk}}$$

Where:
Y_{jk} was response variable being (Weight gain/loss);
μ was overall mean common to all observations;
T_k is effect of the kth treatment (k=1, 2, 3, 4, 5);
ε_{jk} are the random residuals with a distribution of $N\sim(0;\sigma^2_E)$.

RESULTS

Chemical composition

The nutritional composition of supplements and rangeland biomass is presented in Table 3. The diets were significantly different on crude protein content as well as acid detergent fibre. BSM had higher total digestible nutrient (TDN) compared to other farm based feed supplements. Forage supplements had higher calcium and phosphorus content compared to BSM and urea-treated maize stover.

Table 3. Chemical composition of experimental diets

Experimental diet	% nutrient composition						
	DM	CP	ADF	NDF	P	Ca	Ash
Beef survival meal (BSM)	88.0	11.0	ND	ND	0.5	0.85	ND
Urea-treated Stover (UTS)	95.7	15.7	42.7	71.8	0.3	0.6	9.7
L. Leucocephala meal (LLM)	90.3	21.7	29.6	33.2	0.7	2.1	8.1
Mixed forage meal (MFM)	96.3	13.3	31.5	43.1	1.1	0.8	7.5
Natural pasture (NP)	97.6	3.2	48.2	76.2	0.6	0.3	8.6

Calving rates and weight changes trends

The trend in weight changes is presented in Tables 4, 5 and 6. In year one there was no significant weight loss in all the treatment groups. All cows maintained weight during the period. However, significant (P<0.05) weight loss was observed in animals that were not supplemented. It was also observed that animals with non-conventional protein and non-nitrogen supplement compared well with animal supplemented with commercial diet BSM. It was also observed that animals supplement with BSM and LLM had 83.3% calving rate compared to 66.7% for UTS and MFM whilst animals that were not supplement had 50% calving rate during the study period. No animal loses were recorded during the three year study period though it was observed that animal loss weight as from April but stabilises with the weaning period in June and then start losing weights in August up to the beginning of rain season.

DISCUSSION

In the first year of the experiment the least significant mean weight for experimental animals were not significantly different. The animals in the natural pasture NP were receiving molasses and salt at the same level as those on LLM, UTS, MFM and BSM. Therefore, there was no significant mean weight differences observed. Molasses can be used as an energy source for livestock, particularly in situations where grains are unavailable (Chaudhary *et al.*, 2001). Moreover, it is a rich source of calcium (about 0.9 % DM) due to the addition of calcium hydroxide during processing (Blair, 2007). Studies showed equal or greater animal performance when molasses was substituted for maize grain (Morales *et al.*, 1989). Hence in low amounts of molasses in a roughage-based diet NP stimulate rumen fermentation and the rumen cellulolytic potential is maintained or improved with low quality forage diets. However, molasses fed in large amounts is toxic to livestock (Pérez, 1995; Mesfin and Ledin, 2004). Sodium in the salt improve herbage digestibility and promoted growth of bacteria that digest fibres in the rumen. An increase in sodium recycling in the saliva reduces rumen acidity, the cow will spend more time ruminating. Consequently, if there is enough herbage on the farm animals consuming salt will be able to perform well during the dry season.

Ruminants consuming low-quality forage during the dry period (NP) deficient in CP encounter the problem of inadequate ruminal N that hinders microbial growth and as a result, decreases ruminal fermentation and the quantity of potentially absorbable N presented to the small intestine (Bohnert *et al.*, 2002a; Mupenze *et al.*, 2009). The dry season causes nutritional stress (Matenga *et al.*, 2003; Ngongoni *et al.*, 2007) and consequently decreases animal productivity (Souza *et al.*, 2010; Allam *et al.*, 2012). The decline in growth rates of cattle in dry season delays the attainment of slaughter weight and adversely affects the quality and quantity of beef (Muchenje *et al.*, 2008). Supplemental degradable intake protein (LLM, UTS, MFM and BSM) provides the main source of N for the growth of ruminal microorganisms.

Table 4. Least square mean weight for experimental animals during the first year feeding period in kilograms (mean ±se)

Year 1	Treatment					
	NP	LLM	UTS	MFM	BSM	LSD
Weight 1	243.2a±13.2	239.0a±13.2	244.6a±13.2	244.3a±13.2	250.2a±13.2	54.8
Weight 2	247.0a±13.1	246.7a±13.1	249.5a±13.1	248.5a±13.1	257.7a±13.1	54.5
Weight 3	250.5a±12.9	253.5a±12.9	252.8a±12.9	252.8a±12.9	267.3a±12.9	53.9
Weight 4	252.7a±12.8	261.3a±12.8	261.5a±12.8	258.8a±12.8	274.0a±12.8	53.3
Weight 5	255.5a±12.3	267.7a±12.3	268.0a±12.3	264.8a±12.3	278.3a±12.3	51.5
Weight 6	261.0a±11.9	275.8a±11.9	274.2a±11.9	271.2a±11.9	285.2a±11.9	49.6
Weight 7	268.3a±11.4	281.2a±11.4	279.7a±11.4	276.2a±11.4	293.2a±11.4	47.3
Weight 8	278.7a±11.5	287.8a±11.5	284.7a±11.5	278.2a±11.5	298.7a±11.5	47.7
Weight 9	286.3a±10.5	294.3a±10.5	291.5a±10.5	286.7a±10.5	305.0a±10.5	43.5
Lactating cows	6	6	6	6	6	
Number of calves	6	6	6	6	6	

Different lower case letters in the same row indicate significant different at P<0.05

Table 5. Least square mean weight for experimental animals during the year 2 feeding period in kilograms (mean ±se)

Year 2	Treatment					
	NP	LLM	UTS	MFM	BSM	LSD
Weight 1	261.2b±15.9	320.3ab±15.9	268.7ab±15.9	301.0ab±15.9	328.7a±15.9	65.9
Weight 2	244.8c±14.8	317.2ab±14.8	265.2bc±14.8	297.5abc±14.8	343.2a±14.8	61.5
Weight 3	239.0c±14.1	310.3ab±14.1	267.2bc±14.1	286.8abc±14.1	333.8a±14.1	58.5
Weight 4	232.3c±13.5	305.0ab±13.5	257.0bc±13.5	284.0abc±13.5	336.0a±13.5	56.0
Weight 5	231.5c±12.8	310.2ab±12.8	263.3bc±12.8	280.7bc±12.8	335.7a±12.8	53.3
Weight 6	233.5c±13.9	308.3ab±13.9	272.8bc±13.9	304.3ab±13.9	340.3a±13.9	57.9
Weight 7	247.0c±13.4	324.5ab±13.4	291.2bc±13.4	309.0b±13.4	365.5a±13.4	55.7
Weight 8	267.7c±13.9	340.5ab±13.9	300.7bc±13.9	335.7ab±13.9	375.8a±13.9	57.5
Weight 9	284.8c±13.9	349.3ab±13.9	314.0bc±13.9	340.8abc±13.9	388.2a±13.9	57.5
Lactating cows	3	5	4	4	5	
% calving rate	50	83.3	66.7	66.7	83.3	

Different lower case letters in the same row indicate significant different at P<0.05

Table 6: Least square mean weight for experimental animals during the year 3 feeding period in kilograms (means ±se)

Year 3	Treatment					LSD
	NP	LLM	UTS	MFM	BSM	
Weight 1	276.3a±13.8	304.2a±13.8	286.0a±13.8	294.5a±13.8	307.8a±13.8	57.3
Weight 2	268.0a±14.0	290.7a±14.0	280.8a±14.0	294.3a±14.0	307.0a±14.0	57.9
Weight 3	264.8a±14.2	283.5a±14.2	274.2a±14.2	287.3a±14.2	292.8a±14.2	58.8
Weight 4	247.7a±10.8	289.2a±10.8	265.5a±10.8	283.0a±10.8	270.3a±10.8	44.9
Weight 5	231.2c±8.8	304.8a±8.8	281.2ab±8.8	274.8ab±8.8	266.3bc±8.8	36.7
Weight 6	240.7c±9.1	321.7a±9.1	291.5ab±9.1	297.0ab±9.1	272.0bc±9.1	37.9
Weight 7	260.0b±8.7	330.0a±8.7	299.7a±8.7	304.8a±8.7	313.3a±8.7	36.1
Weight 8	274.2b±9.3	348.0a±9.3	320.2a±9.3	324.3a±9.3	329.5a±9.3	38.4
Weight 9	286.3b±9.3	358.3a±9.3	331.3a±9.3	337.0a±9.3	339.2a±9.3	38.4
Lactating cows	3	4	4	4	4	
% calving rate	50	66.7	66.7	66.7	66.7	

Different lower case letters in the same row indicate significant different at P<0.05

Leucaena is rich in protein, minerals and vitamin content and it enhances voluntary feed intake, nutrient digestibility, rumen fermentation and microorganisms (Lazzarini *et al.,* 2010; Kang *et al.,* 2012). Protein level ranges around 292 g/kg CP in the leaf meal and 220.3 g/kg CP in forage (Garcia *et al.,* 1996; Gusha *et al.,* 2013; Chakoma, 2012). Moreover, it contains condensed tannin content of 10.1–10.5 g/kg that can protect protein from rumen microbial degradation and reduce methane production (Kang *et al.,* 2012). The effect of supplementing with LLM was the same as using the conventional BSM in the study. Therefore, Leucaena can be used as a substitute to beef survival meal to supplement ruminants in the dry period because it is a low-cost technology.

The animals were only supplemented 2 days a week and a significant difference was noted because decreasing the frequency of protein supplementation to ruminants consuming low-quality forage has been shown to result in acceptable levels of performance with only minimal impacts on nutrient intake and digestibility (Bohnert *et al.,* 2002b). Legume supplements supply ruminal microorganisms with a readily available source of nitrogen (N) that enables them to breakdown basal diets efficiently (Edwards *et al.,* 2012). Therefore, animals supplemented on LLM and MFM treatments performed better than the animals on the natural pasture (NP) alone in the second and third year. UTS supply a readily degradable source of N to rumen microbes resulting in increased fermentation and hence increased utilization of the basal diet (Lazzarini *et al.,* 2010;

Yousuf *et al.,* 2007). The UTS provides the capacity to maintain N efficiency which is a consequence of the N recycling ability of ruminants (Mahesh and Mohini, 2013; Mesfin and Kebede, 2011).

Calving rates were significantly different despite weight changes being not significantly different in year one. This indicates that molasses and salt alone cannot stimulate oestrus cycle hence low conception rate. Animals on NP had fewer calves born over the three year, which shows the low productivity of animal without supplements due to non availability of supplements or high cost of supplements. All the non-conventional and non-nitrogen supplements used demonstrated their potential in addressing nutritional changes in resources constrained framers.

CONCLUSION

An average weight gain in the first year was lowest in the group fed on pasture, though they were not significantly different to those in supplemented groups. However, in the subsequent years the performance of the animals improved greatly. The performance in all the different supplementation diets was comparable to the commercial beef meal; hence the adoption and use of the non-conventional feed supplements such as Leucaena and urea-treated stover can be promoted in periods of feed deficit.

REFERENCES

A.O.A.C. (Association of Official Analytical Chemists), 1990. Official Methods of Analysis. 15th edition 2200 Wilson Boulevard Arlington Virginia 22201. U.S.A.

Allam, A.M.,Bakhashwain, A.A., Nagadi, S.A. and Sallam, S.M.A., 2012. Nutritive value assessment of some sub-tropical browses grown in arid region using *in vitro* gas production technique. Journal of Food, Agriculture & Environment 10 (2), 1339-1343

Baloyi, J.J., Hamudikuwanda, H. and Ngongoni, N.T., 2009. Estimation of true intestinal digestibility of dry matter, nitrogen and amino acids of cowpea and silverleaf desmodium forage legumes and Brachystegia spiciformis (musasa) browse legume, African Journal of Range and Forage Science, 26(2), 51-57.

Blair, R., 2007. Nutrition and feeding of organic pigs. Cabi Series, CABI, Wallingford, UK.

Bohnert, D. W., Shauer, C. S. C., Bauer, M. L. and DelCurto, T., 2002b. Influence of rumen protein degradability and supplementation frequency on steers consuming low-quality forage: I. Site of digestion and microbial efficiency. Journal Animal Science. 80, 2967–2977.

Bohnert, D.W., Schauer, C. S., Falck, S. J. and DelCurto, T., 2002a. Influence of rumen protein degradability and supplementation frequency on steers consuming low-quality forage: II. Ruminal fermentation characteristics. Journal of Animal Science, 80, 2978-2988.

Chakoma, C.I., 2012. Sustainable forage production strategies for small scale livestock production in Zimbabwe. International Journal of Agriculture Innovations and Research, 1(3), 85-90.

Chaudhary, L. C., Sahoo, A., Agarwa, N.l; Kamra, D. N. and Pathak, N. N., 2001. Effect of replacing grain with deoiled rice bran and molasses from the diet of lactating cows. Asian-Australian Journal of Animal Science, 14 (5), 646-650. http://www.feedipedia.org /node/2877.

Danovaro, R., 2009. Total organic carbon, total nitrogen and organic phosphorus in marine sediments: In Methods for the study of deep-sea sediments, their functioning and biodiversity. Taylor and Francis Group, LLC.

Edwards A., Mlambo V., Lallo C.H.O., Garcia G.W., Diptee M., 2012. *In Vitro* Ruminal Protein Degradability Of Leaves From Three Tree Species Harvested At Two Cutting Intervals. Online Journal of Animal Feed Research, 2(2), 224-230.

Garcia, G.W., Ferguson, T.U., Neckles, F.A., Archibald, K.A.E., 1996. The nutritive value and forage productivity of Leucaena leucocephala. Animal Feed Science and Technology 60, 29–41.

Goering, H.K. and Van Soest, P. J., 1970. Forage Fiber Analysis USDA. Agricultural Handbook 379, USDA, Washington, DC.

Gusha, J., Katsande, S., Zvinorova, P.I. and Tavengwa, I. 2013a. Calving period affects cow and calf performance in semi-arid areas in Zimbabwe. IOSR Journal of Agriculture and Veterinary Science, 3(3), 61-65.

Gusha J., Ngongoni NT .and Halimani, T.E. 2013b. Nutritional composition and effective degradability of four forage trees grown for protein supplementation. Online Journal of Animal Feed Research, 3(4), 170-175.

Gusha J., Manyuchi, C.R., Imbwayaro-Chikosi, I.V., Hamandishe, V., Katsande, S. and Zvinorova, I.P. 2014. Production and economic performance of f1-crossbred dairy cattle fed non-conventional protein supplements in Zimbabwe Tropical Animal

Health and Production, 46 (1), 229-234.

Kang, S., Wanapat, M., Pakdee, P., Pilajun, R., and Cherdthong, A., 2012.Effects of energy level and Leucaena leucocephala leaf meal as a protein source on rumen fermentation efficiency and digestibility in swamp buffalo. Animal Feed Science and Technology 174, 131– 139.

Kaur, D., 2007. Development of a cheap and rapid method to determine calcium in milk fraction in an industrial environment, Master of Applied Science Thesis Auckland University of Technology, Auckland, New Zealand.

Lazzarini I., Paulino R.P.V. and Queiroz, A.C., 2010. Intake and digestibility in cattle fed low-quality tropical forage and supplemented with nitrogenous compounds. Tropical Animal Health and Production, 42, 1471-1479.

Mahesh, M.S. and Mohini, M., 2013. Biological treatment of crop residues for ruminant feeding: A review. African Journal of Biotechnology, 12(27), 4221-4231.

Matenga, V.R., Ngongoni, N.T., Titterton, M., and Maasdorp, B.V., 2003. Mucuna seed as a feed ingredient for small ruminants and effect of ensiling on its nutritive value. Tropical and Subtropical Agroecosystem, 1(2-3), 97-105

Mavedzenge, B.Z., Mahenehene, J., Murimbarimba, F., Scoones, I. and Wolmer, W., 2005. Changes in the Livestock Sector in Zimbabwe Following Land Reform: The Case of Masvingo Province. A Report of a Discussion Workshop. IDS: Brighton

Mesfin, R. and Kebede, G., 2011. On-farm experience of feeding urea-molasses treated barley straw to crossbreed dairy cows in Jeldu District, highlands of Ethiopia. Livestock Research for Rural Development, 23, Article #41. August 12, 2013, available at http://www.lrrd.org/lrrd23/2/mesf23041.htm

Mesfin, R. and Ledin, I., 2004. Comparison of feeding urea-treated teff and barley straw based diets with hay based diet to crossbred dairy cows on feed intake, milk yield, milk composition and economic benefits. Livestock Research for Rural Development, 16, Article. # 104. (August 9, 2013), available at http://www.lrrd.org/lrrd16/12/mesf16104.htm

Morales, J. L., Van Horn, H. H. and Moore, J. E., 1989. Dietary interaction of cane molasses with source of roughage: intake and lactation effects. Journal of Dairy Science, 72 (9): 2331-2338.

Muchenje, V., Dzama, K., Chimonyo, M., Raats, J. G. and Strydom, P. E. 2008. Meat quality of Nguni, Bonsmara and Angus steers raised on natural pasture in the Eastern Cape, South Africa. Meat Science 79:20-28.

Mupenzi, M., Karenzi, E., Kanani, T. and Lussa Birasa, A., 2009. Uses of supplement levels of *Stylosanthes scabra* (Stylo) leaf meal on milk yield of A*nkole cows.* Livestock Research for Rural Development. Volume 21, Article #63. (December 26, 2013), available at http://www.lrrd.org/lrrd21/5/muti21063.htm

Ngongoni, N.T., Mapiye, C., Mupeta, B., Mwale, M. and Chimonyo, M., 2007. Potential of farm-produced crop residues as protein sources for small-medium yielding dairy cows. African Journal of Agricultural Research, 2, 309-317.

Ngongoni, N.T., Mapiye, C., Mwale, M. and Mupeta, B. 2006. Factors affecting milk production in the smallholder dairy sector of Zimbabwe. Livestock Research for Rural Development, 18, Article #89. (August 20, 2013), available at http://www.lrrd.org/lrrd18/6/ngon18089.htm.

Pérez, R., 1995. Molasses. In: Tropical Feeds and Feeding Systems, First FAO Electronic Conference, 1995). http://www.fao.org/ag/AGA/AGAP/FRG/ECONF95/HTML/MOLASSES.HTM

SAS Institute, 2010. SAS/STAT user's guide. 9 ed. (SAS Institute, Cary).

Sibanda, S. and Khombe, C.T., 2006**.** Livestock research and development In Rukuni, M, P. Tawonezvi, C. Eicher, with M. Munyuki-Hungwe and P. Matondi 2006. Zimbabwe agricultural revolution revisited, University of Zimbabwe, Zimbabwe

Souzza M.A., Detmann E., Paulino M.F., Sampaio, C.B., Lazzarini I. and Filho S.V., 2010. Intake, digestibility and rumen dynamics of neutral detergent fibre in cattle fed low quality tropical forage and supplement with nitrogen and/or starch. Tropical Animal Health and Production, 42, 1299-1310.

Yousuf, M.B., Belewu, M.A., Daramola, J.O. and Ogundun, N.I., 2007. Protein supplementary values of cassava-, leucaena- and gliricidia-leaf meals in goats fed low quality *Panicum maximum* hay. Livestock Research for Rural Development, 19, Article #23, (October 8, 2013) available at http://www.lrrd.org/lrrd19/2/yous19023.htm

19

EFFECT OF DIETARY INCLUSION OF PURSLANE (*Portulaca oleracea L.*) ON YOLK OMEGA-3 FATTY ACIDS CONTENT, EGG QUALITY AND PRODUCTIVE PERFORMANCE OF RHODE ISLAND RED HENS

Esther Evaris[1], Luis A. Sarmiento-Franco[1*], Jose C. Segura-Correa[1], Concepción M. Capetillo-Leal[1]

[1]Departamento de Nutrición Animal, Facultad de Medicina Veterinaria y Zootecnia, Universidad Autónoma de Yucatán, Carretera Mérida-Xmatkuil km 15.5. Apdo. 4-116 Itzimná, C.P. 97100, Mérida, Yucatán, México.
Email: luis.sarmiento@uady.mx
** Corresponding author*

SUMMARY

This study aimed to determine the effects of dietary inclusion of 100 g/kg and 200 g/kg purslane meal on yolk ω-3 fatty acids content, egg quality and productive performance in Rhode Island Red laying hens. Sixty-four hens of 30 weeks of age, with an average body weight of 2100 g, were randomly allocated into four treatments: 0 g/kg (negative control), 100 g/kg and 200 g/kg purslane meal, and 14.7 g/kg flaxseed oil (positive control) in the diet. Hens were individually housed in cages equipped with feeders and drinking troughs. They received food and water *ad libitum* and 18 h of light during 56 days. The inclusion of 200 g/kg purslane and 14.7 g/kg flaxseed oil in laying hens diet showed higher ($P < 0.05$) content of ω-3 fatty acids (α-linolenic acid and docosahexaenoic acid) in egg-yolks compared with those of the groups that received 0 and 100 g/kg purslane. Treatments did not influence ($P>0.05$) parameters of egg quality such as egg longitudinal and transverse diameter, albumin and yolk diameter, yolk height, and yolk weight. Hens that received 200 g/kg purslane in their diet increased ($P < 0.05$) feed intake, egg production, egg mass, and feed conversion compared with those fed with the negative control. It was concluded that the dietary inclusion of both 100 g/kg and 200 g/kg purslane increases yolk ω-3 fatty acids content without any adverse effect on egg quality traits and productive performance of Rhode Island Red hens.

Keywords: ω-3 fatty acids; egg quality; purslane; laying hens; flaxseed oil.

INTRODUCTION

The ω-3 fatty acids ALA, EPA and DHA are essential nutrients from the standpoint of human health (Harris et al., 2009), they have been proven effective in preventing cancer (Pandalai et al., 1996) and cardiovascular problems and lowering blood triglycerides level. These three fatty acids are important in human diet; that is why Simopoulos et al. (1999) recommended a minimum intake of 2.22 g/d ALA; 0.65 g/d EPA and DHA for adults. These amounts should be higher in infants and elderly people. Nowadays, the consumption of egg yolk enriched with ω-3 fatty acids could be a good alternative to increase consumption of these essential fatty acids. Several studies showed that the inclusion of ingredients rich in ω-3 fatty acids in laying hens diet had lead to the incorporation of these nutrients in egg-yolks (Zotte et al., 2005; Betancourt and Díaz, 2009).

Purslane (*Portulaca oleracea* L.) is one of the most widespread herbaceous in the world. It yields up to 20 tons of fresh matter/ha (Cros et al., 2007). Besides of being a non-toxic plant (Gao et al., 2010), purslane is one of the richest plants in ω-3 fatty acids content (Mohamed and Hussein, 1994; Simopoulos 2004). In addition, purslane is a good source of vitamins and antioxidants. One hundred grams of fresh purslane leaves contain approximately 0.3-0.4 g ALA; 0.0122 g α-tocopherol (vitamin E); 0.0266 g ascorbic acid (vitamin C); 0.0019 g β-carotene (precursor of vitamin A) and 0.0148 g glutathione (Simopoulos et al., 1992). The abundance of nutrients in purslane plant makes it an important resource to be used in laying hens diet. The aim of this study was to determine the effects of dietary inclusion of either 100 or 200 g/kg purslane meal on egg yolk ω-3 fatty acids content, egg quality and productive performance of Rhode Island Red hens.

MATERIALS AND METHODS

Animals and diets

The study was conducted for 56 days at the Faculty of Veterinary Medicine and Animal Science (FMVZ) of the University of Yucatán (UADY), in Yucatán, Mexico. Sixty-four Rhode Island Red hens of 30 weeks of age and 2100 g body weight were randomly allocated into four dietary treatments: 0 g/kg (as negative control), 100 g/kg and 200 g/kg of purslane meal and 14.7 g/kg flaxseed oil (positive control). Each treatment consisted of 16 hens. Hens were individually housed in cages with individual feeding and drinking troughs; they received food and water *ad libitum* and 18 h of light. Fresh purslane leaves were collected at the FMVZ-UADY in a flat landscape (Rendzinas and Lithosols soils), dried at 60 °C for 72 h, ground (3 mm mesh) and added in proportions of 100 and 200 g/kg to the diets. Flaxseed oil was purchased from a Canadian provider. Diets were formulated to be isonitrogenous with 160 g/kg of crude protein and isocaloric with 11.7 MJ/kg of metabolizable energy. Ingredients, calculated composition, and fatty acids composition of experimental diets are shown in Table 1.

Samples preparation and oil extraction

During days 27-28 (sample one) and 55-56 (sample two) of the experiment, 24 fresh eggs from each treatment and sampling period (a total of 192 eggs) were randomly collected for fatty acids analyses; from eggs were clustered into 6 samples of 4 egg yolks each, and homogeneously mixed. From this mixture, 10 g were taken, and oil was extracted according to the method proposed by Bligh and Dyer (1959). The lipids extracted were kept at 4 °C until analysis.

Fatty acids composition

FAMEs were prepared according to the method proposed by Joseph and Ackman (1992) and quantified by gas chromatography using a Varian 450-GC gas chromatographer fitted with a flame ionization detector and a TR-FAME capillary column (120m, × 0.25 mm i.d., 0.25 μm film thickness thermo P/N:260M166L Thermo Scientific, Asheville, NC, USA). The conditions were as follow: detector temperature, 280 °C; injector temperature, 250 °C; column temperature, 150 °C for 0.5 minute at 10 °C/minute increased to 180 °C, at 1.5 °C/minute increased to 220 °C, and at 30 °C/minute increased to 260 °C and held at the final time of 46.50 minutes. Helium was used as the carrier and make-up gas at 1.3 μL/minute and 25 μL/minute, respectively. The injection volume was 1μl with a split ratio of 1:5. FAMEs were identified by comparison with retention times of standards Supelco® 37 Component FAME Mix (Sigma-Aldrich 47885; Saint Louis, MO, USA), Fluka α- Linolenic acid (Sigma-Aldrich 62160, Saint Louis, MO, USA), Fluka Eicosapentaenoic Acid Methyl Ester (Sigma-Aldrich 17266; Saint Louis, MO, USA), and Fluka Docosahexaenoic Acid Methyl Ester (Sigma-Aldrich 05832; Saint Louis, MO, USA). Galaxie data integration system was used to integrate peak areas. Yolk ω-3 fatty acids content were calculated using Heptadecanoic acid (C17:0) as an internal standard.

Table 1. Ingredients, calculated composition and fatty acids composition (g/kg of total fatty acids, as dry matter) of the experimental diets.

	Experimental diets (g/kg)			
	Purslane			Flaxseed oil
Ingredients	**0**	**100**	**200**	**14.7**
Sorghum	694.2	632.5	561.9	645.4
Flaxseed oil[a]	0	0	0	14.7
Purslane[bc]	0	100	200	0
Soya meal	200.7	175.3	144.8	196.2
Wheat bran	11	0	0	51.3
Corn oil	1.6	1.7	4.8	0
Salt	2.5	2.5	2.5	2.5
Premix minerals[d]	0.5	0.5	0.5	0.5
Premix vitamins[d]	0.5	0.5	0.5	0.5
Calcium carbonate	89	87	85	88.9
Calculated Composition				
Crude protein	160	160	160	160
Metabolizable energy (MJ/kg)	11.7	11.7	11.7	11.7
Crude fiber	29.2	28.3	28.1	28.2
Lysine	7.5	7.5	7.7	76
Methionine + Cystine	59	59	59	59
Methionine	0.32	0.33	0.33	0.32
Calcium	35	35	35	35
Total phosphorus	0.30	0.30	0.27	0.27
Linoleic acid	18.4	15	13.4	19.1
Linolenic acid	0	4.1	8.2	8.3
Fatty Acids Composition (g/kg fat)				
Palmitic	190.1	170.4	164	175.7
Stearic	32.9	34.5	34.3	45.8
Oleic	336.2	290.5	285.4	298.9
Linoleic	415.8	460.1	448.2	348.2
α-Linolenic	24.9	44.5	64.1	129.7

[a]Fatty acids profile (g/L FAMEs) of flaxseed oil: C16:0, 50.5; C17:0, 42.1; C18:0, 34.3; C18:1(ω-9c), 158.8; C18:2(ω-6c), 161.9; α-C18:3(ω-3), 549.7; C20:0, 1.7; C22:0, 0.7; C22:6(ω-3), 0.3.
[b]Fatty acids profile (g/L FAMEs) of dry purslane: C16:0, 240.6; C18:0, 72.4; C18:1(ω-9c), 110; C18:2(ω-6c), 220; C18:2(ω-6t), 9.5; α-C18:3(ω-3), 312.3; C20:0, 15.6; C22:0, 13.4; C22:6(ω-3), 6.1.
[c]Proximate analysis (g/kg) of purslane as dry matter : dry matter, 89.7; ash, 213; crude protein, 192; crude fiber 62.6; ether extract, 29.4; nitrogen free extract, 503.
[d]Minerals and Vitamins Premix: Content kg^{-1} of diet: Manganese, 65 mg; iodine, 1 mg; iron, 55 mg, copper, 6 mg; zinc, 55 mg; selenium, 0.3 mg; [b]: Content kg^{-1}: vitamin A, 8000 UI; vitamin D, 2500 UI; vitamin E, 8 UI; vitamin K, 2 mg; vitamin B12, 0.002 mg; riboflavin, 5.5 mg; pantothenate of calcium, 13 mg; niacin, 36 mg; choline, 500 mg; folic acid, 0.5 mg; thiamine, 1 mg; pyridoxine, 2.2 mg; biotin, 0.05 mg.

Egg quality

From each treatment, 16 eggs were collected to determine the effect of the experimental diets on egg quality traits such as egg longitudinal and transverse diameter (cm), albumin diameter (cm), yolk diameter (cm), and yolk height (cm) measured with a Vernier and yolk weight (g) measured with an electrical scale.

Productive performance

Egg production was recorded daily and calculated at the end of each week. Daily, an electronic scale was used to weigh the eggs individually according to their replication and treatment. Egg mass was estimated as grams of eggs produced by hens per day. Feed intake was calculated by the difference between offered and refused feed, weighing the refused feed every 72 h.

Feed conversion was determined weekly as grams of feed consumed for a gram of egg produced. Hens' body weight was estimated as weight at the end of the experiment.

Statistical analysis

The statistical replicates for yolk's ω-3 fatty acids content, egg quality and productive performance were 6 samples, 16 eggs, and 16 hens by treatment, respectively. Data were analyzed using analysis of variance (ANOVA) followed by Tukey multiple comparison test. P-values <0.05 were considered as statistically significant. It should be noted that because of no statistically significant difference between samples one and two, analyzed data were the average of both samples. All statistical tests were performed with the SAS program (SAS, 2000).

RESULTS

Yolk ω-3 fatty acids composition

The effects of experimental diets on yolk's *ω*-3 fatty acids content are shown in Table 2. Yolk ALA's content increased as purslane in the diet increased. Eggs from hens that were fed with the flaxseed oil diet were richer in ALA than those produced by hens that received the other treatments. The difference between treatments was significant. On the other hand, DHA content increased (P=0.001) in egg-yolks produced by hens that fed the diet supplemented with 200 g/kg purslane meal compared with those from hens that received the diets with 0 or 100 g/kg purslane meal. Flaxseed oil increased significantly DHA in egg-yolks. As for EPA, the difference between treatments was not significant.

Egg quality

The experimental diets did not affect (P>0.05) the parameters used to evaluate egg quality such as egg longitudinal and transverse diameter, albumin diameter, yolk diameter, yolk height and yolk weight (Table 2).

Productive performance

Results for egg production, egg weight, egg mass, feed intake, feed conversion and body weight are shown in Table 3. The diet supplemented with 200 g/kg purslane increased (P<0.05) egg production, feed intake, and feed conversion compared with the negative control diet. Moreover, egg mass increased (P=0.001) for hens fed with 100 and 200 g/kg purslane compared with those that received the negative control diet. Experimental diets did not affect egg weight and hens body weight (P>0.05).

Table 2. Effects of experimental diets on yolk *ω*-3 fatty acids content (g/kg of total fatty acids)[a] and egg longitudinal diameter (ELD), egg transverse diameter (ETD), albumin diameter (AD), yolk diameter (YD), yolk height (YH) and yolk weight (YW) of Rhode Island Red laying hens[b]

	Dietary levels (g/kg) of					
	Purslane			Flaxseed oil		
Yolk *ω*-3 fatty acids	**0**	**100**	**200**	**14.7**	**SEM**	***P-Value***
ALA	3^{w}	4.5^{x}	7^{y}	28.5^{z}	2.16	0.001
DHA	10.6^{w}	11.2^{w}	16.04^{x}	30.7^{y}	1.73	0.001
EPA	0	0	0.04	0.05	0.01	0.001
Egg quality parameters						
ELD (cm)	5.65	5.64	5.76	5.74	0.03	0.460
ETD (cm)	4.29	4.39	4.35	4.37	0.04	0.169
AD (cm)	8.50	8.35	8.60	8.64	0.10	0.738
YD (cm)	4.05	4.09	4.15	4.15	0.02	0.427
YH (cm)	1.65	1.66	1.63	1.65	0.01	0.742
YW (g)	15.86	16.03	16.13	16.9	0.16	0.133

Means with different letters (w, x, y, z) within a row are significantly different at P<0.05.
[a]Results are means and standard errors of the means of 6 samples/treatment.
[b]Results are means and standard errors of the means of 16 eggs/treatment.

Table 3. Effects of experimental diets on productive performance of Rhode Island Red laying hens[a].

	Dietary levels (g/kg) of					
	Purslane			Flaxseed oil		
Parameters	**0**	**100**	**200**	**14.7**	**S.E.M**	***P- Value***
Egg production (%)	65.28^{w}	73.61^{wx}	77.77^{x}	70.80^{wx}	1.36	0.005
Egg weight (g)	56.11	60.16	60.82	57.90	1.18	0.497
Egg mass (g/hen/day)	36.55^{w}	43.82^{xy}	47.23^{y}	40.90^{wxy}	1.06	0.001
Feed intake (g/hen/day)	127.94^{w}	144.81^{wx}	147.23^{x}	135.30^{wx}	2.74	0.039
Feed conversion	3.27^{w}	3.06^{wx}	2.71^{x}	3.10^{wx}	0.07	0.049
Body weight (g)	2168.62	2171.25	2172.62	2166.15	25.89	0.988

Means with different letters (w, x, y) within a row are significantly different at $P<0.05$.
[a]Results are means and standard errors of the means of 16 laying hens/treatment fed for 56 days.

DISCUSSION

In this study, ALA and DHA on egg yolk increased significantly with 200 g/kg purslane in the diet. Similar results were reported by Zotte et al. (2005). However, in the current study, yolks ALA and DHA contents were higher than those reported by these researchers. The increase of ALA in the yolks could be due to the increase of purslane in diet, for other studies have shown that the level of yolk ALA increases with the amount of ALA rich materials in hens' diets (Caston and Leeson, 1990; Van Elswyk, 1997). The increase of DHA in the yolks could be due to the desaturation and elongation of ALA in hen's liver, since it has been shown that laying hens can convert ALA to DHA and deposit it in the egg yolk (Kralik et al., 2008). The few amount of EPA in the yolks can be explained by the fact that laying hens have the unusual ability of rapidly converting ALA into DHA in significant quantities, and into EPA in lesser amounts (Farrell, 2010).

The increase of egg production found in this study could arise because of the increase of feed intake of the hens fed with the diet supplemented with 200 g/kg purslane. The significant increase of egg mass can be justified by the fact that hens fed with 100 and 200 g/kg purslane had the highest egg production and egg weight. The results for feed intake in this study disagree with those reported by Aydin and Dogan (2010). The high feed intake reported in this study could be attributed to a lower concentration of dietary metabolizable energy since it has been reported that hens tend to increase feed intake when there is a low energy concentration in the diet (NRC, 1987; Ferket and Gernat, 2006). However, in this study, a trial was not performed in order to verify the metabolizable energy of the experimental diets.

CONCLUSION

The dietary inclusion of purslane meal up to 200 g/kg increased yolk's ω-3 fatty acids content without any adverse effect on egg quality parameters and productive performance of Rhode Island Red hens.

REFERENCES

Aydin, R., Dogan I., 2010. Fatty acid profile and cholesterol content of egg yolk from chickens fed diets supplemented with purslane (Portulaca oleracea L.). Journal of the Science of Food and Agriculture 90, 1759-1763.

Betancourt, L., Díaz, G., 2009. Egg enrichment with omega-3 fatty acids by means of flaxseed supplement (Linum usitatissimum) in the diet. Revista MVZ Córdoba 14, 1602-1610.

Bligh, E.G., Dyer, W.J., 1959. A rapid method for total lipid extraction and purification. Canadian Journal of Biochemistry and Physiology 37, 911–917.

Caston, L., Leeson, S., 1990. Research note: Dietary flax and egg composition. Poultry Science 69, 1617-1620.

Cros, V., Sánchez, J.J.M., Franco, J.A., 2007. Good yields of common purslane with a high fatty acid content can be obtained in a peat-based floating system. HortTechnology 17, 14-20.

Farrell, D., 2010. The role of poultry in human nutrition. The omega-3 fatty acids. Food and Agriculture Organization of the United Nations. Poultry Development Review. School of Land, Crops and Food Sciences, The University of Queensland, St. Lucia 4072, Queensland, Australia.

http://www.fao.org/docrep/013/al709e/al709e00.pdf (last checked 06/19/2013).

Ferket, P.R., Gernat, A.G., 2006. Factors that affect feed intake of meat birds: a review. International Journal of Poultry Science 5, 905-911.

Gao, D., Li, Q., Fan, Y., 2010. Hypoglycemic effects and mechanisms of Portulaca oleracea L. in alloxan-induced diabetic rats. Journal of Medicinal Plants Research 4, 1996-2003.

Harris, W.S., Mozaffarian, D., Lefevre, M., Toner, C.D., Colombo, J., Cunnane, S.C., Holden, J.M., Klurfeld, D.M., Morris, M.C., Whelan, J., 2009. Towards establishing dietary reference intakes for eicosapentaenoic and docosahexaenoic acids. Journal of Nutrition 139, 1-16.

Joseph, J.D., Ackman, R.G., 1992. Capillary column gas chromatography method for analysis of encapsulated fish oil and fish oil ethyl esters: collaborative study Journal of AOAC International 75, 488-506.

Kralik, G., Škrtić, Z., Suchý, P., Straková, E., Gajčević, Z., 2008. Feeding fish oil and linseed oil to laying hens to increase the n-3 PUFA of egg yolk. Acta Veterinaria Brno 77, 561-568.

Mohamed, A.I., Hussein, A.S., 1994. Chemical composition of purslane (Portulaca oleracea). Plant Foods for Human Nutrition 45, 1-9.

NRC, 1987. National Research Council. Estimating feed intake of laying hens, ed. National Academy Press, Washington D.C., pp. 44-45.

Pandalai, P.K., Pilat, M.J., Yamazaki, K., Naik, H., Pienta, K.J., 1996. The effects of omega-3 and omega-6 fatty acids on in vitro prostate cancer growth. Anticancer Research 16, 815-820.

SAS, 2000. SAS/STAT User's guide for Personal Computers, Version 6 Vol. 2. SAS Institute Inc. Cary, NC, USA.

Simopoulos, A.P., Norman, H.A., Gillaspy, J.E., Duke, J.A., 1992. Common purslane: a source of omega-3 fatty acids and antioxidants. Journal of the American College of Nutrition 11, 374-382.

Simopoulos, A.P., Leaf, A., Salem, N.J.R., 1999. Workshop on the essentiality of and recommended dietary intakes for omega-6 and omega-3 fatty acids. Journal of the American College of Nutrition 18, 487-489.

Simopoulos, A.P., 2004. Omega-3 fatty acids and antioxidants in edible plants. Biological Research 37, 263-277.

Van Elswyk, M.E., 1997. Nutritional and physiological effects of flax seed in diets for laying fowl. World's Poultry Science Journal 53, 253-264.

Zotte, A.D., Tomasello, F., Andrighetto, I., 2005. The dietary inclusion of Portulaca oleracea to the diet of laying hens increases the n-3 fatty acids content and reduces the cholesterol content in the egg yolk. Italian Journal of Animal Science 4, 157-159.

FEED RESOURCES QUALITY AND FEEDING PRACTICES IN URBAN AND PERI-URBAN DAIRY PRODUCTION OF SOUTHERN ETHIOPIA

Girma Chalchissa[1]*, Yoseph Mekasha[2] and Mengistu Urge[3]

[1]Adami Tulu Agricultural Research Center, P.O. Box 35, Ziway, Ethiopia
[2]International Livestock Research Institute, P.O. Box 5689, Addis Ababa, Ethiopia.
[3]Haramaya University, School of Animal and Range Sciences,
P.O. Box 138, Dire-Dawa, Ethiopia
**Corresponding author*

SUMMARY

The study was conducted in urban and peri-urban areas of Southern Ethiopia to assess the quality of available feed resources and feeding practices in the area. Shashamane city was considered as urban, while Kerara Filicha, Kuyera and Arsi Negele were considered as peri-urban production system. Structured questionnaire, secondary data sources and field observations were employed to generate data. A total of 60 dairy farmers from urban and 60 from peri-urban (Kerara Filicha =20, Kuyera = 20 and Arsi Negele = 20) were selected for the study. Average herd size per farm in urban area was 5.08±0.35 out of which small and medium scale farms had 2.6±0.2 and 7.5±0.29 crossbred cattle, respectively. In peri-urban areas average herd size per farm was 4.7±0.34 crossbred cattle out of which 2.3±0.16 were in small and 7.06±0.24 in medium scale farms. Fifteen major feed types used by dairy farms were identified in the area and categorized into five classes: grazing, green feeds, crop residues, agro-industrial by-products and non-conventional feeds. The result of the study indicated that wheat straw and *teff* straw were the main basal diets in the study area. Laboratory analysis of major feed resources indicated that crop residues had CP, digestibility and metabolizable energy contents of 2.9-5.9%, 46.6% and 5.9-8.7MJ/kg DM, respectively. About 50% of small and 51.7% of medium scale farms use roughage feeds as a major feed source. Therefore, from the current study it was concluded that the quality of available basal roughage feed is generally low and needs strategic supplementation with protein and energy rich feeds.

Key words: Feed quality; feeding practices; urban; peri-urban; dairy production; Ethiopia.

INTRODUCTION

In Ethiopia, market-oriented urban and peri-urban milk productions are emerging as main suppliers of milk and dairy products to the cities. They possess both indigenous and crossbred dairy animals ranging from 50% to high grade Friesian breeds in small, medium and large sized farms. They depend on purchased roughage and concentrate feeds with limited grazing for milk production (Kelay, 2002; Sintayehu *et al.*, 2008, Azage *et al.,* 2013). However, the productivity of these animals is very low due to lack of appropriate nutritional management, health care, marketing system, extension and training (Azage and Alemu, 1998). Lack of appropriate nutritional management indicates that inadequacy of feed in terms of quality and quantity. Livestock feed resources in Ethiopia are mainly derived from natural pasture/grasslands, crop residues, improved pasture, forage crops, agro-industrial by-products and non-conventional feeds (CSA, 2012). The contribution of these feed resources, however, depends up on the agro-ecology, the types of crop produced, accessibility and production system (Seyoum *et al.*, 2001; Ahimed *et al.*, 2010). Seasonal variations in quantity and quality of the forages are a major concern especially during the dry season. The crude protein (CP) content of pastures is lower than the forage crude protein content of 7 percent which would cover the maintenance requirements of ruminants (McDonald *et al.*, 1995). Deficiencies in nutritive value of natural pasture stress the importance of supplementation with energy and nitrogen especially during the dry season. Different research works point out different percentage on the contribution of various livestock feed resources. This may be due to the wide range of ecological variation between different localities in the country and also variation in time which in turn results with variation in crop species and cropping intensity. So location and time specific feed resource assessment is required in order to know the feed gap between feed supply and feed requirement within specified animal production level.

MATERIALS AND METHODS

Description of the study area

Location: The study was conducted on private urban and peri-urban dairy farms in and around Shashamane city, which is one of the potential areas for milk production in Southern Ethiopia. The area is located at a distance of 230 to 260km south of the capital Addis Ababa. The area relieves an annual rain fall of 800- 1300mm; characterized with different altitude ranges of 1900 and 2200 meters above sea level and average minimum and maximum temperature of 12 and 27°C, respectively.

Sampling methods

A preliminary visit was conducted in the study area to get general picture of the study sites and to identify the target farms. Thus, dairy farms in Shashamane and three associated peri-urban areas were purposively identified based on their potential to produce milk and supply to Shashamane city. For the purpose of this study, the farms were stratified into urban and peri-urban sub-systems. Each production sub-system was further stratified into two based on herd size: small holders (farms with less than four cows) and medium level (farms with 4-10 cows); Ike (2002).

There were a total number of 593 dairy farms keeping crossbred cattle in the study area, of which about 69% were urban and 31% were peri-urban farms. For this study a total of 120 dairy farms, of which 60 were from urban and the remaining 60 were from peri-urban dairy systems, 30 from each farm scale (small and medium size) were purposively selected. Following identification of dairy farms, semi-structured questionnaires were developed and pre tested before commencement of the actual survey. Information was gathered by interviewing the household heads and farm managers.

Feed sampling and chemical analysis

Representative samples of feed resources commonly employed by the farms during the study period were collected monthly and samples of the same feed type were bulked together. They were thoroughly mixed and sub sampled. Wet feed materials were allowed to lose moisture under shed before transportation. The air dried samples were taken to Hawassa University Animal Nutrition Laboratory and dried in an oven at 60°C for 48 hrs and ground in Willey mill to pass through 1mm sieve and allowed to equilibrate at room temperature for 24 hrs. The ground samples were kept in air tight containers pending analysis for chemical composition.

Feed samples were analyzed for DM and Ash using the standard procedures of AOAC (1990). Nitrogen (N) content was determined by Kjeldahl method and Crude Protein (CP) was calculated as N*6.25 (AOAC, 1995). The organic matter was calculated as difference between 100 DM and ash content. Acid Detergent Fiber (ADF), Acid Detergent Lignin (ADL), Neutral Detergent Fiber (NDF), and *In vitro* Digestible Organic Matter in the Dry Matter (IVDOMD) was determined by the modified Tilley and Terry method (Van Soest and Robertson, 1985). Metabolizable Energy (ME) content of a particular feed was estimated from IVDOMD and as per the following equations: ME (MJ/kg DM) =

0.015*IVDOMD (g/kg) (MAFF, 1984). The study was conducted from August to December 2012.

Statistical analysis

Both quantitative and qualitative data collected during the survey were analyzed using Statistical Analysis System software (SAS, 2004). Descriptive statistics such as means, percentages, standard error of mean and frequency distributions were used to describe the qualitative and quantitative variables in the production system.

RESULTS

Occupational status of households

Overall, about 29.2% of interviewed dairy farms owners were owned by dairy cow producers (those whose livelihood depend mainly on dairy cow rearing/that engaged only in dairy production), followed by farmers (21.7%), traders (19.2%), household wives (10.8%), government employees (10.8%), and retired persons (8.3%). Across the two farm scales, 48.3% of medium scale farms and 10% of small scale farms were owned by households that engaged only in dairy production (Table 1). Among the production systems, about 5.3% of urban and 43.3% of peri-urban households were engaged in both dairy production and crop farming activities.

Inventory of feed resources and feeding practices

In this study, fifteen different feed types were identified and categorized into five classes: grazing, green feeds (purchased green grass, green maize stover and elephant grass), crop residues (wheat straw, *teff* straw, haricot bean straw and maize stover), agro-industrial byproducts (wheat bran, linseed cake, brewery waste and mixed concentrates) and non conventional feeds (*atela* and mill byproducts). Two major feeding systems were identified in the study area: semi-intensive and stall feeding/zero grazing. Non conventional feeds, particularly, local liquor residue (*atela*) was highly used as major feed source in peri-urban farms.

Chemical composition of commonly used feed stuffs

The dry matter (DM) content of green feeds varied from 17% in elephant grass to 30.61% in green maize stover. The crude protein (CP) content of different green feeds was 8.2%, 12% and 14% for purchased green grass, green maize stover and elephant grass, respectively. The mean *in vitro* digestible organic matter in the dry matter (IVDOMD) for green feeds was about 54.9%.

The mean DM content of crop residues was 92.3%, which corresponds with Solomon (2004) and Sisay (2006). The CP content of crop residues varied from 2.98% in wheat straw to 5.97% in haricot bean straw. The energy content of crop residues ranged from 5.96 MJ/kg DM in wheat straw to 8.79 MJ/kg DM in maize stover.

Table1: Major occupational status of interviewed households in urban and peri-urban dairy farms in Shashamane milk shed.

	Household head occupation (%)					
Variables	Housewife	Trader	Dairy farming only	Farmer	Civil servant	Pensioner
Production system						
Urban (n=60)	16.7	23.3	28	5.3	15	11.7
Peri-urban (n=60)	5	15	25	43.3	6.7	5
Karara Filicha (n=20)	-	5	25	60	-	10
Kuyera (n=20)	5	25	20	40	10	-
Arsi Negele (n=20)	10	15	30	30	10	5
Herd size						
Small (n=60)	15	26.7	10	23.3	15	10
Medium (n=60)	6.6	11.7	48.3	20	6.7	6.7
Overall (n=120)	10.8	19.2	29.2	21.7	10.8	8.3

n= number of household heads

Table 2: Feeding systems and major feed types used by urban and peri-urban dairy producers in Shashamane milk shed.

Variables	Production sub-systems				
	Urban		Peri-urban		
	Small (n=30)	Medium (n=30)	Small (n=30)	Medium (n=30)	Overall (n=120)
Feeding systems (% Farms)					
Stall feeding (zero grazing)	100	100	86.7	96.7	95.8
Semi intensive	0	0	13.3	3.3	4.2
Main Feed types (% Farms)					
Grazing/foraging	0	0	3.3	0	0.8
Green feeds	10	6.7	6.7	6.7	7.5
Crop residues	36.7	43.3	50	46.7	44.2
Agro-industrial byproducts	36.7	40	6.7	13.3	24.2
Katikala atela and mill byproduct	16.6	10	33.3	33.3	23.3
Type of mineral supplements					
Common salt	96.7	93.3	93.3	96.7	95
Multi nutrient block	3.3	3.3	0	0	1.7
Boji	0	3.4	6.7	3.3	3.3

n= number of households, *atela* = residue of locally produced beverage, *Boji*= locally available mineral salt.

Table 4: Chemical composition and nutritive value of commonly used feed resources in the study area.

Feedstuff		Chemical composition (% DM)						Nutritive values	
	DM (%)	Ash	OM	NDF	ADF	EE	Lignin	CP	IVDOMD (%)	ME *
Green feeds										
Purchased green	28.02	10.7	89.3	75.8	41	1.3	6.43	8.2	51.58	7.73
Green maize stover	30.61	7.6	92.4	62.3	66.55	1.2	6.24	12	61.7	9.25
Elephant grass	17	16.2	83.8	51.9	57.2	1.5	7.2	14	51.4	7.71
Crop residues										
Wheat straw	92.45	8.8	91.2	78.6	58.1	0.7	12.6	2.98	39.74	5.96
Tef straw	93.9	7.84	92.16	81.5	46.8	1.12	9.23	4.34	43.35	6.5
Haricot bean straw	89.86	7.58	92.42	76.1	63.32	0.9	17.02	5.97	44.57	6.68
Maize stover	93.6	9.29	90.71	82.13	51.72	0.64	9.54	4.56	58.65	8.79
Agro-industrial byproducts										
Linseed cake	93.23	8.5	91.5	25.1	9.52	9.13	5.14	29.8	67.1	10.06
Nouge seed cake	92.2	10.92	89.08	34.75	26.73	5.94	6.98	32.1	63.12	9.46
Wheat bran	90.06	4.28	95.72	38.19	9.39	4.3	2.52	16.98	77.7	11.65
Commercial mix	91.4	6.92	93.08	36.86	25.7	4.5	6.38	22.7	67	10.05
Brewery waste	23.5	4.83	95.17	77.2	38.8	8	9.9	27	61.25	9.18
Non conventional Feeds										
Katikala Attela	17.4	4.9	95.1	58.8	23.5	3.6	10.3	20.6	78.4	9.76
Mill by-products	91.73	10.2	89.8	39.2	35.6	9.2	5.52	19.2	69.02	10.35

Katikala Atela = by-product of local beverage.
*MJ/kg DM

The neutral detergent fiber (NDF) content of crop residues ranged from 76.1% in haricot bean straw to 83.8% in maize stover. Acid detergent fiber (ADF) contents of crop residue varied from 46.8% in *teff* straw to 63.32% in haricot bean straw. Metabolizable energy (ME) content of commonly used energy supplements such as wheat bran, commercial mix, mill byproducts and *atela* varied from 9.18 to 10.05 MJ/kg DM. *Atela* had relatively lower energy content compared to wheat bran and mill by-products. Among the protein supplements, *atela* had slightly lower CP (20.6%) than nouge seed cake (32.1%), linseed cake (29.8%) and brewery waste (27%). Wheat bran and mill by-product had lower CP content of 16.98% and 19.2%, respectively, compared to other supplementary feeds.

Herd size and composition

The average herd size of crossbred cattle per household in the urban area was 5.08±0.35 with medium farms having average of 7.5±0.29 and small farms 2.6±0.2 cattle. In peri-urban area the average herd size was 4.7±0.34 cattle with medium farms having 7.06±0.24 and small farms 2.3±0.16 cattle. In all the systems, farmers possessed dairy cows followed by heifers as their future herd replacement animals. Distributions of cows on their productive state showed that among the total number of cows in urban farms about 37.4% were pregnant and milked, 28.6% milked and non pregnant, 31.7% were dry pregnant and 2.3% were dry and non pregnant, whereas, the values in peri-urban farms were 34.3% pregnant and milked, 33.7% milked and non pregnant, 26.4% dry pregnant and 5.6% were dry and non pregnant. In both cases, higher numbers of cows were pregnant and lactating.

Table 5: Herd size and composition of crossbred cattle in Shashamane milk shed by production sub-system and herd size.

Variables	Production sub-systems					
	Urban			Peri-urban		
	Small (n=30)	Medium (n=30)	Mean ±SE	Small (n=30)	Medium (n=30)	Mean ±SE
Herd size (mean ±SE)	2.6±0.2	7.5±0.29	5.08±0.35	2.3±0.16	7.06±0.24	4.7±0.34
Herd composition (mean ±SE)						
Cows	1.53±0.09	4.1±0.23	2.8±0.21	1.64±0.1	4.3±0.19	2.9±0.2
Milking cows	1.06±0.08	2.6±0.13	1.85±0.12	1.3±0.08	2.9±0.17	2.1±0.14
Pregnant milking (%)	30.4	40	37.4	22	39	34.3
Non pregnant milking (%)	39.2	24.8	28.6	56	25	33.7
Dry cows	0.47±0.09	1.5±0.14	0.95±0.1	0.34±0.08	1.5±0.21	0.9±0.13
Pregnant dry (%)	26	33.6	31.7	16	30.5	26.4
Non pregnant dry (%)	4.4	2.3	2.3	6	5.5	5.6
Heifers (1-3 years age)	0.43±0.09	1.5±0.17	0.9±0.12	0.3±0.08	0.96±0.14	0.6±0.09
Pregnant (%)	38.5	45.7	44	33.3	41.4	39.5
Non pregnant (%)	61.5	54.3	56	66.7	58.6	60.5
Heifer calves (1-11 months)	0.44±0.09	1.2±0.15	0.8±0.09	0.26±0.09	0.8±0.13	0.5±0.08
Male calves (1-18 months)	0.2±0.07	0.4±0.09	0.3±0.06	0.1±0.07	0.8±0.13	0.46±0.09
Breeding bulls	0.0	0.3±0.08	0.15±0.01	0.0	0.2±0.03	0.1±0.02
Dominant cattle breeds (% Farms)						
Friesian crosses	100	96.7	98.3	93.3	90	91.7
Friesian crosses and local zebu	0.0	3.3	1.7	6.7	10	8.3
Presence of other species of animals (sheep, poultry and donkey) (% Farm)	16.7	100	58.3	26.7	100	63.3

n= number of respondents, SE= standard error

DISCUSSION

The involvement of farmers in dairy production in the current report was higher than 13.7% reported for Sebeta Awsa area (Dereje, 2012). Higher involvement of farmers in dairy production in the current report might be related to attractive market access in the area. Generally, the result of this study indicates that dairying is the major occupation for the majority of the interviewed household head dairy producers in area.

Stall feeding practice in urban area might be related to shortage of grazing lands in urban areas. Similar result was reported for Addis Ababa milk shed (Yoseph, 1999) and Dire Dawa city (Emebet, 2006) in which grazing/foraging was not practiced by urban dairy farms.). The lower use of agro-industrial byproducts by peri-urban farms may be related to less availability of these feed types and more availability of local liquor residue (*atela*) in the area. Opposing to the current report, agro-industrial by-products were used as main feed source by 25.8% and 24.2% of peri-urban small and medium farms respectively, compared to 20.9% and 22.6% of small and medium farms, respectively, in urban farms of Sebeta Awas areas (Dereje, 2012). Higher use of non conventional feed in peri-urban farms may be related to high production potential of *Katikala (Areki)* in the area (particularly Arsi Negele and Kuyera areas).

The result of the current study was in agreement with 7.96 MJ/kg DM energy content of green feeds in Sinana area (Solomon, 2004). Crude protein value for wheat straw in the current study was lower than 4.1% reported in Sinana area (Solomon *et al.*, 2008). All crop residues evaluated had lower CP contents than the minimum level of 7% CP required for optimum rumen microbial function (Van Soest, 1982). The mean *in vitro* digestible organic matter in the dry matter (IVDOMD) for crop residues was 46.6%, which is lower than the minimum level of 50% required for quality roughages (Mosi and Butterworth, 1985; Seyoum and Fekede, 2008). Lower crude protein contents of crop residues may be compensated with supplementation of feeds rich in protein to improve cattle performance. The NDF and ADF contents of crop residues in the current report was higher than the range of 66.5% to 77.2% for NDF and 42.4% to 54.2% for ADF reported for central Ethiopia (Seyoum and Fekede, 2008). Neutral detergent fiber (NDF) contents of more than 55% limits dry matter intake (Van Soest, 1982). Hence, the crop residues assessed in the current study area could affect feed intake of dairy cattle unless supplemented with feeds having better nutritional content.
The energy contents for crop residues was within the range reported by Zewdie (2010) for central highlands, but higher than reported by Solomon (2004) for Bale highlands. Differences might be due to difference in management practices, soil fertility and/or crop variety used. The energy content of a*tela* (9.76 MJ/kg DM) reported in this study was lower than 11 MJ/kg DM reported by Nega *et al.* (2006) which could be due to differences in the chemical composition and type of grains used as a raw material to produce this by-product.

Similar herd size per farm was reported in Hawassa city (Haile *et al.*, 2012) where the average herd size in the urban areas was 5.48± 0.49 cattle. Similarly, Ike (2002) reported average herd size of 6.23±4.39 cattle in peri-urban areas of Hawassa. Higher herd size of 3.2 and 11.3 cattle was reported for small and medium scale farms, respectively, in Dire Dawa city (Emebet and Zeleke, 2007). Higher proportions of pregnant milking cows in urban and peri-urban farms might be related to long lactation length in the area. The proportion of dry cows in this study was higher than the ideal recommended value of 17% set by Radostits *et al.* (1994). Keeping these herds require extra expenses for their feeding and other management. Compared to the current study, high percent (14%) non pregnant dry cows were reported by Mohamed *et al.* (2003) in Harar milk shed.

CONCLUSION

Therefore, from the current study it was concluded that the quality of available basal roughage feeds is generally low and strategic supplementation of protein and energy rich feeds should be required. In relation with the increasing market price of concentrate and roughage feeds, other optional feeds like fruit waste and other non-conventional feed resources should be further considered.

Acknowledgements

The authors are grateful to the Oromia Agricultural Research Institute for financially supporting this study.

REFRENCES

Ahmed, H., Abule, E., Mohammed, K. and Treydte, A.C. 2010. Livestock feed resources utilization and management as influenced by altitude in the Central Highlands of Ethiopia. Livestock Research for Rural Development. http://www.cipav.org.co/lrrd/lrrd19/12/cont1912.htm, (Accessed on January 5, 2012).

AOAC (Association of Official Analytical Chemists) 1990. Official methods of analysis. 13th edition. AOAC, Washington DC, USA.

AOAC (Association of Official Analytical Chemists), 1995. Official Methods of Analysis. PP.5-13. (16th edition), Washington DC.

Azage Tegegne and Alemu Gebrewold, 1998. Prospects for peri-urban dairy development in Ethiopia. In: Fifth national conference of ESAP (Ethiopian Society of Animal Production). ESAP, Addis Ababa, Ethiopia.

Azage Tegegne, Berhanu Gebremedhin, Dirk Hoekstra, Berhanu Belay and Yoseph Mekasha, 2013. Smallholder dairy production and marketing systems in Ethiopia: IPMS experiences and opportunities for market-oriented development. Improving Productivity and Market Success of Ethiopian Farmers Project (IPMS) working paper 31. International Livestock Research Institute (ILRI), Addis Ababa, Ethiopia.

CSA (Central Statistical Agency), 2012. Agricultural Sample Survey 2011/12. Report on Livestock and Livestock Characteristics, Volume II, Statistical Bulletin, 532. Addis Ababa, Ethiopia.

Dereje Shibru, 2012. Evaluation of the Performance of Crossbred Dairy Cows and Heifer Calves in Urban and Peri-urban Dairy Systems in Sebeta Awas Woreda, Oromia, Ethiopia. An MSc. Thesis, Haramaya University, Dire Dawa, Ethiopia. 114p.

Emebet Moreda, 2006. Reproductive Performance of Dairy Cows under Urban Dairy Production Systems in Dire-Dawa. MSc. Thesis, Alemaya University, Dire Dawa, Ethiopia. 82p.

Emebet Moreda and Zeleke Mekuriaw, 2007. Characteristics and constraints of crossbred dairy cattle production in lowland areas of Eastern Ethiopia. Livestock Research for Rural Development, Volume 20, Article #.57http://www.lrrd.org/lrrd20/4/mure20057.htm.

Haile Welearegay, Zelalem Yilma and Yosef Tekle-Giorgis, 2012. Challenges and opportunities of milk production under different urban dairy farm sizes in Hawassa City, Southern Ethiopia. African Journal of Agricultural Research Vol. 7(26), pp. 3860-3866.

Ike, A., 2002. Urban dairying in Awassa, Ethiopia. MSc thesis, University of Hohenheim, Stuttgart-Hohenheim, Germany. 113p.

Kelay Belihu, 2002. Analysis of Dairy Cattle Breeding Practices in Selected Areas of Ethiopia, PhD. Dissertation, Humboldt University of Berlin, Germany.164p.

MAFF (Ministry of Agriculture Fisheries and Food), 1984. Energy allowances and feeding systems for ruminants. Reference Book 413 HMOs, London, 85p.

McDonald, P., R.A Edwards, J.F. Greenhalgh, and C.A. Morgan, 1995. Animal Nutrition. (Fifth Edition). Longman Group, Harlow, United Kingdom. 607p.

Mohamed Ahmed Mohamed, A., Ehui, S. and Yemesrach Asefa, 2003. Dairy Development in Ethiopia. Paper presented at the InWent, IFPRI, CTA, NEPAD Conference Paper No. 6, December 1-3, 2003, Pretoria.

Mosi, A.K. and M.H., Butterworth, 1985. The voluntary intake and digestibility of diets containing different proportions of *tef* (*Eragrostis tef*) straw and *Trifolium tembense* hay when fed to sheep. Tropical animal production 10, 19-22.

Nega Tolla, P. Vijchulata, P. Chairatanayuth and S. Swsdiphanich, 2006. Comparative Performances of Holstein-Friesian Cows under Smallholder and Large Scale Farmers' Management in Central Rift Valley, Ethiopia. Kasetsart Journal. (Natural Sciences) 40 : 58 – 68.

SAS, 2004. Statistical Analysis System software, Version 9.0, SAS Institute, Inc., Cary, NC, USA.

Seyoum Bediye and Zinash Sileshi, 1988. Feeding value of some Ethiopian feedstuffs. PP.175-185. IAR Proceedings, Second National Livestock Improvement Conference. 24-26 February.1988. Addis Ababa, Ethiopia.

Sintayehu Yigrem, Fekadu Beyene, Azage Tegenye and Berhanu Gebremedhin, 2008. Dairy production, processing and marketing systems of Shashemene, Dilla area, South Ethiopia. IPMS Improving Productivity and Market Success) of Ethiopian Farmers Project Working Paper 9. ILRI (International Livestock Research Institute), Nairobi, Kenya. 61p.

Solomon Bogale, 2004. Assessment of Livestock Production Systems Feed Resource base in Sinana Dinsho district of Bale highlands,

Southeast Oromya, An MSc. Thesis, Alemaya University, Dire Dawa, Ethiopia.

Van Soest, P.J. 1982. Nutritional Ecology of the Ruminants: Ruminant metabolism, Nutritional strategies, the cellulolytic Fermentation and the Chemistry of Forages and Plant Fibers. Ithaca, New York. 373p.

Van Soest, P.J. and Robertson, J.B. 1985. Analysis of Forages and Fibrous Foods. A Laboratory Manual for Animal Science 613. Cornel University, Ithaca. New York, USA, 202p.

Yoseph Mekasha. 1999. Impact of feed resources on productive and reproductive performance of dairy cows in the urban and peri-urban dairy production system in the Addis Ababa milk shed and evaluation of non- conventional feed resources using sheep. MSc. Thesis, Alemaya University, Dire Dawa, Ethiopia. 197p.

Zewdie, W. 2010. Livestock Production Systems in Relation with Feed Availability in the Highlands and Central Rift Valley of Ethiopia. MSc. Thesis, Haramaya University, Dire Dawa, Ethiopia. 160p.

21

EFFECT OF MUCUNA (*Mucuna pruriens*) ON SPERMIOGRAMS OF WEST AFRICAN DWARF BUCKS

James O. Daramola[1]*, John A. Abiona[1]., Olusiji F. Smith[1], Olubukola A. Isah[2], Olusiji S. Sowande[3], Moshood O. Olaniyi[4], Jimoh A. Olanite[5]

[1]*Department of Animal Physiology, Email: daramolajames2003@yahoo.com;*

[2]*Department of Animal Nutrition*
[3]*Department of Animal Production*
[4]*College of Veterinary Medicine*
[5]*Department of Pasture & Range Management*
Federal University of Agriculture Abeokuta, PMB 2240, Nigeria
**Corresponding author*

SUMMARY

The study was carried out to investigate the potential effect of *Mucuna pruriens* seed powder on spermiograms of West African Dwarf bucks (WAD). Twenty bucks were randomly assigned to five treatments consisting of four animals each receiving 0, 25, 50, 75, 100mg/Bw of *Mucuna pruriens* seed powder administered via oral route for 30 days consecutively. Following treatment, spermiograms of the animals were determined. The results showed that exposure of bucks to *Mucuna pruriens* powder led to a significant increase in weight gain, scrotal width, scrotal circumference, testicular weight, gonadosomatic index, sperm motility and testicular parenchyma of animals compared to the control and the improvement in these parameters was much pronounced in bucks that received 75mg *Mucuna pruriens* seed powder ($P<0.05$). Higher ($P<0.05$) sperm concentration was observed in 100mg inclusion of *Mucuna pruriens* seed powder compared to other treatments. The results showed a decline ($P<0.05$) in the level of testosterone with increasing level of *Mucuna pruriens* seed powder administration. The findings showed that oral administration of *Mucuna pruriens* seed powder to the bucks for 30 days resulted in improved testicular measurements and sperm quality without any adverse effect on structure of reproductive physiology of the bucks.

Keywords: Bucks; *Mucuna pruriens;* spermiograms; sperm quality.

INTRODUCTION

The Problem of idiopathic male infertility is a major constraint in livestock production. Restriction is being imposed on the use of synthetic hormones and antioxidants because of their carcinogenicity (Osei-bonsu, 1998), hence the need for natural antioxidants therefore become imperative and desirable. In the vast resource of phytoproducts in South Western Nigeria, there are various plants that are claimed to contain antioxidants and phytohormones. *Mucuna pruriens* is a tropical legume known as velvet bean, found in Africa, India and the Caribbean. *Mucuna pruriens* is indigenous to and commonly found in South Western Nigeria. Roots, leaves and seeds of the plant are commonly used in the treatment of impotence, snake bite, diabetes, cancer and Parkinsonism (Alo *et al.* 2012). *Mucuna pruriens* has also shown to exhibit neuroprotective effect by increase brain mitochondrial complex-I activity and significantly restoring dopamine and norepinephrine levels in Parkinsonism animal model (Manyam *et al.* 2004). Seeds of *Mucuna pruriens* possess antioxidant, neuroprotective activities and bioactive substances (Sharma *et al.* 1978; Misra and Wagner, 2007). With the increasing acceptance of natural source of substances as an alternative form of health care (Sharma *et al.* 1978; Misra *et al.* 2007), the screening of medicinal plants for active compounds has become very important because these may serve as promising sources of novel hormone and antioxidant prototypes. Presently, no available information in literarure on the effect of the *Mucuna pruriens* on spermiograms of West African Dwarf bucks. The research was therefore conducted to investigate the potential effect of *Mucuna pruriens* seed powder on spermiograms of West African Dwarf bucks.

MATERIALS AND METHODS

The experiment was carried out at the Small Ruminant Unit, Federal University of Agriculture, Abeokuta, located in the tropical rain forest zone of Nigeria within 7^0 10'N and 3^0 2'E in line with the regulations of the Institute of Food Security, Environmental Resources and Agriculture Research (IFSERAR) of Federal University of Agriculture Abeokuta, Nigeria. Twenty (20) West Africa Dwarf bucks of ages ranging between 8-12 months and weighting 8.355 $\pm$ 0.8674kg were used. The bucks were kept and maintained under semi intensive method and uniform nutritional regime. The animals were fed with concentrate at 300g/body weight (Bw) while *Panicum maximum* and water was given to them *ad libitum*. The animals were divided into 5 groups of 4 animals each and each group was randomly assigned to one of experimental treatments consisting of 0, 25, 50, 75, 100mg/BW of *Mucuna pruriens* seed powder administered via oral route to the animals for 30 days consecutively.

Data Collection

The body weight, scrotal circumference, scrotal length and scrotal width of the animals were recorded weekly for 65 days. The scrotal circumference was measured at the widest point of paired testes.Scrotal length was measured as the length of the testis when still in the scrotum while scrotal width was measured as the distance between the two sides of the broadest part of the scrotum using the thread and the thread measured on meter rule (Notter *et al.* 1985). Following castration under local anesthesia, testicular parameters (the testis weight, length, width of the testis and semen characteristics) and testicular histomorphometry were determined at the 65th day of the trial. The paired testes were milked out, weighed and recorded in grams after the epididymis was trimmed off. Gonadosomatic index (%) was calculated (GSI = [testis weight/body weight] x 100). The length and width of each of the testis was measured using a pair of caliper. The volume of each testis was recorded using Archimedes' principle of water displacement by the amount of water displaced in a 500 cm^3 measuring cylinder filled with water at $300cm^3$. The tunica albuginea (peeled-off of the testis) was weighed individually and recorded in grams. The testicular parenchyma was then calculated as the difference between the whole testis and weight of tunica albuginea, while the testicular index was calculated for each buck as testis length multiplied by testis width.

Progressive sperm motility, sperm concentration and morphological examinations were carried out in line with Bearden and Fuquay (1997). Progressive sperm motility was determined using a phase-contrast microscope (400 x magnifications). A drop of semen was placed on a microscopic slide and sperm cells moving straight forward over the microscopic field were recorded. For each sample, at least five microscopic fields were examined. The mean of the five successive evaluations was recorded as the final motility score. The concentration was determined by the use of an improved Neubauer haemocytometer. Morphological examination of the sperms was carried out and primary abnormalities of the sperm cells located in the head, midpiece and tail were observed under a microscope.

Statistical Analysis

Data obtained were subjected to analysis of variance in a completely randomized design and means separated by Duncan Multiple Range Test (Duncan, 1955) in SPSS version 16 using the following model:

$Y_{ijk} = \mu + A_i + L_j + \sum_{ijk}$

Where,

Y_{ijk}= Dependent variables

μ = Population mean

A_i= effect due to i^{th} *Mucuna pruriens* treatments

L_j = effect due to j^{th} level of inclusion, j = 0, 25, 50, 75, 100

$\sum_{ijk}$= Experimental error

RESULTS

The means of spermiograms, of bucks exposed to *Mucuna pruriens* seed powder are presented in Table 1. Exposure of bucks to *Mucuna pruriens* seed powder led to a significant increase in weight gain, scrotal width, scrotal circumference, testicular weight, gonadosomatic index and testicular parenchyma of animals compared to the control ($P<0.05$). The improvement in weight gain, testicular weight, testicular length, testicular width, tunica albuginea, testicular parenchyma and testicular index were much pronounced in bucks that received 75mg *Mucuna pruriens* seed powder while much pronounced improvement of gonadosomatic index were observed in bucks that received both 75mg and 100mg of *Mucuna pruriens* seed powder ($P<0.05$). Total testicular semen volume and total sperm cells followed similar pattern with much pronounced higher (P<0.05) values observed in 25mg and 50mg inclusion of *Mucuna pruriens* seed powder compared to other treatments.

The means values of sperm characteristics, testosterone and physic-clinical parameters of bucks exposed to *Mucuna pruriens* seed powder are presented in Table 2. The results showed that bucks treated with Mucuna pruriens seed powder had higher sperm motility compared to the control and the improvement was much pronounced in bucks that received both 75mg and 100mg of *Mucuna pruriens* seed powder ($P<0.05$). Treatment also led to maintenance of sperm abnormality at low level particularly at 75mg levels of inclusion. Higher (P<0.05) sperm concentration was observed in 100mg inclusion of *Mucuna pruriens* seed powder compared to other treatments. The results showed a decline (P<0.05) in the level of testosterone with increasing level of *Mucuna pruriens* seed powder administration. The results showed that rectal temperature and pulse rate were similar in the treated groups and the control.

Table1. Means (SEM) of spermiograms of bucks exposed to *Mucuna pruriens* seed powder

Parameter	0mg	25mg	50mg	75mg	100mg	SEM
Weight gain (g)	2.00^{e}	3.50^{c}	4.00^{b}	4.50^{a}	3.10^{d}	0.743
Change in scrotal width (cm)	0.13^{d}	0.73^{b}	0.45^{c}	0.98^{b}	1.87^{a}	0.760
Change in scrotal circumference (cm)	1.00^{d}	3.20^{c}	5.10^{b}	3.50^{c}	6.00^{a}	1.090
Total testicular weight (g)	69.80^{d}	80.30^{b}	73.90^{c}	88.20^{a}	71.90^{c}	12.90
Total testicular length (cm)	13.09^{b}	13.66^{b}	13.08^{b}	15.27^{a}	12.15^{b}	0.330
Total testicular width (cm)	3.25^{c}	3.67^{b}	3.09^{c}	5.08^{a}	3.23^{c}	3.230
Total testicular semen volume (ml)	4.05^{bc}	4.26^{a}	4.26^{a}	4.10^{b}	3.69^{c}	3.690
Gonadosomatic index (%)	0.72^{d}	0.82^{b}	0.75^{c}	0.90^{a}	0.81^{a}	0.810
Tunica albuginea (g)	2.02^{b}	2.13^{b}	2.15^{b}	2.95^{a}	1.84^{c}	0.173
Testicular parenchyma (g)	32.88^{d}	37.99^{b}	34.84^{c}	42.01^{a}	34.88^{c}	4.606
Testicular index (cm^2)	10.63^{b}	12.65^{ab}	10.12^{b}	16.06^{a}	9.83^{b}	1.492

[abcd] Means within row with different superscripts differ significantly (P<0.05); SEM = Standard Error of Mean

Table 2. Means (SEM) of sperm characteristics, testosterone and physico-clinical parameters of bucks exposed to *Mucuna. Pruriens* seed powder

Parameter	0mg	25mg	50mg	75mg	100mg	SEM
Sperm motility (%)	70.00^{c}	87.00^{ab}	80.00^{b}	91.00^{a}	90.00^{a}	12.850
Sperm concentration ($x10^9$/ml)	0.19^{b}	0.19^{b}	0.19^{b}	0.18^{b}	0.21^{a}	3.700
% Abnormality	6.00^{a}	3.00^{b}	2.80^{b}	1.04^{d}	2.40^{c}	0.051
Total Sperm cells ($x10^9$/ml)	0.76^{bc}	0.81^{a}	0.81^{a}	0.74^{c}	0.78^{b}	3.695
Testosterone (ng/ml)	7.60^{a}	7.00^{a}	4.60^{b}	1.60^{c}	0.30^{d}	0.021
Rectal temperature (°C)	39.90	39.50	39.40	39.40	39.30	0.100
Pulse rate (beat/minute)	69.50	66.40	63.80	65.50	66.30	2.620

[abcd] Means within row with different superscripts differ significantly (P<0.05); SEM = Standard Error of Mean

DISCUSSION

Measurement of scrotal circumference and mean diameter of testis give good indication of semen production (Knight, 1977). The increase in scrotal and testicular measurements coupled with improved spermatozoa indices suggested viable spermatozoa per unit of testes and agrees with Orji and Steinbach (1976), Knight (1977) and Notter *et al.* (1985) that sperm production correlates highly with the testicular size. The increase in scrotal and testicular measurements without the incidence of any depressive effect on sperm characteristics coupled with comparative values of rectal temperature and pulse rate indicated that the treatment did not have any adverse effect on structure of the reproductive physiology of the animal. According to Ezekwe (1998), Hamilton and Stark (1997) and Perry and Petterson (2001), testes size is a good indicator of present and future sperm production. The improved scrotal and testicular indices in animals treated with *Mucuna pruriens* seed powder indicated its possible positive effect on the spermatogenic potential of the buck treated with this antioxidant-rich plant. *Mucuna pruriens* supplementation in infertile men has been associated with increased sperm count and motility (Shukla *et al.* 2009) indicative of its potential effect on male fertility (Shukla *et al.* 2010; Gupta, *et al.* 2011). The present results however contradicted the observation of Paul and Joseph (2001) that studied the effects of seeds of *Mucuna pruriens* on the gonads and sex accessory glands of male guinea-pigs and showed the presence of potential male antifertility agent in *Mucuna pruriens*.

Moreover, the decline in the level of testosterone with increasing level of *Mucuna pruriens* powder administration was in contrast to previous studies in men and rats. Testosterone level in the treated groups compared to control could probably be due to mechanism of action through levodopa content, in which the increase in serum dopamine antagonizes (works against) prolactin's suppressive effect on libido and testosterone (Shukla *et al.* 2009). Testosterone has been increased in infertile men without any impairment in seminal parameters following 5g of Mucuna Pruriens extract over 3 months (Shukla *et al.* 2009). Testosterone was also increased in the seminal experimental groups (those with low sperm motility or count), and to a more significant degree. Moreover, in rats with type II diabetes, increase in testosterone has been recorded with an oral dose of 200mg/kg bodyweight *Mucuna Pruriens* (Suresh and Prakash, 2010). The finding of the present study in contrast to these previous reports suggests that *Mucuna pruriens* action on increased testosterone could be much observed in infertile or ill healthy animals. *Mucuna pruriens* therefore appears not to trigger the action of testosterone in normal fertile bucks in spite of its effective action on sperm indices. The improved spermiograms observed in this study might be due to the presence of L-DOPA in *Mucuna pruriens* which serve as a precursor of neurotransmitter, acting as a nerviness tonic that prevents male sterility.

The effectiveness of using *Mucuna* seed powder over synthetic L-DOPA has been established by clinical trials (Hussain and Manyam, 1997). Kumar *et al.* (1994) reported that *Mucuna pruriens* is rich in L-DOPA, a precursor to the neurotransmitter dopamine, a marker of sexual pleasure (Molloy *et al.* 2006), besides having several other alkaloids and flavonoids. Therefore, reduction in abnormality following administration of *Mucuna pruriens* seed powder could be linked to high L-DOPA content of this antioxidant- rich plant (Sato *et al.* 1996). The improvement in sperm indices following treatment with *Mucuna pruriens* seed powder might probably be due to protective action of antioxidant in this plant against oxidative stress during spermatogenic process. *Mucuna pruriens* seed contains many bioactive constituents, including alkaloids, coumarins, flavonoids and alkylamines etc. which play an important role in increasing the antioxidant capacity (Misra and Wagner, 2007). Seeds of *Mucuna pruriens* also possess antioxidant, hypoglycemic, lipid lowering and neuroprotective activities (Sharma *et al.* 1978). Moreover, treatment with *Mucuna pruriens* might also contribute to proper functioning of male genital system and facilitate sperm transport (Fait *et al.* 2001). The results of the present study showed that oral administration of *Mucuna pruriens* seed powder to the bucks for 30 days did not only result in improved testicular measurements, sperm motility and sperm concentration but also led to reduction in sperm abnormality. Although the amount of nutrient intake was not measured in this study but the administration of *Mucuna pruriens* resulted in better sperm parameters measured. Muinga*et al.* (2003) reported that supplementation of diets with mucuna improved both dry matter intake and average daily gain of ruminant animals. In contrast, Chikagwa*et al.* (2009) reported that Mucuna supplementation in sheep diet reduced dry matter, nitrogen retention and crude protein intake. The difference in observation could be due to species of animal, methods of administration or processing of this plant material. The results of the present study indicated that constituents of *Mucuna pruriens* did not adversely affect reproductive functions of the bucks.

CONCLUSION

The findings from this study showed that oral administration of *Mucuna pruriens* to the bucks for 30 days resulted in improved testicular measurements and sperm motility but also maintained sperm abnormality at a low level. On the basis of results of the present study, it may be concluded that *Mucuna pruriens* improves semen quality without any adverse effect on structure of the reproductive physiology of the bucks.

Acknowledgement

The research was financially supported by Tertiary Education Trust Fund (TETFund) through the Institute of Food Security, Environmental Resources and Agriculture Research (IFSERAR) of Federal University of Agriculture Abeokuta, Nigeria under the grant number IRG48/2010. The Authors are grateful to Laboratory Technologists, Department of Animal Physiology, Federal University of Agriculture Abeokuta for their technical assistance.

REFERENCES

Alo, M.N., Okeh, O.C., Anyim, C., Orji, J.O. 2012. The effects of ethanol extract of *Mucuna pruriens* leaves on aspartate aminotransferase, alanine aminotransferase and alkaline phosphatase in albino rats. Journal Natural Production Plant Resource 2 (4), 465-470.

Bearden, H.J., Fuquay, J.W. 1997. Semen evaluation. In: Applied animal reproduction. 4th ed. New Jersey: Prentice Hall; pp. 158-169.

Chikagwa-Malunga, S.K., Adesogan, A.T.,Szabo, N.J., Littell, R.C., Phatak, S.C., Kima, S.C., Arriola, K.G.,Huisden. C.M., Deana, D.B., Krueger, N.A. 2009. Effect of replacing soybean meal with Mucuna on intake, digestibility,N balance and microbialprotein synthesis in sheep. Animal Feed Science and Technology 148, 107–123.

Duncan, B. D. 1955. Multiple Range Test and Multiple F-test, Biometrics 11, 1-42.

Ezekwe, A.G. 1998. Gonadal and extragonadal sperm reserve and testicular histometry of post pubertal Muturu bulls. Nigeria Journal of Animal Production 25, 106-110.

Fait, G., Vered, Y., Yogev, L., Gamzu, R., Lessing, J.B., Paz, G. 2001. High levels of catecholamines in human semen: a preliminary study. Andrologia 33, 347–50.

Gupta, A. Mahdi, A.A., Ahmad, M.K., Shukla, K.K., Bansal, N., Jaiswer, S.P., Shankhwar, S.N. 2011. A proton NMR study of the effect of *Mucuna pruriens* on seminal plasma metabolites of infertile males. Journal of Pharmaceutical and Biomedical Analysis 15, 55 (5):1060-6.

Hamilton, T., Stark, D. 1997. Beef bull fertility, Ministry of Agriculture and Food. Fact Sheet. Queens Printer for Ontario ag.infor@omf.gov.on.ca, pp: 1-7.

Hussain, G., Manyam, B. V. 1997. *Mucuna pruriens* proves more effective than L-Dopa in Parkinson's disease animal model. Phytotherapy Research 11, 419-423.

Knight, T.W. 1977. Methods of the indirect estimate of testes weight and sperm numbers in Merino and Romney rams. New Zealand Journal of Agricultural Research 20, 291 – 296.

Kumar, K.V.A., Srinivasan, K.K., Shanbhag, T. Rao, S.G. 1994. Aphrodisiac activity of the seeds of *Mucuna pruriens*. Indian Drug 31, 321–7.

Manyam, B.V., Dhanasekaran, M., Hare, T. 2004. Neuroprotective effects of the antiparkinson drug Mucuna pruriens. Phytotherapy Research 18, 706-712.

Misra, L., Wagner, H. 2007. Extraction of bioactive principle from *Mucuna pruriens* seeds. Indian Journal Biochemistry and Biophysics 44, 56–60.

Molloy, S.A., Rowan, E.N., O'Brien, J.T., McKeith, I.G., Wesnes, K., Burn, D.J. 2006. Effect of levodopa on cognitive function in Parkinson's disease with and without dementia and dementia with Lewy bodies. Journal of Neurology, Neurosurgery and Psychiatry 77, 1323–8.

Muinga, R.W., Saha, H.M., Mureithi, J.G., 2003. The effect of *Mucuna* (*Mucuna pruriens*) forage on the performance of lactating cows. Tropical and. Subtropical. Agroecosystems 1 (2–3), 87–92.

Notter, D.R., Lucas, J.R., Mcclaughertly, F.S., Copenhaver, J.S. 1985. Breed group differences in testicular growth patterns in spring-born lambs. Journal of Animal Science 60, 622-631.

Orji, B. I., Steinbach, J. 1976. Postnatal development of testis and epididymis in the Nigerian dwarf sheep. Proceedings of the 8th International Congress on Animal Production and Artificial Insemination 3, Krakow 73-76.

Osei-bonsu, P. 1998. On-farm trials of *Mucuna* species in Ghana. In: Cover crops in West Africa; contributing to a sustainable agriculture. Ottawa, Canada, IDRC, Ibadan, Nigeria, IITA, p. 201-202.

Paul, U., Joseph, E. 2001. Effect of *Mucuna pruriens* (horse eye bean) on the gonads of male guinea-pigs. Phytotherapy Research 15, 99-102.

Perry, G., Petterson, D. 2001. Determining Reproductive Fertility in Herd Bulls. University of Missiouri Agricultural Publication G, USA.

Sato, Y., Suzuki, N., Horita, H., Wada, H., Shibuya, A., Adachi, H. 1996. Effect of long term psychological stress on sexual behavior and brain catecholamine levels. Journal Andrology 17, 83–90.

Sharma, M.L., Chandhoke, N., Ray Ghatak, B.J., Jamwal, K.S., Gupta, O.P., Singh, G.B.1978. Pharmacological screening of Indian medicinal plants. Indian Journal of Experimental Biology 16, 228–35.

Shukla, K.K., Mahdi, A.A., Ahmad, M.K., Shankhwar, S.N, Rajender, S., Jaiswar, S.P. 2009. *Mucuna pruriens* improves male fertility by its action on the hypothalamus-pituitary-gonadal axis. Fertility and Sterility 92(6), 1934-40.

Shukla, K.K., Mahdi, A.A., Ahmad, M.K., Jaiswar, S.P., Shankwar, S.N., Tiwari, S.C. 2010. *Mucuna pruriens* Reduces Stress and Improves the Quality of Semen in Infertile Men. Evidence Based Complement Alternative Medicine 7(1),137-44.

Suresh, S., Prakash, S. 2010. Effect of *Mucuna pruriens (Linn.)* on sexual behavior and sperm parameters in streptozotocin-induced diabetic male rat. Journal of Sexual Medicine 9(12), 3066-78.

22

FECAL N-ALKANES VARIATION IN LACTATING DAIRY COWS GRAZING A TROPICAL PASTURE (*Cynodon nlemfüensis* VANDERYST VAR. NLEMFÜENSIS)

D.E. Oliveira[1*], S.R. Medeiros[2], L.O. Tedeschi[3], S.C. Silva[4], D.P.D. Lanna[4]

[1] *Santa Catarina State University, Lages, SC, Av. Luiz de Camões, 2090 - Conta Dinheiro, 88520-000, Brazil.*
E-mail: dimas.oliveira@udesc.br
[2] *Embrapa Beef Cattle, BR 262 km 4, 79002-970 - Campo Grande, MS - Brazil*
[3] *Texas A&M University, 230 Kleberg Center, College Station, TX 77843-2471, USA*
[4] *USP/ESALQ, C.P. 09, Piracicaba, São Paulo, 13418-900, Brazil*
**Corresponding author*

SUMMARY

The objective of this study was to measure fecal n-alkane concentration variation and herbage intake in grazing lactating dairy cows fed with two sources of fat: conjugated linoleic acid (CLA) or Megalac (Control). Cows were dosed with n-alkanes using a controlled-release capsule. There was no difference in herbage dry matter intake between CLA and Control. In the first period, there were no differences in fecal concentration of both natural and synthetic n-alkanes from the capsules. For the second period, there was a difference in fecal concentration for time of collection for all natural, dosed, and alkanes ratios. In the third period, there was a difference for time of collection for natural n-alkanes and their respective ratios, with higher excretion values recorded in the afternoon. In addition, there was an effect of day of collection for all n-alkanes and ratios, with higher excretion values recorded on days 2 and 4. Overall, our results highlight the importance of strict control of grazing management and sward structure in studies where external markers are used. Monitoring herbage, sward and animal managements might have an influence on n-alkanes excretion, causing a change in the ratio of their fecal concentration. These modifications may result in erroneous estimates of intake.

Key words: Diet Composition; Grazing Animals; Intake; Markers)

INTRODUCTION

Grazing animals modify their diet composition by selecting different plant species or plant parts (e.g., leaves and stem), increasing the difficulty in the determination of herbage intake and composition of the consumed herbage, hence interfering with the accuracy and planning of feeding programs. The most common method for estimating herbage intake in grazing ruminants is the calculation of fecal output (FO) using an external marker (e.g., chromium oxide) and herbage dry matter (DM) digestibility using *in vitro* procedures (Mayes and Dove, 2000). The assumptions of this method are that the external marker is fully recovered in the feces and a single herbage DM digestibility value is applicable to all animals involved. Additionally, the discrete infusion of markers has shown variations on diurnal and/or daily excretions on fecal concentration of markers resulting in erroneous estimates of intake. The use of natural to dosed n-alkanes ratio has been proposed as an alternative method in an attempt to overcome these limitations (Mayes et al., 1986) and it has been used to estimate herbage DM intake (Dove and Mayes, 1991; Oliveira et al., 2007; Oliveira et al., 2008). However, some studies with stall fed animals (Dillon, 1989) have shown that the n-alkanes technique may also present some problems such as daily and/or diurnal fecal concentration variations affecting estimates of herbage intake (Aguiar et al., 2013). In spite of that, data on fecal concentration variation in grazing animals on tropical pastures is scarce.

The main objectives of this study were to measure fecal n-alkane concentration variation and dry matter intake estimations in lactating dairy cows grazing stargrass (*Cynodon nlemfüensis* Vanderyst var. nlemfüensis), supplemented with two sources of fat.

MATERIALS AND METHODS

Animals, treatments, and experimental design

The experiment was designed to test the fecal n-alkanes variation and DM intake in lactating dairy cows fed with two sources of supplemental fat. All procedures regarding treatments, management and feeding animals, concentrate feeds, milk measures and analysis were described by Medeiros et al. (2010) and will not be addressed here. Briefly, the experiment was carried out in Valença, RJ, Brazil (43°42' W, 22°21' S) using thirty lactating Holstein x Gir crossbred cows receiving one of the following two treatments: 150 g/d of Megalac (Dwight & Church, salts of calcium from palm oil = Control) or 150 g/d of CLA-60 (Dwight & Church, salts of calcium of a mixture with 60% of isomers of CLA containing 24% of cis-9, trans-11, 35% of trans-10, cis-12, 15% of cis-8, trans-10, 17% trans-11, trans-13 and 9% of others).

All cows were rotationally grazed on a stargrass (*Cynodon nlemfüensis* Vanderyst var. nlemfüensis) pasture managed according to a 28-day grazing cycle (i.e., 2 days of pasture occupation, 26 days of pasture resting period) on fourteen 0.5 ha-paddocks. The experiment was divided into three 28-day evaluation periods during which estimates of herbage DMI were performed using the double n-alkanes technique (Mayes et al., 1986).

Herbage Sampling and Analysis

Herbage sampling (12 per period) was carried out on paddocks in which animals were grazing. The herbage sampling was performed during the same period in which fecal sample collection was performed to calculate DMI.

Herbage samples from each paddock were collected twice daily (morning and afternoon) using the hand-plucking procedure (Prates, 1974) during the two days of occupation. Samples were harvested during a 30-minute period each day to allow a larger proportion of the paddock area to be covered. Herbage samples were hand-separated into leaf (leaf lamina/blade), stem (leaf sheath + stem), and dead material. Herbage samples were dried at 65 °C in a forced-draught oven until constant weight. The results were used to calculate the percentage of each morphological component in sward herbage mass.

Concurrently, pre- and post-grazing herbage mass were measured by cutting the herbage at 10 cm from ground level using ten 0.25 m^2 metallic frames randomly positioned in each paddock used by animals at the time of sampling. Similarly, samples were dried in forced-draught oven as described above. Before grazing, four areas were randomly selected in each paddock to measure morphological composition of sward herbage mass above the post-grazing height (30 cm). Samples received the same treatment as herbage samples used to estimate morphological composition of sward herbage mass.

The hand-plucked herbage whole sample and components plus concentrate feed were analyzed for DM (method #934.01), ash (method #942.05), ether extract (EE, method #920.29) and crude protein (CP, method #988.05)) content according to AOAC (1997). Determinations of DM were also performed on fecal samples. Herbage samples were analyzed for neutral detergent fiber (NDF), acid detergent fiber (ADF), neutral detergent insoluble protein (NDFIP), acid detergent insoluble protein (ADFIP), and sulfuric lignin according to Van Soest et al. (1991). Because the detergent system is not accurate for animal byproducts (i.e. fish meal), fractions of NDF, ADF, NDFIP, ADFIP, and lignin of the concentrate were estimated based on individual composition of each ingredient and their corresponding proportion in the concentrate mixture. The rations were formulated using the Large Ruminant Nutrition System (LRNS; http://nutritionmodels.tamu.edu/lrns.html). The herbage carbohydrate and protein fractional degradation rates (%/h) were assigned to match those reported by Tedeschi et al. (2002). The chemical composition of the feeds is shown in Table 1. Each animal was individually characterized and data used to compute animal metabolizable energy requirement using the LRNS model. The results of feed analyses were used to estimate metabolizable energy content of feedstuffs used.

Milking management

Cows were milked at 0500 h and individually fed half of their daily concentrate allowance after milking. At 1130 h, animals were removed from the pasture and housed in a sheltered barn until the afternoon milking at 0230 h. After milking, cows were fed the second half of their daily concentrate allowance and at 0400 h they were returned to the pasture.

Herbage intake and n-alkane analysis

Cows were dosed with a MCM Alkanes code 60421 – Captec, NZ controlled-release capsule containing 8 g of n-hexatriacontane (C_{36}) and 8 g of n-dotriacontane (C_{32}) to estimate herbage DMI, in each of the 28-day periods of measurement. The release rate used for the C_{32} marker was 345 mg/d as previously determined by Oliveira et al. (2008). Eight days after dosing the n-alkane capsules, fecal samples were directly taken from the rectum of the animals immediately after the morning (0500 h) and afternoon (0230 h) milking for five consecutive days.

For the first and second periods of evaluation, samples from each cow from five days of collection were pooled for each collection time of the day (morning and afternoon). For the third period, for each animal, samples were analyzed separately for each different collection time and day. All fecal samples were first stored at -20 °C and then dried at 65 °C in a forced-draught oven until constant weight. Dried fecal samples were ground to pass a 1 mm mesh sieve and stored in plastic containers for subsequent n-alkane analyses.

Table 1. Chemical composition and LRNS estimated metabolizable energy for herbage and concentrate

Sample	n[1]	CP	EE	NDF	NDFIP	ADF	ADFIP	LIGN	ASH	ME
Period 1	12									
Herbage		14.8	2.3	64.8	46.8	30.9	14.9	4.6	9.6	1.95
Leaf		19.4	2.9	66.3	55.6	29.7	12.0	4.8	9.3	2.02
Stem		11.3	0.9	71.3	45.4	34.7	22.5	6.3	9.5	1.64
Concentrate		24.7	5.6	11*	6.6*	3.9*	2.5*	3.3*	16.1	2.99
Period 2	12									
Herbage		14.2	1.7	71.1	42.8	34.3	17.4	2.8	9.6	1.91
Leaf		17.0	2.4	75.1	72.1	32.9	15.8	2.3	8.1	2.11
Stem		11.8	1.3	81.8	65.8	40.7	15.7	4.5	10.1	1.66
Concentrate		25.5	5.6	11*	6.6*	3.9*	2.5*	3.3*	16.1	2.79
Period 3	12									
Herbage		14.3	1.9	66.9	41.2	27.5	12.4	5.2	9.9	1.96
Leaf		16.9	2.7	67.4	60.8	28.7	13.0	3.4	9.0	2.26
Stem		11.8	1.3	72.0	41.8	36.1	13.4	5.0	10.1	1.84
Concentrate		25.5	5.6	11*	6.6*	3.9*	2.5*	3.3*	16.1	2.79

* Estimated using individual ingredient composition and their percentage of inclusion in the concentrate. Where: CP = crude protein; EE = ether extract; NDF = neutral detergent fiber; NDFIP = insoluble protein bound to the NDF (%CP); ADF = acid detergent fiber; ADFIP = insoluble protein bound to the ADF (% CP); LIGN = lignin; ASH = ash; ME = metabolizable energy (Mcal/kg DM estimated by LRNS); (DM basis, %).

[1] Sample of three paddocks.

The n-alkane concentrations in herbage and fecal samples, tablets of controlled-release capsules, and concentrate feeds used were determined in duplicates using the method described by Dillon and Stakelum

(1990) and modified by Oliveira (2004). Basically, modifications corresponded to: 1) n-heptane was used as solvent; 2) during the saponification procedure a bath-water was used instead of a dry-block heater and; 3) plastic 5 mL syringes were used to replace the disposable columns (Oliveira and Tedeschi, 2010). The n-alkane determinations and quantifications were made as described by Oliveira et al. (2008). Two tablets from two controlled-release capsules were extracted for calibration purposes and presented an average of 1540 mg (±63 mg) of C_{32} and 1390 mg (±12 mg) of C_{36}. Herbage DMI was computed as proposed by Mayes et al. (1986) using the C_{33}:C_{32} and C_{31}:C_{32} n-alkane ratios.

Statistical analysis

For the first and second periods of study, the variations on fecal concentrations of C_{31}, C_{32}, C_{33}, C_{35} and C_{31}:C_{32} and C_{33}:C_{32} n-alkane ratios in relation to time of sample collection (morning and afternoon) were analyzed considering as sources of variation treatments and time of collection as well as their interactions. For the third period of measurement, analysis of variation was performed as described, but also separately for days of collection (day 1 until day 5). In this case, sources of variation corresponded to treatments, time of collection and days of collection, including their interactions. All analyses were performed using PROC MIXED of SAS (SAS Inst. Inc., 2000) and when the interaction was not significant, it was excluded from the model. The DMI was analyzed and means estimated using LSMEANS. Comparisons, when appropriate, were performed using the F test and 5% probability.

RESULTS

The effects of CLA on the milk yield, composition, fatty acid profile, blood parameters, and reproduction were reported by Medeiros et al. (2010) and will not be addressed. The average pre-grazing herbage mass for the first, second and third periods of study were 3810 (±230), 4290 (±600) and 1800 (±170) kg/ha, respectively. The corresponding post-grazing herbage mass were 1730 (±340), 1606 (±128) and 1310 (±290) kg/ha. The proportion of leaf, stem, and dead material in the pre-grazing herbage mass was 47.9% (±3.1), 51.5% (±3.1), 0.6% (±0.06) for the first period, 32.8% (±7.2), 52.4% (±11.4), 14.9% (±4.8) for the second period and 50.8% (±14.0), 44.1% (±9.3), 5.0% (±6.7) for the third period, respectively.

n-Alkane profile

The n-alkane profile for the whole herbage, its morphological components, and the concentrate are shown in Table 2 (data for even-chained alkanes are not shown). The n-alkanes with carbon chain length varying from C_{22} to C_{35} were quantified. The greatest concentrations were recorded for C_{33}, C_{31} and C_{29} n-alkanes.

Table 2. N-alkane content of the dietary feeds offered to the animals used to estimate the herbage intake

Period/Sample	C_{23}	C_{25}	C_{27}	C_{29}	C_{31}	C_{32}	C_{33}	C_{35}
Period 1								
[1]HP Herbage	6	15	31	39	79	10	114	25
Leaf	12	25	58	90	122	7	133	15
Stem	43	83	118	113	180	29	256	76
Concentrate	2	3	3	3	1	nd	nd	nd
Period 2								
[1]HP Herbage	17	25	37	51	107	10	171	37
Leaf	2	6	21	49	108	5	141	18
Stem	5	10	13	24	82	3	157	39
Concentrate	50	79	54	26	10	8	6	3
Period 3								
[1]HP Herbage	35	55	68	71	106	20	137	27
Leaf	55	88	99	128	127	37	168	24
Stem	7	12	17	31	88	5	163	36
Concentrate	50	79	54	26	10	8	6	3

[nd] n-alkane not detected; [1] HP herbage sample by the hand-plucked method; n-alkane content (mg/kg of DM) was determined in duplicate.

Intake estimates of herbage

Grazing period 1

The average DM herbage intake (CLA= 11.2 ±1.9, Megalac= 11.1 ±2.6 kg/d, $p>0.05$) was not different between treatments. In the investigation of the concentration of individual and/or n-alkanes ratio in fecal samples, there was no effect of treatment neither interaction between treatment and time of collection on fecal concentrations of individual C_{31}, C_{32}, C_{33}, C_{35} alkanes as well as C_{31}:C_{32} and C_{33}:C_{32} ratios. There was an effect of time of collection (morning and afternoon) on the individual concentration of C_{31}, C_{33} and C_{35} alkanes in the herbage in which the greatest values of fecal concentrations were recorded during the afternoon period (Table 3).

Grazing period 2

There was no treatment effect on the herbage DMI (CLA= 9.0 ±2.1, Megalac= 8.2 ±2.1, kg/day, $p>0.05$). There was no effect of treatment on the individual C_{31}, C_{32}, C_{33}, C_{35} fecal concentrations, but there was an effect of treatment on C_{31}:C_{32} and C_{33}:C_{32} ratios. There was an effect of time of collection on all individual n-alkanes and n-alkane ratios (Table 3).

Grazing period 3

There was no treatment effect on the herbage DMI (CLA= 11.8 ±2.4, Megalac= 12.6 ±2.4, kg/day, $p>0.05$). The least-square means for treatments, time of collection, days of collection, and their interactions on individual n-alkanes and fecal concentration ratios of n-alkanes are listed in Table 4. There was an effect of day of collection on C_{31} fecal concentration, which also varied with time of collection with greater concentration found in the afternoon. The fecal concentration of C_{32} comprises both the C_{32} coming from pasture and from the slow-release capsule. The fecal concentration of C_{33} varied with time of collection (morning= 323.1 vs. afternoon= 367.6 mg, $p<0.01$) and an effect of days of collection was observed with the greatest concentration occurring on day two (467.4 mg/kg of DM). Fecal concentration of C_{35} varied with time of collection (morning= 67.1 vs. afternoon= 76.8 mg, $p<0.01$), and during days of collection ($p<0.01$). The fecal concentration ratios of C_{31}:C_{32} and C_{33}:C_{32} varied with time and days of collection ($p<0.01$).

Table 3. Least square means of individual and fecal ratios concentration for the n-alkanes (mg/kg DM) of C_{31}, C_{32}, C_{33}, C_{35}, and C_{31}:C_{32} and C_{33}:C_{32} ratios on the first and second periods

Alkane/ Ratio	P1[1]		p[3]			P2[2]		p[3]		
	M[4]	A[5]	Tr[6]	TC[7]	TrXTC[8]	M[4]	A[5]	Tr[6]	TC[7]	TrXTC[8]
C_{31}	217.8	240.8	ns	*	ns	253.7	275.7	ns	**	ns
C_{32}	89.9	93.3	ns	ns	ns	123.5	137.1	ns	**	ns
C_{33}	236.4	261.9	ns	**	ns	420.8	453.0	ns	**	ns
C_{35}	42.4	47.5	ns	**	ns	92.3	99.2	ns	**	ns
C_{31}:C_{32}	2.43	2.64	ns	ns	ns	2.14	2.32	**	**	ns
C_{33}:C_{32}	2.64	2,88	ns	ns	ns	3.56	3.84	**	**	ns

*($p<0.05$); **($p<0.01$); [ns](non significant); P1= grazing period 1; P2= grazing period 2; p= probability; M= morning; A= afternoon; Tr= treatment; TC= time of collection; Tr x TC= interaction between treatment and time of collection; mg/kg DM.

Table 4. Least square means of individual and fecal ratios concentration for the n-alkanes (mg/kg DM) between collection time and days of collection on the grazing period 3

Marker	Treatment		Collection Time		Day of Collection				
	CLA	Meg	M	A	1	2	3	4	5
C_{31}	258.3^a	249.2^a	237.4^b	270.1^a	250.8^b	344.1^a	222.4^c	231.5^c	219.8^c
C_{32}	117.0^a	116.5^a	114.0^a	119.5^a	129.8^b	160.1^a	100.6^c	99.4^c	93.8^c
C_{33}	349.9^a	340.8^a	323.1^b	367.6^a	296.4^d	467.4^a	319.2^{bc}	336.9^b	306.9^{cd}
C_{35}	73.4^a	70.5^a	67.1^b	76.8^a	57.6^d	92.0^a	67.4^c	75.4^b	67.4^c
C_{31}:C_{32}	2.29^a	2.18^a	2.15^b	2.33^a	1.99^b	2.23^a	2.25^a	2.35^a	2.37^a
C_{33}:C_{32}	3.14^a	3.01^a	2.96^b	3.19^a	2.35^d	3.04^c	3.24^{ab}	3.43^a	3.32^a

CLA= conjugated linoleic acid; Meg= megalac; M= morning; A= afternoon; Different small letters between treatments within each marker, within each collection time and days of collection are significantly different ($p<0.01$).

DISCUSSION

The results of n-alkane profile of our study are in agreement with the results from other studies with tropical pasture species (Oliveira et al., 1997; Delgado et al., 2000; Genro et al., 2001). There was a variation in the concentration of C_{31}, C_{33}, and C_{35} in the whole herbage samples as well as in the leaf and stem fractions between periods of evaluation. Similar results were reported by Laredo et al. (1991) for *Pennisetum glaucum* and *Sorghum* sp, Oliveira et al. (1997) for *Pennisetum purpureum* cv Napier, and Genro et al. (2001) for *Brachiaria brizantha* cv Marandu, *Panicum maximum* cv Mombaça and *Pennisetum purpureum* cv Cameroon.

The first period of study indicated the basic assumption of no variation in the n-alkanes ratios, using the pair C_{33}:C_{32} to estimate DMI as suggested by Mayes et al. (1986). The effects of time and day of collection on fecal concentration of individual n-alkanes and on the n-alkane ratios were probably caused by the management of animals. They were retrieved from paddocks at 1130 h to a sheltered place, returning to pasture after the afternoon milking, around 0230 h. Therefore, grazing was more intense during the afternoon period and at night than during the morning period, certainly resulting in variations in rumen fill. A larger rumen fill after the more intense grazing period might have caused larger excretions and greater fecal concentrations of n-alkanes in the afternoon period of the following day once concentrate amounts were constant and only forage DMI varied. Therefore, differences in rates of passage of both solid and liquid phases of the digesta throughout the day might have resulted in larger diurnal variation in fecal concentration of the markers. Moreover, changes in sward structure (e.g. leaf to stem ratio) may have resulted in consumption of different plant parts, altering fecal concentration of the markers used. Dillon (1989) worked with lactating cows and found diurnal variations for the pairs C_{31}:C_{32}, C_{33}:C_{32} and C_{35}:C_{36} of n-alkanes, pointing out that feeding management might have contributed to those variations.

The changes in fecal concentration of alkanes with day of collection were likely a consequence of variations in morphological composition of the pasture. As grazing progressed, the effect on fecal concentration of the marker could be related to the depletion of leaf in sward herbage mass, a fact reinforced by the smaller proportion of the leaf component in the following paddocks of the grazing sequence (Figure 1).

In the second period, the difference between treatments was due to differences in rumen metabolism of fatty acids, which might have influenced fiber digestion by altering passage rate and modifying fecal concentration of the markers. According to Mayes et al. (1986), smaller fecal concentrations of natural C_{27} and C_{29} alkanes were found when animals were receiving a synthetic source of C_{28} and C_{32} n-alkanes with stearic and palmitic acids mix. On the other hand, Ohajuruka and Palmquist (1991) measured the intake of lactating dairy cows using external markers and found no effect of added fat on the concentration and fecal recovery of C_{31} and C_{32} n-alkanes.

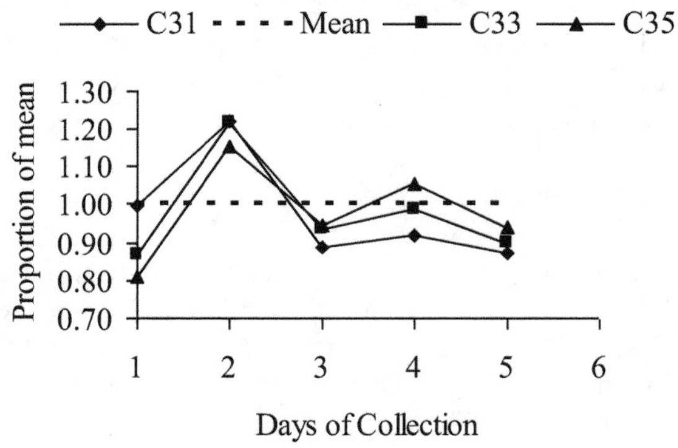

Figure 1 - Mean variation in the fecal concentrations for C_{31}, C_{33} and C_{35} (mg/kg of DM), expressed as proportions of the mean concentration for each cow over five days.

Although it was not in the scope of this study, monitoring and maintenance of swards under control is important because as grazing occurs, changes in sward structure take place throughout the pasture growing season and could affect intake and performance of animals by altering the amount and composition of the herbage consumed. Vulich et al. (1993) suggested that pasture should be sampled frequently because of variations in n-alkanes concentration between days and weeks.

Fecal concentration of C_{33} changed between morning and afternoon, but because this n-alkane had the same concentration in both leaf and stem fractions, fecal concentration was maintained by the increase in intake of the stem component as grazing progressed (Table 4).

Fecal concentration of C_{35} might represent an increase in consumption of stems relative to leaves because the leaf to stem ratio of sward herbage mass decreased during grazing and this n-alkane appeared in larger concentrations in the stem relative to the leaf fraction (Table 4).

Although fecal concentration of C_{33} varied over time, variation in the ratios of C_{31} and C_{33} relative to C_{32} must have been caused by variation in fecal concentration of the C_{32} n-alkane (Figures 2 and 3), since the dose was the same each day, being associated with a decreasing intake of C_{32} from pasture because this n-alkane was found in higher concentrations on leaves relative to stems. This finding is in agreement with those of Dove et al. (1991). Those authors reported variation in individual fecal concentrations of C_{29} n-alkane from the diet, C_{28} and C_{32} from the dosed slow-release capsule and C_{36} fed via paper pellet.

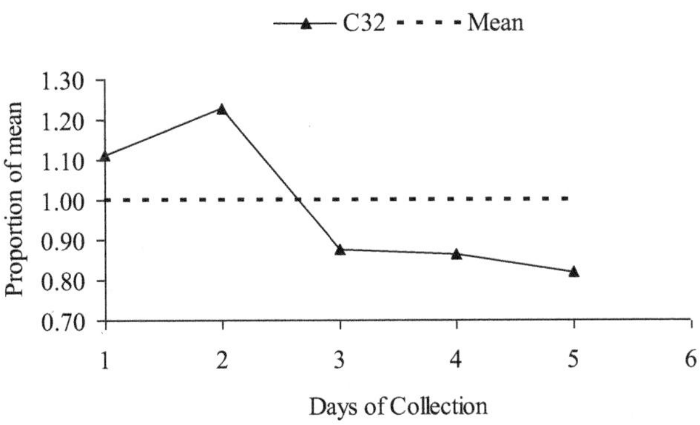

Figure 2 - Mean variation in the fecal concentrations of the C_{32} (mg/kg of DM), expressed as proportions of the mean concentration for each cow over five days.

Dove et al. (1992) found an interaction between herbage intake and feeding level on fecal concentration of the n-alkanes used in their study. It is possible that in this study the changes in sward herbage mass between the two grazing days (occupation days) of each sampling paddock produced a similar effect characterized by high feeding level on the first day and low feeding level on second day, influencing the fecal concentration of n-alkanes. Stakelum and Dillon (1990) used lactating dairy cows to test different patterns (hours) and feeding levels and found a marked daily variation in fecal concentration ratio of the pairs C_{31}:C_{32}, C_{33}:C_{32} and C_{35}:C_{36}. These diurnal and/or daily systematic variations in fecal concentrations of n-alkanes can also be explained by changes in grazing patterns caused by changes in sward structure and selective grazing.

As pointed out by Dove and Mayes (1991) and Mayes and Dove (2000), variations in fecal concentration of individual n-alkanes may not interfere with estimates of herbage intake provided that variation in fecal concentration ratio of the pair of n-alkanes used in the estimation process remains stable.

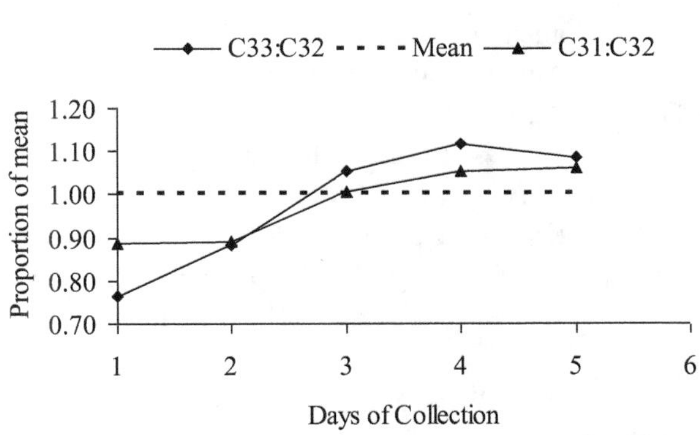

Figure 3 - Mean variation of the ratio in the feces of C_{31} and C_{33} odd- chain (diet) n-alkane concentration to C_{32} even-chain (dosed) n-alkane concentration (mg/kg of DM) expressed as proportions of the mean concentration for each cow over five days.

CONCLUSIONS

Overall, our results highlight the importance of strict control of grazing management and sward structure in studies where external markers are used. We concluded that monitoring herbage and sward and animal managements might have an influence on n-alkanes excretion, causing a change in the ratio of their fecal concentration. These modifications may result in erroneous estimates of intake.

Conflict of interest

The authors declare that they have no conflict of interest.

Acknowledgment

We thank the Fundação de Amparo à Pesquisa do Estado São Paulo (FAPESP) for the financial support to this study and a scholarship for the first author.

REFERENCES

Aguiar, A.D., Forbes T.D.A., Rouquette Jr., F.M., Tedeschi, L.O., and Randel, R.D. 2013. Evaluating the statistical variation in estimating forage dry matter intake of grazing Brahman bulls using n-alkanes. Journal of Agricultural Science, Cambridge, 151: 129-140.

AOAC, 1997. Official methods of analysis. 16th edn. Gaithersburg.

Delgado, D.C., Oliveira, M.C., Navarro, A., 2000. Composition of cuticular n-alkanes in tropical plants. Their potential as markers to estimate consumption and selection of grazing ruminants. Cuban Journal of Agricultural Science 34: 147-152.

Dillon, P.G., 1989. The use of dosed and herbage n-alkanes to estimate herbage intake with dairy cows. MSc. Thesis, National University of Ireland, Cork.

Dillon, P.G., Stakellum, G., 1990. The analysis of n-alkanes in faeces and herbage. In: Proceedings European Grazing Workshop, Wageningen, 7.

Dove, H., Mayes, R.W., 1991. The use of plant wax alkanes as marker substances in studies of the nutrition of herbivores: a review. Australian Journal of Agricultural Research 42: 913-952.

Dove, H., Mayes, R.W., Lamb, C.S., Ellis, K.J., 1991. Evaluation of an intra-ruminal controlled-release device for estimating herbage intake using synthetic and plant cuticular wax alkanes. In: Proceedings of the International Symposium Nutrition Herbivores, 3, p. 82.

Dove, H., Siever-Kelly, C., Leury, B.J, Gatford, K.L., Simpson, R.J., 1992. Using plant wax alkanes to quantify the intake of plant parts by grazing animals. In: Proceedings of Nutrition Society Australian 13, Melbourne. p. 149.

Genro, T.C.M., Prates, E.R., Thiago, L.R.L.S., Herrero, M., Sabatel, V. Identification and quantification of n-alkanes in three tropical grasses, 2001. In: Proceedings of the International Grassland Congress, 19, Águas de São Pedro. pp. 402-403.

Laredo, M.A., Simpson, G.D., Minson, D.J., Orpin, C.G., 1991. The potential for using n-alkanes in tropical forages as a marker for determination of dry matter by grazing ruminants. Journal of Agricultural Science, Cambridge, 117: 355-361.

Mayes, R.W., Lamb, C.S., Colgrove, P.M., 1986. The use of dosed and herbage n-alkanes as markers for determination of herbage intake. Journal of Agricultural Science, Cambridge, 107: 161-170.

Mayes, R.W., Dove, H., 2000. Measurement of dietary nutrient intake in free-ranging mammalian herbivores. Nutritional Research Reviews 13: 107-138.

Medeiros, S.R., Oliveira, D.E., Aroeira, L.J.M., Mcguire, M.A., Bauman, D.E., Lanna, D.P.D., 2010. Effects of dietary supplementation of rumen-protected conjugated linoleic acid to grazing cows in early lactation. Journal of Dairy Science 93: 1126-1137.

Ohajuruka, O.A., Palmquist, D.L., 1991. Evaluation of n-alkanes as digesta markers in dairy cows. Journal of Animal Science 69: 1726-1732.

Oliveira, D.E., 2004. Determinação de alcanos. Manual de extração e análise cromatográfica em forragens, concentrados e fezes. FEALQ, Piracicaba. 30p.

Oliveira, D.E., Prates, E.R., Peralba, M.C.R., 1997. Identificação e quantificação de n-alcanos presentes nas ceras de plantas forrageiras. Revista Brasileira de Zootecnia. 26: 881-886.

Oliveira, D.E, Medeiros, S.R., Tedeschi, L.O., Aroeira, L.J.M., da Silva, S.C., 2007. Estimating forage intake of lactating dual-purpose cows using chromium oxide and n-alkanes as external markers. Scientia Agricola. 64: 103-110.

Oliveira, D.E, Manella, M.Q., Tedeschi, L.O., da Silva, S.C., Lanna, D.P.D., 2008. N-alkanes to estimate the voluntary forage intake of cattle using controlled-release capsules. Scientia Agricola. 65: 230-238.

Oliveira, D.E, Tedeschi, L.O., 2010. Comparing plastic syringes and disposable columns to extract n-alkanes in forage and concentrate feeds. Ciencia Rural, 40, 8: 1844-1847.

Prates, E.R. 1974. Nutritional evaluation of pensacola bahiagrass pasture by animal and laboratory techniques. PhD Thesis, University of Florida, Gainesville, Florida.

SAS Institute Inc. 2000 SAS/STAT Guide for personal computers. SAS Institute Inc., Cary, North Carolina.

Stakelum, G., Dillon, P., 1990. Dosed and herbage alkanes as feed intake predictors with dairy cows: The effect of feeding level and frequency of perennial ryegrass. In: Proceedings of European Grazing Workshop, 7, Wageningen, 1990, 1v.

Tedeschi, L.O., Fox, D.G., Pell, A.N., Lanna, D.P.D., Boin, C., 2002. Development and evaluation of a tropical feed library for the Cornell Net Carbohydrate and Protein System model. Scientia Agricola. 59: 1-18.

Van Soest, P.J., Robertson, J.B., Lewis, B.A., 1991. Methods for dietary fiber, neutral detergent fiber, and nonstarch polysaccharides in relation to animal nutrition. Journal of Dairy Science. 74: 3583-3597.

23

ADAPTING A BITE CODING GRID FOR SMALL RUMINANTS BROWSING A DECIDUOUS TROPICAL FOREST

P.G. González-Pech[*a1,2], **J.F.J. Torres-Acosta**[1], **C.A. Sandoval-Castro**[1]

[1]*Facultad de Medicina Veterinaria y Zootecnia, Universidad Autónoma de Yucatán.*
[2]*Centro Multidisciplinario de Educación, Ciencia y Cultura.*
Km 15.5 Carretera Mérida-Xmatkuil, Apdo. 4-116 Itzimná. C.P. 97100, Mérida, Yucatán, México. Emails: pedro.gonzalez@uady.mx, tacosta@uady.mx, carlos.sandoval@uady.mx
**Corresponding author*

SUMMARY

The aim of this study was to build a bite coding grid that considers the diversity of plant parts consumed by sheep and goats in the heterogeneous vegetation of a deciduous tropical forest. The architecture of plant species and the form and size of the plant parts consumed by sheep and goats grazing in Yucatán, México (20°52'N, 89°37'W) were recorded. This information was used to create a bite coding grid with 33 bite categories (BC). This new grid was validated using the direct observation method with 3 sheep and 3 goats during the dry and rainy seasons. The coding grid enabled the classification of all the plant architectures consumed (grasses, herbs, leafs of woody mono- and dicotyledoneans, bipinnate leafs and creeper plants) as well as other parts of plants consumed by the animals (round/oblong leafs, elongated leafs, steams, vines, flower, fruits, pods). It also considered the mixture of dry leafs and pods consumed directly from the ground. The size (length) of the BC varied from < 1 cm to up to 20 cm. The new grid incorporated BC specific for the deciduous tropical forest of Yucatán, such as plants with bipinnate leafs, creepers' leafs and vines, as well as fallen pods collected from the ground. Of the 33 BC possible in the new coding grid, 28 and 32 BC were used during dry and rainy seasons respectively. Thus, the adapted bite coding grid can be used together with the direct observation method to investigate feeding behavior of sheep and goats browsing in the tropical forests.

Key words: Feeding behaviour; direct observation; bite categories; sheep; goats; bite categories.

INTRODUCTION

Rangelands are heterogeneous sources of food with multi-stratified distribution of feed resources and seasonal variation in terms of nutritional quality and the quantity of available biomasses (Meuret et al., 1991). Previous exposure to grazing by animals could influence the range of plant species and plant materials that are considered a feed resource (Provenza, 2003). In the case of the low deciduous tropical forests of the Yucatan, the vegetation is composed of more than 260 plant species, including a wide variety of bushes, trees, creepers, climbers, and herbs (Flores et al., 2006). This variability reflects the typical circumstance in which animals are fed. The direct observation of the feeding behaviour is a methodology that was developed and validated for the identification of the feeding behaviour either *in situ* (Reppert, 1960; Neff, 1974; Bourbouze, 1980) or under simulated free grazing conditions (Meuret, 1988). The improved direct observation method developed by Agreil and Meuret (2004) allowed for the continuous observation of the feeding behaviour of sheep and goats. This method has been used for conducting research on feeding behavior and identify the diversity and variability of part plants consumed by ruminants in grasslands and shrubby rangelands (Agreil et al, 2005, Pontes et al, 2010) of temperate regions. Those results showed that small ruminants were capable of adjusting the size of their bites, as well as the plant species they consumed.

The feeding behavior of free grazing animals under tropical conditions is scarce and have focused on preference studies (Becker and Lohrmann, 1992; Kronberg and Malechek, 1997). In the Yucatan, the pioneering work on this topic was developed in the first half of the 1980's for hair sheep (Ortega-Reyes et al., 1985) and for goats (Rios and Riley, 1985). Both studies reported qualitative information for the plants consumed, but lacked quantitative measurements for the ingested plant materials (plant species and plant parts), as well as the size and mass of the bites, which may be used to characterize the harvesting strategies of sheep and goats in this tropical forest. This information is needed for the sustainable management of the vegetation and could also be applied to investigate supplementation strategies adapted to the conditions of the tropical forests of Yucatan. To achieve this, it is essential to adapt a tool for the conditions of the heterogeneous vegetation in the tropical forests. The objective of this study was to build a bite coding grid that considers the diversity of plant parts consumed by sheep and goats in the heterogeneous vegetation of a deciduous tropical forest.

MATERIALS AND METHODS

Study site

The study was conducted in the central zone of Yucatán, 15.5 km South of Mérida, México (20°52' 7.14" N and 89°37' 24.04" W). The prevailing climate is tropical warm sub-humid with summer rainfall (AW_0). The study focused on the vegetation of a low deciduous tropical forest (112 ha) (Flores et al., 2006). The observations were conducted from January to May 2012, and from July to August covering the dry and rainy season.

Plant architecture, plants species, and plant parts consumed

The harvesting behaviour of sheep and goats was observed to determine the plant architectures, plant species, and plant organs consumed during the biting action or collected from the fallen leafs on the ground. Observations were performed over three consecutive days at two different time points during January and February 2012 (four observations points). Observations were performed ten days apart. A flock of 30 sheep and 70 goats was accompanied on their normal grazing circuits, which lasted for four hours. The consumption of all the plant materials was recorded for three sheep and three goats previously habituated to the presence of the human observers. Data was recorded from the different plant materials harvested to obtain information concerning the plant and bite arquitecture:

(i) The architecture of plants. The shape/form of the harvested plants was classified according to the following categories: (a) grasses, (b) herbs, (c) ramified woody plants, (d) woody plants of bipinnate leafs, and (f) climber or creeper plants.

(ii) The botanical identification of the plant materials consumed. The botanical identification of the plants consumed by sheep or goats was conducted by the staff of the herbarium at the Faculty of Veterinary and Animal Science (UADY).

(iii) The architecture of bites. A bite was defined as the material that was harvested with the mouth, either directly from the plants or collected from the floor of the tropical forest. The bites included a variety of harvesting procedures, including: cutting, ripping out, plucking, stripping, or picking up with the animal's mouthparts. The bites varied in size (from the very small leafs removed one by one, to the large branches separated from the plant in a single bite). The definition of a bite did not consider either mastication or deglution. Thus, the architecture of the bites was the form of the materials removed by the biting

action, and is categorised as round or oblong leafs, elongated leafs, steams, vines, leafs with steams, leafs with vines, flowers, leafs and/or steams with flowers, leafs and/or vines with flowers, and pods. The materials were measured (cm) from simulated bites performed on plants similar to those harvested (Wallis De Vries, 1995).

Adapting the bite coding grid for a tropical forest

The information obtained from the preliminary observations on the plant species, the different plant architectures, the architecture of the observed bites, and the range of sizes observed (cm) was used to adapt the coding grid of Agreil and Meuret (2004) for deciduous tropical vegetation. The original version consisted of 41 possible BC, describing (in cm) the part clipped on plants from grasses (12 BC), shrubs and ramified dicotyledonous herbaceous plants (seven BC), tree foliages (eight BC), creeper plants (four BC), flowers (five BC), and fruits (five BC). This coding grid had been used to identify a total of 105 different plant species harvested by sheep and goats in different temperate ecosystems containing shrubby and grassland vegetation (Agreil and Meuret, 2004). Due to the size of the leaves and stems of some tropical plants, such as the bipinnate, it was difficult to estimate the BC at 1 cm of precision. Therefore, the new grid only took into consideration larger sizes ranges (i.e. 1–3 cm) rather than the BC included in the original BC grid (i.e., 1 cm, 2 cm, and 3 cm). The grass plants of different tiller (with few or many stems) were included as one classification only. Therefore, the new grid did not require six BC for grass plants. The shrubs, trees plants, and fallen leafs are included in the same BC. Thus, three BC from the original coding grid were not used. The new grid did not use the eight BC from the original grid to describe flowers and fruits, but rather it considered one BC for flowers and one BC for fruits, since their size is closely related to each plant species, which were always identified during the BC data collection. Finally, nine additional BC were designed to describe materials removed from creepers (two new BC), woody bipinnate plants (six BC), and pods collected from ground (one BC).

Validation of the adapted bite coding grid using sheep and goat in a tropical forest

The improved direct observation of feeding behaviour (Agreil and Meuret, 2004) was implemented in the study site using the new BC and the adapted grid. We used the same flock used during the identification of the plant architectures and plant species harvested. Three sheep and three goats (already familiarised to the constant presence of human observers) were observed from March to May 2012 and from July to August covering the dry and rainy season. For each animal, the feeding behaviour was observed during three consecutive days using the focal sampling technique (Altman, 1974). Behaviour was observed at three different times every day: (i) early morning (7:00–8:20 AM), (ii) mid-morning (8:20–9:40 AM), and (iii) late-morning (9:40–11:10 AM). Animals were observed according to a Latin square design, allowing for observations at different time periods during grazing. The procedure accounted for the heterogeneity of available plants due to grazing area, environmental conditions, and the change of the shepherd. Animals were closely followed while observers registered on audio devices the plant species and the corresponding codes from the adapted BC grid during browsing. Every 10 days, a three day observation period was performed, and a total of four complete days of grazing were recorded for every experimental animal.

RESULTS

Plant architecture, plants species, and plant parts consumed

Thirty-four different plant species were identified as effectively harvested by sheep and goats during the dry and rainy seasons. The leaves of woody and woody bipinnate plants accounted for 42.4% (n = 14) of the plants species harvested, followed by herbs (33.3%; n = 11), creepers (15.2%; n = 5). The grass represented only 9% (n = 3) and 11% (n = 4) of the species harvested in the dry and rainy seasons respectively (Table 1). The architecture of the bites included: round, oblong leafs, bipinnate leafs, elongated leafs, steams, vines, flowers, fruits, pods, and fallen dry leafs consumed from the ground. The size of the harvested materials ranged from less than 1 cm to 20 cm (Table 1).

Description of the adapted BC grid

Figure 1 represent the adapted bite coding grid which consisted of 33 BC descriptors. The description of each BC correspond to the following aspects:

The ideograms. The ellipse schematises the mandible. The circles and straight lines illustrate the leafs and stems. The small triangles connected by a sinuous line schematise the BC for creepers plants. The small, oblong, parallel circles represent the leaves of bipinnate plants.

The columns. From left to right, the first column classifies the harvest, consisting of single rounded or oblong leafs on herbs, woody shrubs and trees, and fallen leafs and creeper plants. The second column classifies the same plants, but for those cases when the bite included more than one leaf and a portion of the stems or branches of plants. The third column classifies the bites on creeper plants when they include leafs and vines. The fourth column classifies the bites of woody plants of bipinnate leafs (steam, leaf, and branch included). The fifth column categorise the bites on leafs and stems of grasses, as well as herbs with straight elongated leafs.

The rows. Every row corresponds to a limited range of bite size depending on the length of the harvested material (not stretched or compressed). The size is indicated in cm at the left of the grid.

The code names of every BC. Mono and bi-syllabic codes placed above the icons identifies every BC. The new grid includes thirty BC codes (five columns and six rows) and three codes that were not associated with a particular size: "flo" identified the harvest of flowers, "fru" was used for the fruits, and "vai-sol" (the only bi-syllabic code) was used for the pods collected from the ground. These three codes could be used as single BC or in addition to the any other BC when the harvesting included more than one of these structures.

These codes were designed to be easy to learn and to dictate after the name of the plant species during the feeding observation of small ruminants. Thus, the first two or three words in the codes correspond to especific columns and to the particular architecture of the leaves, stems, and branches of the plants. The last vowel of the code correspond to the rows and their alphabetical order is determined by the size of the corresponding BC.

Validation of the adapted BC grid for sheep and goat in the tropical forest

Of the 33 BC available in the adapted BC grid, 28 and 32 categories were used during the observations at dry and rainy season (Table 2). The largest number of plant species harvested by sheep and goats in the tropical forest corresponded to the first column of the BC: (a) single rounded or oblong leafs of herbs, (b) leafs of woody shrubs and trees, and (c) leafs of creepers plants. The second column of the grid contained the second most common types of harvested plant parts (steams with leafs). The third most common type of plant parts corresponded to woody bipinnate plants (4th column of the BC). The third and fifth columns of the coding grid were used unfrequently and included a small number of plant species such as vines of creepers (3th column) and grasses (5th column).

Irrespective of season, the BC with highest frequency was oba (n = 28311), followed by obe (n=12712), roba (n = 8651), and robe (n = 5986) BC. The lower frequency corresponded to the obu (n = 66), ro (n = 62), fru (n = 44), and ru (n = 10) (Table 2).

DISCUSSION

Plant architecture, plant species, and plant parts consumed

The wide range of plant architectures (five columns, Table 1), as well as the sizes of bites (six rows, table 1; from less than 1 to 20 cm), corresponded well with the expected variability of materials available for harvesting in the heterogeneous vegetation of the tropical forest. The botanical diversity and the heterogeneous arrangement of the vegetation in multi-stratified levels are characteristic of the vegetation in the tropical deciduous forest (Flores et al., 2006, White and Hood, 2004). Thus, a considerable number of combinations of plant architecture, bite architecture, and size of bites were possible. While the grass plants only provided the elongated leaf form of bite architecture, the herbs, woody bipinnate, and creeper plants provided more types of bite architectures, including round or oblong leafs, steams, vines, flowers, fruits, and pods in a large range of sizes (Table 1). Thus, the adapted grid was flexible enough to be used in a variety of harvest possibilities found in the tropical forests.

The plant species identified in the present study were similar to the plant species harvested by goats in a similar period (from January to May) reported in an earlier study performed in the central zone of Yucatan (Rios and Riley, 1985). These results also suggest that, in spite of the different methodologies employed, both studies were able to identify similar plant species.

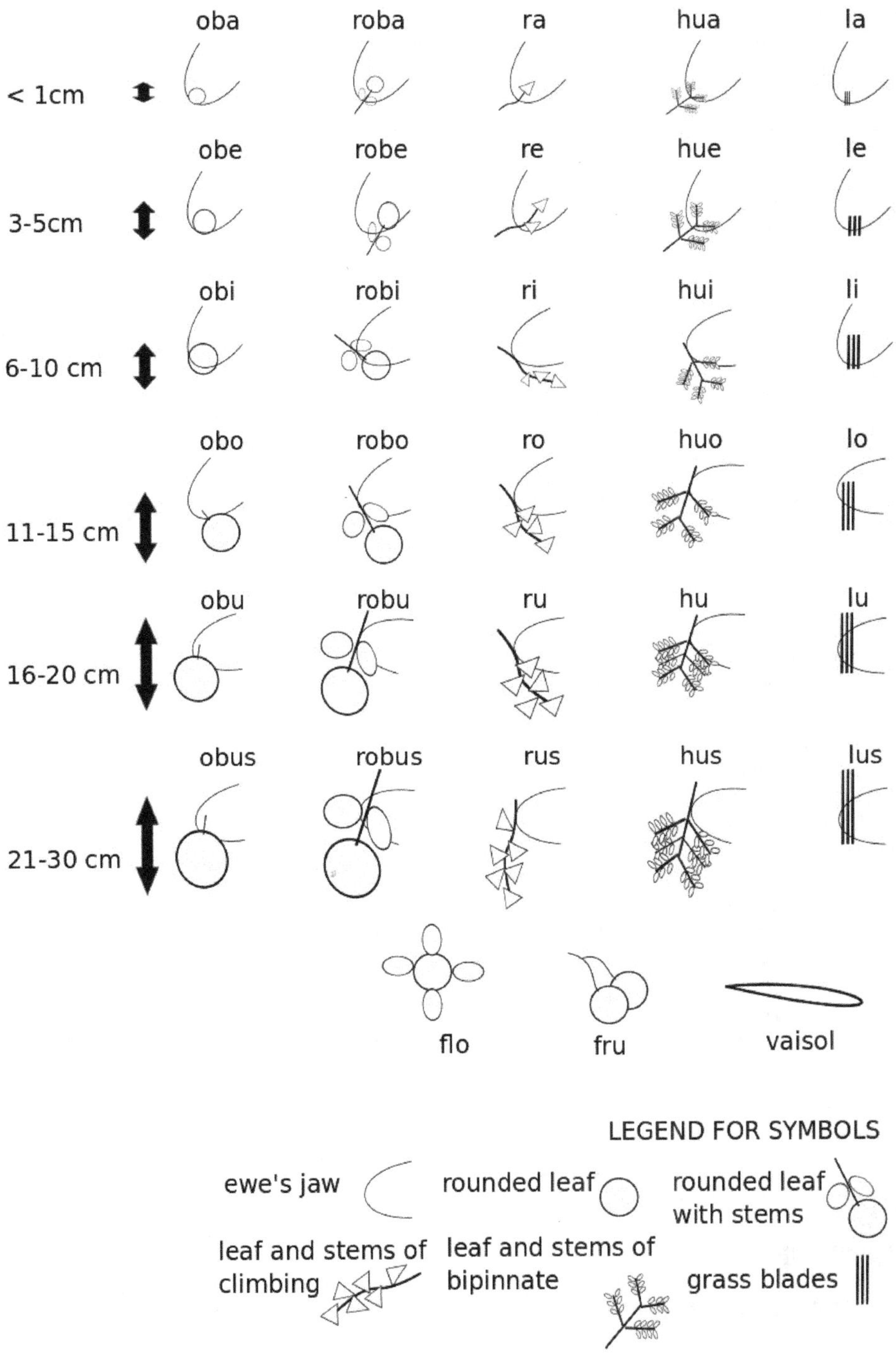

Figure 1. Bites categories (BC) adapted to the deciduous tropical forest vegetation. The codes used during the observation are given above the icons. The length of removed materials is indicated in cm (left of the grid). The BC from left to right (columns 1 to 5) were: (1) when bites included single rounded or oblong leafs (herbs, woody shrubs, trees, and climbers or creepers); (2) the same plants when bites included more than one leaf and a portion of the stem or branch; (3) when bites included climbers and creeper plants with leafs and vines; (4) when bites included woody plants of bipinnate leafs (with steam, leafs, and branches); and (5) when bites included leafs and stems of grass, as well as of herbs of straight, elongated thin leafs. The codes "flo" and "fru" were used to identify the harvest of flowers and fruits, and the code "vai-sol" was used for the fallen pods consumed from the ground. The last three codes could be used as single BC or in addition to the other codes when the bite also included some of these structures.

Table 1. Plant species, plant architecture, and size of the plant organs harvested by sheep and goats obtained from preliminary observations of feeding behaviour during January and February 2012.

Plant species	Plant architecture	Bite architecture							Size of bite (cm)
		Round/oblong leaf	Elongated leaf	steam	Vine	flower	fruit	pod	
Chloris inflata	grass		✔						< 1–10
Eragrostis ciliaris	grass		✔						< 1–20
Eragrostis amabilis	grass		✔						< 1–15
Althernatera flavescens	herb	✔		✔					< 1–5
Blechum pyramidatum	herb	✔		✔					< 1–5
Cnidoscolus aconitifolius	herb	✔		✔					3–20
Bourreria pulchra	herb	✔		✔					< 1–20
Parthenium hysterophorus	herb	✔		✔		✔			< 1–15
Sida acuta	herb	✔		✔					< 1–10
Solanum trydiamum	herb	✔		✔					< 1–20
Viguiera dentata	herb	✔		✔		✔	✔		< 1–20
Waltheria indica	herb	✔		✔		✔			< 1–15
Morinda royoc	herb	✔		✔					< 1–15
Tetramerium nervosum	herb	✔				✔			3–10
Bunchosia swartziana	woody	✔		✔		✔			< 1–20
Caesalpinia gaumeri	woody	✔							< 1–15
Cordia alliodora	woody	✔		✔					< 1–20
Dyospirus anisandra	woody	✔		✔					< 1–15
Gymnopodium floribundum	woody	✔		✔		✔	✔		< 1–20
Neomillspaughia emarginata	woody	✔		✔					< 1–15
Piscida piscipula	woody	✔		✔				✔	< 1–20
Randia aculeata	woody	✔		✔					< 1–15
Acacia collinsi	woody bipinnate	✔							< 1–20
Senegalia gaumeri	woody bipinnate	✔							< 1–10
Acacia pennatula	woody bipinnate	✔						✔	< 1–5
Leucaena leucocephala	woody bipinnate	✔				✔			< 1–20
Lysiloma latisiliquum	woody bipinnate	✔							< 1–20
Mimosa bahamensis	woody bipinnate	✔				✔			< 1–10
Bahuinia divaricate	creeper	✔		✔	✔				< 1–15
Ipomea crinicalyx	creeper	✔		✔	✔	✔			< 1–20
Ipomea nill	creeper	✔		✔	✔	✔			< 1–15
Passiflora foetida	creeper						✔		3–5
Cardiospermum halicacabum L.	creeper	✔		✔					< 1–10
Mix a: mixture of pods of *L. leucocephala, L. latisiliquum, A. pennatula*	woody bipinnate	✔							< 1–15
Mix b : mixture of dry leafs of *C. alliodora, D. anisandra, G. floribundum, P. piscipula*	woody	✔							< 1–15

BC grid adapted to the deciduous tropical forest

Seventeen BC from the grid used by Agreil and Meuret (2004) were not included in the new grid adapted to the conditions of deciduous tropical forests. Nevertheless, those codes could be required to describe the harvest under tall perennial rainforest (e.g. in silvopastoral systems) or areas with trees or conifers (highlands). The large diversity of creepers in the study site required the addition of 2 new BC as well as 6 new BC for the specific bipinnate plants which are endemic for the deciduous tropical forest (Flores et al., 2006). Finally, one BC was added to identify the pods collected from the ground. The latter was frequently observed in the harvest of sheep and goats along the dry season (n = 1505 bites; Table 2).

Validation of the adapted bite coding grid using sheep and goats in a tropical forest

Of the 33 BC from the new grid, 32 were used during the direct observation of feeding behaviour performed on sheep and goats during the dry and rainy season. Although there were a lower number of BC reported in the present trial (n = 33) as compared to the work reported by Agreil and Meuret (2004) (n = 41), the new grid did not require additional codes when it was implemented in the tropical forest. All of the materials consumed by sheep and goats were easily classified using their corresponding BC code (Table 2). Also, the last row of the grid (for BC of very large size; Figure 1) was not used in the dry season (penultimate row; Table 2), but was included since it could be required during the wet season. The materials collected from the ground, such as the mixture of pods and dry leafs, were easy to classify in their corresponding BC during the direct observations.

CONCLUSION

The new bite coding grid adapted for the vegetation of the deciduous tropical forest includes 33 BC, of which nine new BC were added to describe the diversity of the bites observed on creeper plants and the tree leafs from the fabaceae family (legume trees with bipinnate leafs). The adapted BC grid was suitable to identify the plant materials harvested by sheep and goats in the dry and rainy seasons. The grid can be easily used to study the feeding behaviour of sheep and goats using the direct observation method. The plant species harvested by sheep and goats in tropical forests provide include a wide range of materials, such as leafs, steams, vines, flower, fruit, and pods in a large range of sizes. Combinations of these materials and those collected from the ground (i.e. pods and fallen leafs) can also be classified using this grid.

Table 2. Bite categories (BC) used during the direct observation of feeding behaviour performed with the adapted BC grid for sheep and goats. The frequency of the observed bites is indicated in the parenthesis above the BC and the plants species involved are indicated by the numbers.

Bite Size	Single round or oblong leafs	Stems + single rounded or oblong leafs	leafs + vines	bipinnate leafs	Elongated leafs + stems
< 1 cm	**oba** (n= 28311) 4, 5, 7, 8, 9, 10, 11, 12, 13, 15, 16, 17, 18, 19, 20, 21, 29, 30, 31, 33, Mix b	**roba** (n= 8651) 4, 5, 8, 9, 10, 11, 12, 16, 17, 18, 19, 20, 21, Mix b	**ra** (n= 613) 29, 30, 31, 33	**hua** (n= 4001) 22, 23, 24, 25, 26, 27, 28	**la** (n=2922) 1, 2, 3,
3–5 cm	**obe** (n= 12712) 5, 6, 7, 8, 9, 10, 11, 12, 13, 14, 15, 16, 17, 18, 19, 20, 21, 29, 30, 31, 33, Mix b	**robe** (n= 5986) 4, 8, 9, 10, 11, 12, 13, 15, 16, 17, 18, 20, 21, Mix b	**re** (n= 569) 29, 30, 31,	**hue** (n= 2533) 22, 23, 24, 25, 26, 27, 28	**le** (n= 1905) 1, 2, 3,
6–10 cm	**obi** (n= 2718) 6, 7, 8, 9, 10, 11, 12, 13, 14, 15, 16, 17, 18, 19, 20, 21, 29, 30, 31, 33, Mix b	**robi** (n= 1784) 8, 9, 11, 12, 13, 16, 17, 18, 20, 21, Mix b	**ri** (n= 210) 29, 30, 31	**hui** (n= 946) 22, 23, 25, 26, 27, 28	**li** (n= 834) 1, 2, 3
11–15 cm	**obo** (n= 813) 7, 10, 11, 12, 13, 15, 16, 17, 18, 19, 20, 21, 29, 30, 31, Mix b	**robo** (n= 971) 6, 7, 8, 11, 12, 16, 17, 18, 20, 21, Mix b	**ro** (n= 62) 30	**huo** (n= 76) 22, 25, 26, 28	**lo** (n= 151) 2, 3
16–20 cm	**obu** (n= 66) 7, 10, 11, 16, 20, 30, Mix b	**robu** (n= 1430) 7, 11, 12, 15, 16, 18, 20, Mix b	**ru** (n= 10) 30	**hu** (n= 215) 22, 25, 26,	**lu** (n= 150) 2
21–30 cm	**obus** (n= 0) 30*, 9*, 19*, 10*	**robus** (n= 0) 30*, 11*, 18*, 11*, 12*	**rus** (n= 0)	**hus** (n= 0) 25*	**lus** (n= 0) 34* 2*
			flo (n= 130) 8, 11, 12, 14, 15, 18, 26, 28, 30, 31	**fru** (n= 44) 18, 32	**vaisol**(n= 1985) 20, 24, Mix a

Grass	Herbs	Woody	Woody bipinnate
1 C. inflata	*4 A. flavescens*	*15 B. swartziana*	*22 A. collinsi*
2 E. ciliaris	*5 B. pyramidatum*	*16 C. alliodora*	*23 S. gaumeri*
3 E. amabilis	*6 C. aconitifolius*	*17 D. anisandra*	*24 A. pennatula*
34 P. maximum	*7 B. pulchra*	*18 G. floribundum*	*25 L. leucocephala*
	8 P. hysterophorus	*19 N. emarginata*	*26 L. latisiliquum*
Creepers	*9 S. acuta*	*20 P. piscipula*	*27 M. bahamensis*
29 B. divaricata	*10 S. trydanum*	*21 R. aculeata*	*28 C. gaumeri*
30 I. crinicalyx	*11 V. dentata*		
31 I. Nill	*12 W. indica*	* Only observed during rainy season	
32 P. foetida	*13 M. royoc*		
33 C. halicacabum	*14 T. nervosum*		

Mix a: mixture of pods of *L. leucocephala, L. latisiliquum,* and *A. pennatula*
Mix b: mixture of dry leafs of *C. alliodora, D. anisandra, G. floribundum,* and *P. piscipula*

Acknowledgement

This work received the financial support from the project CB-2008-01/106146 "La relación herbívoro–tanino: adaptaciones fisiológicas de los pequeños ruminates (ovinos y caprinos) expuestos a vegetación rica en taninos" and the Postdoctoral grant No. 290662, both from CONACYT-México. We also thank the financial support of IEPAC project (FEDER Convention No. 31439 - Programme INTERREG IV CARAIBES 20017-2013-No. 41000140). The authors sincerely thank the herbarium staff of the Faculty of Veterinary and Animal Science-UADY, Mexico.

REFERENCES

Agreil, C., Meuret, M., 2004. An improved method for quantifying intake rate and ingestive behavior of ruminants in diverse and variable habitats using direct observation. Small Ruminant Research. 54, 99–113.

Agreil, C., Fritz, H., Meuret, M., 2005. Maintenance of daily intake through bite mass diversity adjustment in sheep grazing on heterogeneous and variable vegetation. Applied Animal Behaviour Science. 91, 35–36.

Altman, J., 1974. Observational study of behavior: sampling methods. Behavior. 9, 227–265.

Becker, K., Lohrmann, J. 1992. Feed selection by goats on tropical semi-humid rangeland. Small Ruminant Research. 8, 285–298.

Bourbouze, A., 1980. Goats grazing on woody rangeland. (in French, with English abstract) Fourrages 82, 121–144.

Flores, S.J., Vermont-Ricalde, M.R., Kantún-Balam, M.J., 2006. Leguminosae diversity in the Yucatan Peninsula and its importance for sheep and goats feeding. In: Sandoval-Castro, C.A., Hovell, F.D. DeB., Torres-Acosta, J.F.J., Ayala-Burgos, A. (Eds.), Herbivores: The Assessment of Intake, Digestibility and the Roles of Secondary Compounds. Nottingham University Press, Nottingham, pp. 291–299.

Kronberg, SL., Malechek, J.C., 1997. Relationship between nutrition and foraging behaviour of free-ranging sheep and goat. Journal Animal Science. 75, 1756–1763.

Meuret, M., 1988. Feasibility of in vivo digestibility trials with lactating goats browsing fresh leafy branches. Small Ruminant Research. 1, 273_290.

Meuret, M, Boza, J., Narjisse, H, Nastis, A., 1991. Evaluation and utilization of rangeland feed by goats. In: Goat nutrition, Morand-Fehr P. (Eds) Centre for Agricultural Publishing and Documentation (Pudoc). Pudoc, Wageningen, Netherlands, pp. 160–171.

Neff, D.J., 1974. Forage preference of trained mule deer on the Beavercreek watersheds. Arizona Game and Fish Department Special Report, 4, 1–61.

Ortega-Reyes, L., 1985. Composición química y digestibilidad de la dieta de ovinos pelibuey bajo condiciones de libre pastoreo en un henequenal de Yucatán, México. Tecnica Pecuaria Mexico. 48, 17–23.

Pontes, L. da S., Magda, D., Agreil, C., Gleizes, B., 2010. Feeding behaviour of sheep on shrubs (*Cytisus scoparius L.*) in response to contrasting herbaceous cover, in Shcnyder et al. (Eds), Grassland in a changing world. Proceedings of the 23rd General Meeting of the European Grassland Federation, Kiel, Germany, 29th August- 2nd September 2010, pp. 910–912.

Provenza, F., 2003. Foraging behaviour: managing to survive in a world of change. Behavioural Principles for Human, Animal, Vegetation, and Ecosystem Management. Utah, USA. Utah State University.

Reppert, J.N., 1960. Forage preferences and grazing habits of cattle at the Eastern Colorado Range Station. Journal Range Management. 13, 58–62.

Rios, G., Riley, J.A., 1985. Preliminary studies on the utilization of the natural vegetation in the henequen zone of Yucatán for the production of goats I. Selection and nutritive value of native plants. Tropical Animal Production. 10, 1–10.

Wallis De Vries, M.F. 1995. Estimating forage intake and quality in grazing cattle: a reconsideration of the hand-plucking method. Journal of Range Management. 48, 370-375.

White, D.A., Hood, C.S., 2004. Vegetation patterns and environmental gradients in tropical dry forest of the northern Yucatan Peninsula. Journal Vegetation Science. 15, 151–160.

24

EFFECT OF VITAMIN E ON SPERM AND OXIDATIVE STRESS PARAMETERS OF WEST AFRICAN DWARF GOAT BUCKS

James O. Daramola*, Ezekiel O. Adekunle, Oyegunle E. Oke, Opeoluwa Ogundele, Ebun-Oluwa A. Saanu and Adebisi J. Odeyemi

Department of Animal Physiology, Federal University of Agriculture Abeokuta, P.M.B 2240, Abeokuta, Nigeria. E-mail:daramolajames2003@yahoo.com

**Corresponding Author*

SUMMARY

The aim of the present study was to determine the effect of supplementation of vitamin E on sperm and oxidative stress parameters in West African Dwarf goat bucks. The bucks were allocated to 4 treatments consisting of 0 mg, 15 mg, 30 mg and 45 mg of vitamin E. At the end of 30 days consecutive administration of vitamin E, semen and blood samples were collected and evaluated for sperm and oxidative stress parameters. The results showed that sperm motility, acrosome integrity, membrane integrity and live spermatozoa; seminal Malondialdehyde (MDA), arginase activity and leukocytes; and serum MDA and testosterone were similar in all the levels of inclusion and comparable to the control. However, higher arginase activities were observed in all the levels of vitamin E compared to the control while 15 mg of vitamin E had higher percentage of sperm abnormality compared to other treatments and the control. The findings indicated that the levels of vitamin E used did not have beneficial effect on the sperm and oxidative stress parameters. Further study of higher levels of vitamin E is therefore necessary to ascertain the appropriate level required for optimal improvement in these parameters.

Keywords: Antioxidant; buck; sperm viability; stress

INTRODUCTION

West African Dwarf (WAD) goats (*Capra hircus*) are widely distributed in the South-Western part of Nigeria with a prevailing tropical climate. The area is particularly noted for unfavourable environment, which is humid and favours high prevalence of diseases. The goat production in this zone makes a major contribution to the agrarian economy. As the demand for animal protein in the sub region is constantly high, the prospects for increasing the numbers and productivity of this breed of goat need to be utilized. Therefore, improvement of this important breed of goat presents a great potential to alleviate the problem of protein malnutrition in the region.

Oxidative stress is one of the major factors that affect male fertility (Agarwal and Prabakaran, 2005) and its effect on sperm malfunctions through induction of lipid peroxidation to bio-membranes has been reported (Arabi and Seidaie, 2008). Goat spermatozoa in particular are sensitive to oxidative stress due to high content of unsaturated fatty acids in phospholipids of plasma membrane and low antioxidant capacity of goat seminal plasma (Donghue and Donoghue, 1997; Watson, 2000). The effect of reactive oxygen species (ROS) on sperm motility and the toxicity of the fatty acid peroxides generated by their attack on the membrane phospholipids were earlier reported (Jones *et al.,* 1978). There are two potential sources of ROS: the leukocytes and the spermatozoa themselves (Aitken *et al.,* 1992). Antioxidant molecules which can counteract the cellular damage due to ROS are present in seminal plasma and spermatozoa. Seminal plasma has been reported to contain a-tocopherol (Moilanen *et al.,* 1993) and enzymatic systems that reduce oxidative damage (Zini *et al.,* 1993). Vitamin E (a-tocopherol) is well known lipid-soluble, chain breaking antioxidant (Burton and Ingold, 1986; Luo *et al.,* 2011) found in cells (Cheeseman *et al.,* 1988) where it contributes to their integrity and function. As a naturally occurring antioxidant, vitamin E is located in biological membranes where it acts to protect the membrane polyunsaturated fatty acid from oxidation and attenuate oxidative damage to cellular membranes (Sugiyama, 1992). Esterbauer *et al.* (1991) reported that vitamin E was effective in preventing lipid peroxidation and other radical-driven oxidative events. Vitamin E plays important role in reproductive function and previous report indicated increase in total sperm output and sperm concentration during dietary supplementation in boars (Brzezinska-Slebodzinska *et al.,* 1995), rabbits (Yousef *et al.,* 2003) and rams (Luo *et al.,* 2004; Yue *et al.,* 2010). Impairment of mammalian fertility was also attributed to vitamin E deficiency. Brzezinska-Slebodzinska *et al.* (1995) observed that supplementation with vitamin E increased the concentration of spermatozoa in semen; an effect possibly linked to the antioxidant properties of this vitamin. However, the chain-breaking antioxidant of vitamin E for preserving the functional competence of spermatozoa has not been reported to be present in WAD goat bucks. Information is scarce on the effect of vitamin E on sperm and oxidative stress parameters of West African Dwarf goat bucks that are managed in this zone associated with less favourable environment.

The present study was therefore designed to evaluate the effect of vitamin E on sperm and oxidative stress parameters of West African Dwarf goat bucks.

MATERIALS AND METHODS

Animal and management

The study was carried out at the Goat Unit of Teaching and Research Farm, Federal University of Agriculture, Abeokuta which falls within 7° 10' N and 3° 2' E and altitude 76 m above sea level. It lies in South-Western part of Nigeria with a prevailing tropical climate, a mean annual rainfall of 1,037 mm and average temperature of 34.7 °C. Twenty (20) WAD goat bucks aged 2.5-3 were used. The animals were kept under an intensive management system and maintained under a uniform and constant nutritional regimen with concentrate feed at 300g/body weight and supplemented with guinea grass (*Panicum maximum*).

Trial and Data Collection

The animals were divided into 4 groups of 5 animals each and each group was randomly assigned one of the experimental treatments consisting of 0, 15, 30 and 45 mg/BW of vitamin E administered via oral route to the animals for 30 days consecutively. Thereafter, semen samples were collected with the aid of artificial vagina. Two replicates each of semen and blood samples were collected from the bucks 3 times at an interval of 15 days post vitamin E administration. Sperm and oxidative stress parameters were determined and values obtained during the 3 times collections were pooled for each parameter.

Evaluation

Subjective microscopic evaluation of sperm motility

Sperm motility was determined as described by Bearden and Fuquay (1997). Briefly, semen was placed in Clifton Water bath (Model: 74178 by Nickel Electro Ltd, Weston-S-Mare Somerset, England) at 37 C and accessed for sperm motility at a magnification of 400× with Celestron PentaView LCD digital microscope. A 5 μL sample of semen was placed directly on a heated microscope slide and overlaid with a 22 x 22 mm cover slip. For each sample, five microscopic fields were examined to observe progressive sperm motility and the mean of the five successive evaluations was recorded as the final motility score.

Sperm concentration

The concentration was determined by the use of a Minitube Photometer [SDM1 Ram 12300/0106, Germany].

Acrosome Integrity

The percentage of spermatozoa with intact acrosomes was determined according to Ahmad *et al.* (2003). Briefly, 50 μL of each semen sample was added to a 500 μL formalin citrate solution (96 mL 2.9 % sodium citrate, with 4 mL 37 % formaldehyde) and mixed carefully. A small drop of the mixture was placed on a microscope slide and a total of 200 spermatozoa were counted in at least three different microscopic fields for each sample at a magnification of 400× with Celestron PentaView LCD digital microscope. Intactness of acrosome characterized by normal apical ridge of spermatozoa was assessed.

Sperm Membrane Integrity

Hypo-osmotic swelling test (HOST) assay as described earlier (Jeyendran *et al.,* 1984) was used to determine sperm membrane integrity and this was done by incubating 10 μL semen in 100 μL hypo-osmotic solution (fructose and sodium citrate) at 37 °C for 30 min, 0.1 mL of the mixture was spread over a warmed slide, covered with a cover slip and observed at a magnification of 400× with Celestron PentaView LCD digital microscope. Two hundred spermatozoa (200) were counted for their swelling characterized by coiled tail, indicating intact plasma membrane.

Sperm abnormality and live sperm

Sperm morphological abnormalities and live sperm were determined as described by Bearden and Fuquay (1997) with the use of eosin-nigrosin smears. A thin smear of mixture of semen and eosin-nigrosin solution was drawn across the slide and dried. Sperm cells that did not absorb eosin-nigrosin solution were recorded as live sperm cells. The percentage of morphologically abnormal spermatozoa with defects in the head, midpiece and tail were observed (400× magnification) with Celestron PentaView LCD digital microscope.

Malondialdehyde (MDA) concentration

Malondialdehyde concentration as index of lipid peroxidation in the semen samples was measured in a thiobarbituric acid reactive substances (TBARS) according to Yagi (1998). For this assay, 0.1 mL of semen sample was incubated with 0.1 mL of 150 mM Tris-HCl (pH 7.1) for 20 min at 37 °C. Subsequently, 1 mL of 10 % trichloroacetic acid (TCA) and 2 mL of 0.375 % thiobarbituric acid were added followed by incubation in boiling water for 30 min. Thereafter, it was centrifuged for 15 min at 3000 g inside the blank tube and the absorbance was read with UV spectrophotometer (SW7504 model by Surgifriend Medicals, England) at 532 nm. The concentration of MDA was calculated as follows: The concentration of malondialdehyde MDA (nmol/mL) = AT – AB/1.56 × 10^5; Where: AT = the absorbance of the sample serum, AB = the absorbance of the blank, 1.56 × 10^5 molar absorptivity of MDA

Seminal leukocytes

Peroxidase test adapted from Endtz (1974) and as recommended by WHO (1992) was used. A stock solution was prepared by mixing 50 mL distilled water with 50 mL 96 % ethanol plus 125 mg benzidine. The working solution was obtained by adding 5 μL 30 % H_2O_2 to 4 mL of stock solution. Twenty microliters of working solution was mixed with 20 μL of semen sample in a small test tube. After incubation for 5 min at room temperature, 20 μL of working solution was mixed with 20 μL of phosphate-buffered saline. Then, 10 μL was placed in a haemocytometer, and peroxidase-positive cells (dark brown round cells) were counted.

Arginase activity

A 0.1 g bovine serum albumin (BSA) as standard in 10 mL of water was used in tubes containing 1 mL alkaline copper reagent (a mixture of copper sulfate reagent, sodium dodecyl sulfate solution, and sodium hydroxide solution (1:2:1) and 0.1 mL semen samples mixed and incubated for 10 min at room temperature (Lowry et al., 1951). After this, 4 mL folin Ciocalteu's phenol reagent was added to the tubes, mixed and incubated for 5 min at 55°C. The absorbance of the samples was recorded at 650 nm in spectrophotometer (UV spectrophotometer, SW7504 model by Surgifriend Medicals, England). The results were expressed as units/mg of protein (specific activity).

Serum testosterone

Blood samples were collected through jugular vein, centrifuged and serum harvested and stored at -20°C for testosterone assay. Testosterone was determined by ELISA (enzyme-linked immunosorbent assay) using testosterone kit, catalog number 2000 from Dialab (UK) and read with Robonic 11 ELISA Reader (UK).

Statistical analysis

Data obtained were subjected to analysis of variance in a completely randomized design and means separated by Duncan Multiple Range Test (Duncan, 1955). The results analyzed using a one-way analysis of variance in SAS 1999 package were expressed as the means ± SEM in SAS 1999 package using the model below:

$$Y_{ijk} = \mu + A_i + L_j + \sum_{ijk}$$

Where, Y_{ijk}= Dependent variables; μ = Population mean; A_i= effect due to i^{th} vitamin E treatments; L_j = effect due to j^{th} level of inclusion, j = 0, 15, 30, 45; $\sum_{ijk}$= Experimental error

RESULTS

The means of sperm characteristics of WAD buck goats administered with different levels of vitamin E are presented in Table 1. The results showed that sperm concentration, sperm motility, acrosome integrity, membrane integrity and live spermatozoa were similar in all the levels of inclusion and comparable to the control. However, 15 mg of vitamin E had increase ($P<0.05$) percentage of sperm abnormality compared to other treatments and the control.

The means of seminal oxidative stress parameters of WAD buck goats administered with different levels of vitamin E are presented in Table 2. The results showed that seminal MDA, arginase activity and leukocytes were similar in all the levels of inclusion and comparable to the control.

The means of serum oxidative stress parameters of WAD buck goats administered with vitamin E are presented in Table 3. The results showed that serum MDA and testosterone were similar in all the levels of inclusion and comparable to the control. However, higher arginase activities were observed in all the levels of vitamin E compared to the control.

Table 1: Means (±SEM) sperm characteristics of WAD buck goats administered with vitamin E

Parameters	0 mg	15 mg	30 mg	45 mg
Sperm concentration ($x10^8$/mL)	22.05±152.43	22.74±210.35	18.43±262.63	19.87±196.43
Motility (%)	94.67±1.28	92.33±4.17	94.67±3.38	96.67±1.20
Acrosome integrity (%)	93.67±2.32^a	84.83±3.15^b	89.50±1.66^{ab}	91.75±1.84^{ab}
Membrane integrity (%)	91.67±2.78	83.58±2.09	89.71±3.32	92.33±1.56
Live sperm (%)	92.08±3.25	87.92±5.12	93.25±2.47	97.50±1.91
Abnormality (%)	1.37±0.22^b	2.66±0.18^a	1.25±0.25^b	1.79±0.22^b

^{a,b} Values within rows with different superscripts differ significantly ($P<0.05$)

Table 2: Means (±SEM) seminal oxidative stress parameters of WAD buck goats administered with vitamin E

Parameters	0 mg	15 mg	30 mg	45 mg
Seminal MDA (nmol/mL)	0.04±0.01	0.02±0.01	0.04±0.02	0.04±0.01
Seminal arginase activity (units/mg protein)	1.13±0.06	1.10±0.21	1.25±0.12	1.22±0.25
Seminal leukocytes (x 10^3/mL)	0.52±0.09	0.63±0.08	0.48±0.06	0.43±0.08

Table 3: Means (±SEM) serum MDA, arginase activity and testosterone of WAD buck goats administered with vitamin E

Parameters	0 mg	15 mg	30 mg	45 mg
Serum MDA (nmol/mL)	0.18±0.02	0.18±0.01	0.18±0.01	0.18±0.01
Serum arginase activity (units/mg protein)	1.44±0.08^b	1.76±0.03^a	1.85±0.16^a	1.75±0.03^a
Serum testosterone (ng/ml)	0.40±0.10	0.40±0.06	0.40±0.06	0.37±0.09

^{a,b} Values within rows with different superscripts differ significantly ($P<0.05$)

DISCUSSION

The present study indicated that the levels of vitamin E used did not have beneficial effect on the sperm and oxidative stress parameters. Pasqualotto et al. (2000) reported an association between high seminal reactive oxygen species levels and reduced sperm count and motility. The protective effects of vitamin E against oxidative damage of sperm cells become more significantly observable when hygienic conditions are poorly controlled such as increased incidence of infections/inflammations of reproductive apparatus (Luo *et al.,* 2011). During inflammation, the antioxidant defence of reproductive system downplays and generates an oxidative stress (Potts and Pasqualotto, 2003), which may impair testis function and affect negatively semen characteristics (O'Bryan *et al.,* 2000). Testicular tissue becomes one of the targets for oxidative stress due to high content of polyunsaturated membrane lipids (Mishra and Acharya, 2004). Furthermore, vitamin E was reported to play a role in association with antioxidant enzymes, for preserving the functional competence of spermatozoa subjected to an oxidative attack (Therond, *et al.,* 1996). Vitamin E has also been observed to increase sperm viability and reduced lipid peroxidation when subjected to oxidative stress inducer (Bansal and Bilaspuri, 2009). However, the animals used in this present study were apparently healthy and were not subjected to oxidative stress; the possible reason why the protective effect of vitamin E was not observed. Moreover, the WAD goat is hardy and thrives in harsh conditions of management (Adeloye, 1998) which is largely possible because of the innate adaptogenic power of the animal to react to a particular quantum stressor (Adeloye and Daramola, 2004). In contrast, high tocopherol concentrations could cause ineffective antioxidant action even becomes a pro-oxidant (free radicals) that precisely reproduce the formation of radicals (Wahjuningsih and Rachmawati, 2012). The type of phenolic antioxidants (such as tocopherol) in excessive concentration will lose its effectiveness as an antioxidant and even to form a pro-oxidant (Wahjuningsih and Rachmawati, 2012). Changes in antioxidant function become pro-oxidant or free radicals cause more unsaturated fatty acids that are subjected to free radicals (Wahjuningsih and Rachmawati, 2012). This situation further accelerates and expands the incidence of lipid peroxidation of sperm plasma membrane damage due to loss of some essential unsaturated fatty acids making up the membrane (Nur *et al.,* 2005). This tendency is reinforced by the fact that the content of MDA as a lipid peroxidation product that is toxic to spermatozoa is found higher at a high dose of tocopherol (Wahjuningsih, 2010). This body of evidence support the present study that the levels of vitamin E used was neither enough for antioxidative effect nor in excess to elicit pro-oxidative effect on sperm and oxidative stress parameters. In contrast, vitamin E as antioxidant (Yue *et al.,* 2010) was observed to prevent membrane damage mediated by free radicals (Gurel *et al.,* 2005). In addition, Kutlubay et al. (2007) reported reduced oxidative stress in the testis following administration of vitamin E. Oxidative stress is one of the major factors that affect male fertility (Agarwal and Prabakaran, 2005). Oxidative stress is known to play a major role in the sperm malfunctions through induction of lipid peroxidation to biomembranes (Arabi *et al.,* 2001]. The difference in observation in the present study and those reported could be due to species of animal or methods of administration. This discrepancy may also be caused by different doses and treatment periods (Chattopadhyay *et al.,* 2001). Although the amount of nutrient intake was not measured in this study but the administration of vitamin E did not adversely affect reproductive functions of the bucks. In a study of effects of feeding different levels of vitamin E in creep feeds on calf creep performance and subsequent 28 days feedlot performance, Gunter et al. (2004) established a linear decrease as the concentration of vitamin E increased in creep feed, and observed that dry matter consumption average was not significant among treatments. Furthermore, studies indicated that vitamin E supplementation had no significant effect on dry matter intake of dairy cows (Bell *et al.,* 2006), dry matter intake and feed intake of Aohan fine wool sheep (Liu *et al.,* 2013).

A positive correlation has been reported between sperm motility and arginase activity in seminal plasma and spermatozoa (Elgun *et al.,* 2000; Eskiocak *et al.,* 2006). The main role of arginase in testis is regulation of nitric oxide (NO) concentration (Nathan, 1997). Increased arginase activity generally results in lower NO concentration and subsequently leads to increased sperm motility (Elgun *et al.,* 2000). This action could be the possible reason for the slight higher serum arginase activity with vitamin E in this study. However, this effect was not reflected in the seminal arginase activity.

Contaminating leucocytes and immature spermatozoa are the major sources of reactive oxygen species in semen (Aitken *et al.,* 1992; Agarwal *et al.,* 2003; Garrido *et al.,* 2004). Excessive reactive oxygen species levels present in sperm cells' environment can either be produced in large amounts by leukocytes or the spermatozoa themselves, the process that results in decreased membrane fluidity of both plasma and organelle membranes and, consequently causes damage to membrane function (Sikka *et al.,* 1996). Sequence to loss of membrane function, spermatozoa loses the ability to function properly and therefore, fertilization is impaired (Riffo and Parraga, 1996). Although contaminating leucocytes can be a critical factor for sperm survival, the phagocytic role of leukocytes in eliminating defective spermatozoa is beneficial to sperm concentration, motility and acrosome reaction and may even stimulate sperm functions through the release of ROS (Kaleli, *et al.*, 2000; Henkel, 2011). Notwithstanding, high concentration of leukocytes, particularly activated leukocytes, are still harmful to sperm functions (Henkel, 2011). The similar concentration of leukocytes following addition of vitamin E in this study further indicated that the levels of vitamin E used in the present study did not play the role of antioxidant in reducing defective spermatozoa or excessive ROS produced by leukocytes.

CONCLUSION

The findings indicated that the levels of vitamin E used did not have beneficial effect on the sperm and oxidative stress parameters. Further study on higher levels of vitamin E is therefore necessary to ascertain appropriate level that can have beneficial effect on these parameters.

Acknowledgements

The animals and some chemicals used in this study were funded by Federal University of Agriculture Abeokuta, Nigeria under the grant number FUNAAB-DGM/01-2012. The Authors are also grateful to the Laboratory Technologists of the Department of Animal Physiology, Federal University of Agriculture Abeokuta for their technical assistance.

REFERENCES

Adeloye, A.A., 1998: The Nigerian small ruminant species. Corporate Office Max: Ilorin, Nigeria. 7-8.

Adeloye, A.A., Daramola, J.O., 2004. Improving living conditions of domestic animals in Nigeria: A Review. In: Studies of Sustainable Agriculture and Animal Science in Sub-Saharan Africa Uche C. Amalu and Franz- Theo Gottwald (eds.). Support Africa International 1, 99-115.

Agarwal, A., Saleh,R.A., Bedaiwy, M.A., 2003. Role of reactive oxygen species in the pathophysiology of human reproduction. Fertility and Sterility 79, 829-843.

Agarwal, A., Prabakaran S.A., 2005. Mechanism, measurement and prevention of oxidative stress in male reproductive physiology. Indian Journal of Experimental Biology 43, 963-974.

Ahmad, Z., Anzar, M., Shahab, N., Ahmad, S., Andrabi, M.H., 2003. Sephadex and sephadex ionexchange filtration improves the quality and freezability of low-grade buffalo semen ejaculates. Theriogenology 59, 1189-1202.

Aitken, R.J., Buckingham, D., West, K., Wu, F.C., Zikopoulos, K., Richardson, D.W., 1992. Differential contribution of leucocytes and spermatozoa to the generation of reactive oxygen species in the ejaculates of oligozoospermic patients and fertile donors. Journal of Reproduction and Fertility 94, 451-462.

Arabi, M., Seidaie, S.R., 2008. Assessment of motility and membrane peroxidation of bull permatozoa in the presence of different concentration of vitamin C. Veterinary Medical Journal 2, 39-46.

Bansal, A.K., Bilaspuri, G.S., 2009. Antioxidant effect of vitamin E on motility, viability and lipid peroxidationof cattle spermatozoa under oxidative stress. Animal Science Paper Reports 27 (1), 5-14.

Bearden, H.J., Fuquay, J.W., 1997. Semen Evaluation. In: Applied Animal Reproduction. 4th ed. New Jersey: Prentice Hall; pp.158-169.

Bell, J.A., Griinari, J.M., Kennelly, J.J., 2006. Effect of safflower oil, flaxseed oil, monensin, and vitamin E on concentration of conjugated linoleic acid in bovine milk fat. Journal of Dairy Science 89, 733-748.

Brzezinska-Slebodzinska, E., Slebodzinski, A.B., Pietras, B., Wieczorek, G., 1995. Antioxidant effect of vitamin E and glutathione on lipid peroxidation in boar semen plasma. Biology and Trace Element Research 47, 69-74.

Burton, G.W., Ingold, K.U., 1986. Vitamin E: application of the principles of physical organic chemistry to the explanation of its structure and function. Accounts of Chemical Research 19, 194-201.

Chattopadhyay, S., Ghosh, S., Debnath, J., Ghosh, D., 2001. Protection of sodium arsenite induced ovarian toxicity by coadministration of L-ascorbate (vitamin C) in mature wistar strain rat. Archives Environmental Contamination and Toxicology 41, 83-89.

Cheeseman, K.H., Emery, S.P., Maddix, Slater, T.F., Burton, G.W., Ingold, K.U. 1988. Studies on lipid peroxidation in normal and tumor tissues. The Yoshida rat liver tumour Biochemical Journal 250, 247-252.

Donghue, A. M., Donoghue, D. J., 1997. Effects of water and lipid soluble antioxidants on turkey sperm in vitro does not improve human sperm motility. Fertility Sterility 72, 484-486.

Duncan, B. D., 1955. Multiple Range Test and Multiple F-test. Biometrics, 11, 1-42.

Elgun, S., Kacmaz, M., Sen, I., Durak, I., 2000. Seminal arginase activity in infertility. Urological Research 28, 20-23.

Eskiocak, S., Gozen, A.S., Taskiran, A., Kilic, A.S., Eskiocak, M., Gulen, S., 2006. Effect of psychological stress on the L-arginine-nitric oxide pathway and semen quality. Brazilian Journal of Medical and Biological Research 39, 581-588.

Esterbauer, H., Dieber-Rotheneder, M., Striegl, G., Waeg, G., 1991. Role of vitamin E in preventing the oxidation of low density lipoprotein. American Journal of Clinical Nutrition 53, 314-321.

Garrido, N., Meseguer, M., Simon, C., Pellicer, A., Remohi, J., 2004. Pro-oxidative and anti-oxidative imbalance in human semen and its relation with male fertility. Asian Journal of Andrology 6, 59-65.

Gunter, S.A., Beck, P.A., Davis, M.E., Wistuba, T.J., 2004. Effects of vitamin e levels in creep feeds for nursing calves on performanceand sudsequent feedlot performance. Proceedings, Western Section, American Society of Animal Science Vol. 55

Gurel, A., Coskun, O., Armutcu, F., Kanter, M., Ozen, O.A., 2005. Vitamin E against oxidative damage caused by formaldehyde in frontal cortex and hippocampus: biochemical and histological studies. Journal of Chemistry and Neuroanatamy 29, 173-178.

Henkel, R. R., 2011. Leukocytes and oxidative stress: dilemma for sperm function and male fertility. Asian Journal of Andrology 13, 43-52.

Jeyengran, R.S., Van der Ven, H.H., Peter-Perez, M., Crabo, B.G., Zaneveld, L.J., 1984. Development of an assay to assess the functional integrity of the human sperm membrane and its relationship to other semen characteristics. Journal Reproduction and Fertility 70 (10), 219-28.

Jones, R., Mann, T., Sherins, R.J., 1978. Adverse effects of peroxidized lipid on human spermatozoa. Proc. R. Soc. Land. B., 201, 413-417.

Kaleli, S., O¨cer, F., Irez, T., Budak, E., Aksu, M.F., 2000. Does leukocytospermia associate with poor semen parameters and sperm functions in male infertility? The role of different seminal leukocyte concentrations. European Journal of Obstetrics and Gynecology and Reproductive Biology 89,185-91.

Kutlubay, R., Oguz, E.O., Can, B., Guven, M.C., Sinik, Z., Tuncay, O.L., 2007. Vitamin E protection from testicular damage caused by intraperitoneal aluminium. International Journal of Toxicology 26, 297-306.

Liu, K., Ge, S., Luo, H., Yue, D., Yan, L., 2013. Effects of dietary vitamin E on muscle vitamin E and fatty acid content in Aohan fine-wool sheep. Journal Animal Science and Biotechnology 4, 21.

Lowry, O.H., Roseenbroug, N.J., Farr, A., Randall, L.J., 1951. Protein measurement with the folin phenol reagent. Journal of Biological Chemistry 193, 265-275.

Luo, H.L., Jia, Z.H., Zhu, S.E., Ding, J.Z., 2004. Effect of vitamin E on the qualities of fresh and frozen thawed ram semen. China Herbivores, 24, 14-16 (in Chinese)

Luo, H., Ge, S., Yue, D., Yan, L., Xu, X., Liu, K., Yuan, F., 2011. Effect of Vitamin E on the Development of Testis in Sheep, Artificial Insemination in Farm Animals, Dr. Milad Manafi (Ed.), ISBN: 978- 953-307-312-5, InTech, Available from: ttp://www.intechopen.com/books/artificial-insemination-in-farmanimals/ effect-of-vitamin-e-on-the-development-of-testis-in-sheep.

Mishra, M., Acharya, U.R., 2004. Protective action of vitamins on the spermatogenesis in lead-treated Swiss mice. Journal of Trace Element and Medical Biology 18 (2), 173-178.

Moilanen, J., Hovatta, O., Lindroth, L., 1993. Vitamin E levels in seminal plasma can be elevated by oral administration of vitamin E in subfertile men. International Journal of Andrology 16, 165- 66.

Nathan, C., 1997. Inducible nitric oxide synthase: what differences does it makes? Journal of Clinical Investigations 100, 2417-2423.

Nur. Z., Dogan, I., Gunay, U., Soylu, M. K., 2005. Relationships between Sperm Membrane Integrity and other Semen Quality Characteristics of the Semen of Saanen Goat

Bucks. Bulletin of the Veterinary Institute in Pulawy 49, 183-187.

O'Bryan, M.K., Schlatt, S., Phillips, D.J., De Kretser, D.M., Hedger, M.P., 2000. Bacterial lipopolysaccharide-induced inflammation compromises testicular function at multiple levels in vivo. Endocrinology 141, 238-246.

Pasqualotto, F.F., Sharma, R.K., Nelson, D.R., Thomas, A.J., Agarwal, A., 2000. Relationship between oxidative stress, semen characteristics, and clinical diagnosis in men undergoing infertility investigation. Fertility and Sterility 73, 459 - 64.

Potts, J.M.. Pasqualotto, F.F., 2003. Seminal oxidative stress in patients with chronic prostatitis. Andrologia 35, 304-308.

Riffo, M.S., Parraga, M., 1996. Study of the acrosome reaction and the fertilizing ability of hamster epididymal cauda spermatozoa treated with antibodies against phospholipase A2 and/or lysophosphatidylcholine. Journal of Experimental Zoology 275, 459-68.

Sikka, S.C., 1996. Oxidative stress and role of antioxidants in normal and abnormal sperm function. Frontier of Bioscience 1, 78-86.

Sugiyama, M., 1992. Role of physiological antioxidants in chromium (VI)-induced cellular injury. Free Radical Biology and Medicine 12, 397-407.

Therond, P., Auger, J., Legrand, A., Jouannet, P., 1996. a-tocopherol in human spermatozoa and seminal plasma:relationships with motility, antioxidant enzymes and leukocytes. Molecular Human Reproduction 2 (10), 739-744.

Wahjuningsih, S., 2010. Study of α-Tocopherol Supplementation on Quality of Frozen Semen of Etawah Crossbreed Goat. Proceeding of The 14th Animal Science Congress of the Asian Australasian Association of Animal Production Societies. Pingtung, Taiwan 23-27 August 2010.

Wahjuningsih, S., Rachmawati, A., 2012. The Effect of α-tocopherol on Plasma Membrane Integrity of Goat Spermatozoa. Journal of Basic and Applied Science Research 2(9), 8857-8860.

Watson, P. F., 2000. The causes of reduced fertility with cryopreserved semen. Animal Reproduction Science 61, 481-92.

WHO, 1992. World Health Organization, Laboratory Manual for the Examination of Human Semen and Sperm–Cervical Mucus Interaction. Cambridge University Press, Cambridge.

Yagi, K., 1998. Simple procedure for specific assay of lipid hydroperoxides in serum or plasma. Free Radical and Antioxidant Protocols 108, 101-106.

Yousef, M.I., Abdallah, G.A., Kamel, K.I., 2003. Effect of ascorbic acid and Vitamin E supplementation on semen quality and biochemical parameters of male rabbits. Animal Reproduction Science 76, 99-111.

Yue, D.B., Yan, L.Y., Luo, H.L., Jin, X.X., Xu, X., 2010. Effect of Vitamin E supplementation on semen quality and the testicular cell membranal and mitochondrial antioxidant abilities in Aohan fine-wool sheep. Animal Reproduction Science 118, 217-222.

Zini, A., De Lamirande, E., Gagnon, C., 1993. Reactive oxygen species in semen of infertile patients: levels of superoxide dismutase and catalase-like activities in seminal plasma and spermatozoa. International Journal of Andrology 16, 183-188.

ANALYSIS OF URBAN HOUSEHOLD DEMAND FOR POULTRY PRODUCTS IN OGBOMOSO NORTH AND SOUTH LOCAL GOVERNMENT AREA OYO STATE, NIGERIA

L.T. Ogunniyi*, A. Ajiboye and W.A. Sanusi

Department of Agricultural Economics and Extension, Ladoke Akintola University of Technology Ogbomoso. E- mail: titiogunniyi@yahoo.com
**Corresponding Author*

SUMMARY

The study analyzed the urban household demand for poultry products in Oyo state, Nigeria. Data were drawn from two local governments on 200 households with the aid of well-structured questionnaire in 2010.These were subsequently analyzed using the Almost Ideal Demand System. The study revealed that there was an inverse relationship between the household expenditure and the budget share of egg, a direct relationship between the price of egg and its budget share while an increase in the price of fish led to an increase in its budget share in the home.

Keywords: AIDS; poultry; elasticity; expenditure.

INTRODUCTION

Since the 1970s, global production, consumption trade of poultry meat has grown faster than that of any other meat. During the 1990s, when demand growth slowed for other meats, including fish, demand growth for poultry meat accelerated and poultry continue to lead the expansion of meat trade. Although demand for poultry meat was strong relative to demand for other meats in developed countries during the 1990s, the rapid global gains in poultry meat supply, demand and trade were led by gains in developing countries. (USDA, 1999). The poultry sector is crucially important in the context of agricultural growth and improvement of diets of people in Nigeria. The sub-sector is particularly important in that it is a significant source for the supply of protein and nutrition, in a household's nutritional intake. It is an attractive economic activity as well, especially to woman and poor population. Livestock farming has remained an essential component of the agricultural sector of the Nigerian economy. This is true, because livestock is a good source of animal protein which is needed for proper and balanced diet. Adeshinwa *et al.*, (1999) observed that the Food and Agricultural Organization (FAO) recommended that an average of 200 g animal protein is required per day for healthy living in the developing countries. This therefore, means that the general wellbeing of the people is directly dependent on the amount of animal protein consumption available to them. Although, FAO (2000) recommended the average protein requirement for healthy living and maintained that the meat protein consumption in most African countries is very low, at a level of 25g. Yet in Nigeria, the consumption is even lower especially in the southern and eastern parts of Nigeria, where in recent times human nutritionist have observed that the production of animal protein has not been high enough to meet the demand of the rapid population growth (Obi, 2003). He asserted that much of the animal protein intake available to the southern and eastern parts of Nigeria, comes from the north in the form of cattle, rams, dogs and goats.

Comparative Statistics by Ademosun (2000) puts Nigeria's total meat production at 810,000 ton for a population of about 110 million, resulting in a meat production index of 22 g per caput, per day. Sonaiya (1982), had rightly envisaged that as consumers become more articulate and organized; their demand for wholesome animal protein will exert a powerful influence upon quality, production method and strategies. He added that recent increases in expendable income of urban dwellers have tendencies to stimulate greater demand, not only for quality but also quantity of meat products. Today, the increasing human population in the face of inelastic production, strategies appears to have widened the demand supply gap and accentuates society of meat products. Regmi (2002) supported this view and noted that, the unprecedented growth that has occurred in the last half-century has created an additional demand for meat and general food in developing countries. Beside the failure of meat production capacity to match with the human population growth, the distribution of livestock in Nigeria is apparently lopsided. Composite transportation cost incurred, coupled with remote distance between major meat producing areas and consuming urban centers together make the value of wholesome beef, mutton, pork, chicken and dog meat often unaffordable (Mdafri and Brorsen, 1993).

As poultry industry is expanding with increasing number households involved in the raising of domestic fowls, turkey, goose, ducks and all the likes; the problems of malnutrition persists. It is very difficult for an average of Nigeria, to consume any of the poultry products either chicken meat or egg, on the scale of international nutritional requirement. This can be traced to the high price of the products which makes poultry products appear infrequently on many of our tables, except during the festive periods. This low level of consumption makes the malnutrition problem to be persistent. Another observation is the fact that the demand for these products is still far higher than the supply, this arises from high pricing and importation of both, chicken and eggs from other countries. Nigeria has the largest population in sub-Saharan Africa. About 47% reside in the urban areas where the population growth rate is estimated at three times that in rural areas (World Bank, 2004). This suggests a shift in increased food demand from the rural to the urban areas. Government policies tend to favor the urban dwellers to the detriment of the rural areas. Over the past three decades, rural households have been significantly poorer than urban households. However, while urban poverty has increased, rural poverty has decreased, especially after the post-adjustment period: 1995 to date (Canagarajah and Thomas, 2001). Urbanization is therefore a key non-income factor explaining the change in demand for animal protein (Ahmed and Gruhn, 1995). Consumption of animal protein has been found to be higher in urban areas than in rural areas (Hussain, 1990). Rising incomes, changing tastes and preferences are likely to shift the demand for meat and meat products. Differences in consumer behavior and demand for meat products, particularly dairy products, exist in Nigeria because the country is diverse and characterized by regional, physical, agro climatic, socioeconomic, and cultural/ethnic differences in food habits (Jabbar and Di Comenico, 1993).

The objective of this study is to fill a gap in the literature by estimating demand functions for urban households in Nigeria. Income and price elasticities are needed to formulate policies, set priorities, and engender investments in the livestock sector. Hence, an understanding of household consumption patterns of meat and meat products, especially the effects of income and price on meat demand and the impact of demographic factors on urban household meat consumption (Young and Hamdok, 1994) could provide important policy insights for Nigeria.

METHODOLOGY

Data was collected from 200 household in two local government areas of the state, using a multistage method. The two local government area selected in Ogbomoso metropolis were, Ogbomoso south and north and these were taken as the sampling unit as the first stage of sampling. At the second stage, five wards were randomly selected at each local government, making a total of ten wards. The last stage involved random selection of twenty household in each ward making a total of 200 respondents.

The study employed the Almost Ideal Demand System (AIDS) developed by Deaton and Muellbauer (1980). The model is flexible enough to allow the assumptions of homogeneity and symmetry to be tested or successfully imposed during empirical analysis. It is easy to estimate, gives arbitrary first order estimation to any demand system, and satisfies the axioms of choice. Many of these good attributes have contributed tremendously to the application of the model to demand equation estimation in many parts of the world.

The AIDS model can be specified as

$$w_i = \alpha_i + \sum_j^n \gamma_{ij} \ln P_j + \beta_i \ln(\frac{x}{p}) + e_i \quad (1)$$

where w_i = budget share of ith commodity defined by P_jQ_i/X

P_j = price of jthcommodity

γ_{ij} =estimated coefficient of prices

β_i =estimated expenditure coefficient

X= total expenditure on all commodities in the system

P=price index

The price index can be further defined as:

$$\ln P = \alpha_0 + \sum_k \alpha_k \ln P_k + \tfrac{1}{2}\sum_j \sum_k \gamma_{jk} \ln P_k \ln P_j \quad (2)$$

The price index makes equation 1 to be non-linear. In order to linearize it, the Stone's index has been incorporated.

$$\ln p = \sum w_j \ln p_j \quad (3)$$

Homogeneity, symmetry, and adding up are respectively imposed on the system through the following parameter restrictions:

$$\sum_j \gamma_{ij} = 0; \gamma_{ij} = \gamma_{ji}; \sum_i a_i = 1; \sum a_{is} = 0; \sum \beta_i = 0; \sum \gamma_{ij} = 0. \quad (4)$$

Following Chalfant (1984) and Ahmed and Shams (1994), the Marshalian and Hicksian elasticities are computed from the estimated parameters of the Linear Approximation AIDS model (LAAIDS) in equation 4 as follows;

Marshalian (Uncompensated)

$$\varepsilon_{ij} = -1 + (\gamma_{ij}/w_i) - \beta_i \text{ (own- price)} \quad (5)$$

$$\varepsilon_{ij} = (\gamma_{ij}/w_i) - \beta_i(w_j/w_i) \text{ (cross-price)} \quad (6)$$

Hicksian (compensated)

$$\delta_{ij} = -1 + (\gamma_{ij}/w_i) + w_i \text{ (own-price)} \quad (7)$$

$$\delta_{ij} = (\gamma_{ij}/w_i) + w_j \text{ (cross-price)} \quad (8)$$

The expenditure elasticity is derived as

$$E_i = -1 + \beta_i / w_i \quad (9)$$

RESULTS AND DISCUSSION

Substitute for Meat

The information in table 1 show that 54.5percent, amounting to more than half of the respondents, took only fish as their close substitute for chicken, while the lesser percentages consume either egg, and cheese as their substitute. This fish was in all people consulted like substitute. This might be due for the easy accessibility of fish, in terms of relative distribution and cost per unit, in comparison with the other available substitute. The cheese is also an important food for people diet with fish (17.5%).

Almost Ideal Demand System

Following Deaton and Muellbauer (1980), the demand equation for egg and chicken were estimated without imposition of any restrictions. The results of the analysis are presented in table 2. From the table, the test for homogeneity was carried out. The results of the tests showed that in both, egg and chicken, homogeneity conditions is significantly violated. This result is in line with the findings of Deaton and Muellbauer (1980), Ahmed and Shams (1994) and Awoyemi *et al.,* (2006).

Table 2 also shows that, all the Durbin-Watson statistics were within the plausible region. In all one can effectively say that the dependent and independent variables, have effectively performed their role.

In the first equation using the budget share of egg as the dependent variable, it was clear that, three of the variables were significant at I percent level of significance. The significant variables are the price of egg, fish and total food expenditure. Expenditure has a negative relationship with budget share for egg. This showed that budget share for egg tends to decrease with increase in household expenditure. There is also a positive relationship between the price of egg and its budget share. As the price of fish increases by 1%, the budget share for egg increases by 0.009%.

For chicken as the second equation, four independent variables were also significant, which are the prices of egg, chicken, fish and total food expenditure respectively at 1% level. There is a positive relationship between the budget share for chicken and the price of egg, chicken and fish. This shows that a 1% increase in price of fish, leads to a 0.011% increase in budget share for chicken in the home.

Marshallian (Uncompensated) Own Price and Cross Price Elasticities

Table 3 presents the full matrices of the uncompensated (Marshallian), own price and cross price elasticities. All the estimates of own price elasticities conform to the law of demand of negative signs. The estimated own price elasticities of poultry products ranged between – 0.274 (egg) and - 0.683 (chicken). It is noteworthy that the own-price elasticity for egg (- 0.274) is the least elastic followed by chicken (-0.683).This probably indicates the preferential importance of chicken in the diet of household in Ogbomoso metropolis, above egg. This study is in line with earlier findings by Ma *et al.,* (2004) who reported a negative and less than one own price elasticity for egg and chicken for urban households in China. Also from the table, all the estimated values for the cross price elasticities have a positive sign, which implies substitution effect. This implies that all products are in substitute type of relationship with poultry products.

Table 1. Distribution of Respondents by Substitute for Chicken

Substitute	Frequency	Percent	Cumul. percent
Fish	109	54.5	54.5
Fish and egg	39	19.5	74.0
Fish, egg and cheese	35	17.5	91.5
Fish, egg, cheese & others (meat)	2	1.0	92.5
Fish, egg & others (meat)	1	0.5	93.0
Fish and cheese	6	3.0	96.0
Egg	3	1.5	97.5
Meat	3	1.5	99.0
No substitute	2	1.0	100.0
Total	200	100.0	

Source: Field survey, 2010

Table 2. Unconstraint Parameter Estimates and Test of Homogeneity

Commodity	Constant	Egg	Chicken	Meat	Fish	Expenditure	R^2	DW
Egg	0.235***	0.059***	-0.004	0.001	0.009***	-0.069***	0.938	1.960
	(10.756)	(20.471)	(-0.019)	(0.072)	(3.461)	(-46.290)		
Chicken	0.413***	0.010**	0.046**	0.002	0.010***	-0.091***	0.987	1.874
	(12.849)	(2.332)	(11.541)	(0.807)	(2.692)	(-93.959)		

Source; Field survey, 2010
Note: Values in parenthesis represents t-values,
*** represent significant at 1% level,
** represent significant at 5% level,

Hicksian (Compensated) Own Price and Cross Price Elasticities

From Table 4, the own price elasticity for the two products are negative, which conform to *a priori* expectation. Egg has the most inelastic own- price elasticity (ε=-0.253) among the two poultry products, considered in the study. This indicates that households in Ogbomoso metropolis, are insensitive to changes in the price of egg. That is, if the price of egg comes down, or there is increase in the per capita income of the household, consumption will not be much affected. Huq and Arshad(2010) calculated compensated own price elasticity for meat at -0.482 and hence in line with this study.

Hicksian cross price elasticities are higher than that of the Marshallian though not up to unity and positive. This implies that Hicksian cross price elasticity for all products are substitutes for each other.

Table 3. Marshallian (Uncompensated) Own Price and Cross Price Elasticities

Commodity	Egg	Chicken	Meat	Fish
Egg	-0.274	0.110	0.194	0.250
Chicken	0.091	-0.683	0.119	0.135

Source: Field Survey, 2010.

Table 4. Hicksian (Compensated) Own Price and Cross Price Elasticities

Commodity	Egg	Chicken	Meat	Fish
Egg	-0.253	0.156	0.251	0.295
Chicken	0.140	-0.572	0.251	0.242

Source: Field Survey, 2010.

Expenditure and Income Elasticities

The estimates of income elasticity are yielded by multiplying the estimated expenditure elasticities with the responsiveness of expenditure on egg and chicken by income change from Engel's curve. The income elasticities for egg and chicken were less than one, which implies that the poultry products are income inelastic. In other words, they are normal and necessity goods. This finding is in line with previous work by Seale *et al.*, (2003) who reported less than one income elasticities for meat for all categories of income groups in Indonesia. Huq and Arshad(2010) calculated expenditure elasticity for meat at 2.461 based on AIDS model and hence could be considered as a luxury. When the expenditure elasticity is less than one, there is weak reference for demand by a particular household. This leads to less consumption of poultry products by such household. This study is in consonance with the findings of Ma *et al.*,(2004) for egg(less than one), while it is not for chicken (greater than one).

Table 5. Estimated Expenditure and Income Elasticities

Product	Expenditure elasticity	Income elasticity
Egg	0.228	0.0048
Chicken	0.550	0.0115

Source: Field Survey, 2010

CONCLUSION AND RECOMMENDATION

It was concluded that as the household expenditure increases, there is a decrease in the budget share to egg and chicken. It shows that, the poor people depend in extreme of egg for their consumption of protein, its means that prices of egg is a very important variable in the household. The public policy should be carefully with price, production and consume of egg. The poverty status of the area should be addressed by the government, this will go a long way in increasing the living standard of the people and change the mentality that poultry meat belong to the few except on festival days.

REFERENCES

Ademosun, A.A.2000. Structural Adjustment and the Nigeria Livestock Industry Stuport in Infancy. Keynote Address delivered at the Nigerian Society for Animal Production. Animal Conference held at University of Agriculture, Makurdi.

Adeshinwa, A.O.K., Famimo, O., Ogunmodebe, B.K., Ogunsanmi, A.O. 1999. Weaner Pigs Fed with Graded Levels of Maize Offal as Replacement for Maize and the Effects on Performance. Tropical Journal of Animal Science. 1: 101-105.

Ahmed, A.U., Shams, Y.S. 1994. Demand Elasticities in Rural Bangladesh; An Application of AIDS Models. The Bangladesh Development Studies. 22: 1-25.

Ahmed, R., Gruhn, P. 1995. Selected issues in the supply and demand for red meat and poultry products in developing countries. In Supply of livestock products torapidly expanding urban populations (ed R.T Wilson). Proceedings of the Joint FAO/WARP/KAAS symposium, Hoam Faculty Club, Seoul National University, Seoul, Korea 16 - 20 May, Food and Agriculture Organization of the United Nations, Rome, Italy.

Awoyemi T.T., Amao, A.O., Fatogun, O.O. 2006. Household Expenditure Pattern on Food in Ibadan Oyo State, Nigeria. Journal of Rural Economics and Development. 15:79-86.

Canagarajah, S., Thomas, S. 2001. Poverty in a Wealthy Economy: The Case of Nigeria. Journal of African Economies. 10: 143 – 173.

Chalfant, J. 1984. A Globally Flexible Almost Ideal Demand System. Journal of Business and Economic Statistics, 5:233-242.

Deaton, A., Muellbauer, J. 1980. An Almost Ideal Demand System. America Economics Review. 70:312-314.

FAO. 2000. Year Book, Food and Agricultural Organization, Rome.

Hussain, M. A. 1990. Nutrition Policy and Urban Poor in Developing Countries. Food Policy. PP.186-192 June.

Huq, A.S.M.A., Arshad, F.M.2010. Demand Elasticities for Different Food Items in Bangladesh. Journal of Applied Sciences. 10:2369-2378.

Jabbar, M. A., Di Comenico, C.M. 1993. Demand for Dairy Products among the Indigenous Population of Southern Nigeria. Indian Journal of Dairy Science 46:363 - 370.

Ma, H., Rae, A., Huang, J.,Rozelle, S. 2004. Chinese Animal Product Consumption in the 1990s. The Australian Journal of Agricultural and Resource Economics. 48:569-590.

Mdarfi, A., Brorsen, B. W. 1993. Demand for Red Meat, Poultry and Fish in Morocco : An Almost Ideal Demand System. Agricultural Systems. 9: 155 – 163.

Obi, C.I. 2003. Game production. An alternative to beef cattle production in southern Nigeria. Nigeria Academic Forum. 4: 36-40.

Regmi, A., Deepak, M.S., James Seale, Jr., Jason, Bernstern,J. 2002: Cross Country Analysis of Food Consumption Patterns. In: Changing Structure of Food Consumption and Trade/ WRS- 01-1. Economic Research Service/USDA. www.citeseerx.ist.psu.edu/

viewdoc/download?doi=10.1.1.143.3253. Accessed on the 27th March,2009

Seale, J., Regmi,A., Bernstein, J.A. 2003. International Evidence on Food Consumption Patterns. Technical Bulletin No. TB1904.

Sonaiya, E.B. 1982. Beef Quality and the Nigerian Consumer. In: Beef Production, Nigeria. Osinowo *et al* (Ed.), Proceeding of Nigeria National Conference of Beef Production, pp. 555-564. www.freepatentsonline.com/ article/European.../260791889.html Accessed on the 3rd 0f June, 2010.

USDA. 1999. United State Department of Agriculture Summary, Layers and Egg Production. Economic Research Services/USDA India poultry sector-Development and prospect/WRS-04-03. .

World Bank. 2004. http://devdata.worldbank.org/ external/CPProfile.asp?PTYPE= CP&CCODE=NGA. Accessed on the 6th of June, 2010

Young, T., Hamdok, A. A.1994. Effects of Household Size and Composition on the Consumption in Rural Households in Matabeleland South, Zimbabwe. Agricultural Economics. 11: 335 – 343.

Permissions

All chapters in this book were first published in TSA, by Universidad Autónoma de Yucatá¡n; hereby published with permission under the Creative Commons Attribution License or equivalent. Every chapter published in this book has been scrutinized by our experts. Their significance has been extensively debated. The topics covered herein carry significant findings which will fuel the growth of the discipline. They may even be implemented as practical applications or may be referred to as a beginning point for another development.

The contributors of this book come from diverse backgrounds, making this book a truly international effort. This book will bring forth new frontiers with its revolutionizing research information and detailed analysis of the nascent developments around the world.

We would like to thank all the contributing authors for lending their expertise to make the book truly unique. They have played a crucial role in the development of this book. Without their invaluable contributions this book wouldn't have been possible. They have made vital efforts to compile up to date information on the varied aspects of this subject to make this book a valuable addition to the collection of many professionals and students.

This book was conceptualized with the vision of imparting up-to-date information and advanced data in this field. To ensure the same, a matchless editorial board was set up. Every individual on the board went through rigorous rounds of assessment to prove their worth. After which they invested a large part of their time researching and compiling the most relevant data for our readers.

The editorial board has been involved in producing this book since its inception. They have spent rigorous hours researching and exploring the diverse topics which have resulted in the successful publishing of this book. They have passed on their knowledge of decades through this book. To expedite this challenging task, the publisher supported the team at every step. A small team of assistant editors was also appointed to further simplify the editing procedure and attain best results for the readers.

Apart from the editorial board, the designing team has also invested a significant amount of their time in understanding the subject and creating the most relevant covers. They scrutinized every image to scout for the most suitable representation of the subject and create an appropriate cover for the book.

The publishing team has been an ardent support to the editorial, designing and production team. Their endless efforts to recruit the best for this project, has resulted in the accomplishment of this book. They are a veteran in the field of academics and their pool of knowledge is as vast as their experience in printing. Their expertise and guidance has proved useful at every step. Their uncompromising quality standards have made this book an exceptional effort. Their encouragement from time to time has been an inspiration for everyone.

The publisher and the editorial board hope that this book will prove to be a valuable piece of knowledge for researchers, students, practitioners and scholars across the globe.

List of Contributors

Jorge Urrutia Morales, José Francisco Cervantes Becerra and María Teresa Rivera Lozano
San Luis Experimental Station, National Institute for Forestry, Agriculture and Livestock Research, San Luis Potosí, México

Héctor Raymundo Vera Ávila and Eugenio Villagómez-Amezcua Manjarrez
National Center for Disciplinary Research on Animal Physiology and Breeding, National Institute of Forestry, Agriculture and Livestock Research, Querétaro, México

F. Duarte-Vera
Campo Experimental Mocochá, Centro de Investigación Regional del Sureste, INIFAP. Km 15 Carr. Mérida-Baca, Yucatán, México
Facultad de Medicina Veterinaria y Zootecnia, Universidad Autónoma de Yucatán, Mérida, Yucatán, México

C. A. Sandoval-Castro, L. A. Sarmiento-Franco and R. Santos-Ricalde
Facultad de Medicina Veterinaria y Zootecnia, Universidad Autónoma de Yucatán, Mérida, Yucatán, México

L. O. Tedeschi
Department of Animal Science, Texas A&M University, College Station, TX 77845-2471, USA

Carolina Fiol
Departamento de Bovinos; Facultad de Veterinaria, Universidad de la República, Lasplaces 1550, Montevideo, Uruguay

Rodolfo Ungerfeld
Departamento de Fisiología; Facultad de Veterinaria, Universidad de la República, Lasplaces 1550, Montevideo, Uruguay

A. Gómez-Brunet, J. Santiago-Moreno, A. Toledano-Diaz and A. López-Sebastián
Departamento de Reproducción Animal, INIA, Madrid, Spain

A. Olmedo-Juáreza, R. Rojo, J. F. Vázquez-Armijo, S. Rebollar-Rebollar and B. Albarran
Centro Universitario UAEM-Temascaltepec, Universidad Autónoma del Estado de México, km 67.5 Carr. Toluca-Tejupilco, Estado de México, 51300, México

A. Z. M. Salem
Facultad de Medicina Veterinaria y Zootecnia, Universidad Autónoma del estado de México Instituto Literario No. 100, Toluca, 50000 Edo. de México, México

J. Lugo
Facultad de Ciencias Laboratorio de Edafología y Ambiente, Universidad Autónoma del Estado de México, Instituto Literario No. 100, Toluca, 50000 Edo. de México, México

Marianna Siegmund-Schultze and Anne Valle Zárate
Hohenheim University, Department of Animal Production in the Tropics and Subtropics, D-70593 Stuttgart, Germany

Girma Abebe
Hawassa University, Department of Animal and Range Sciences, Box - 05, Awassa, Ethiopia

Getahun Legesse
Hohenheim University, Department of Animal Production in the Tropics and Subtropics, D-70593 Stuttgart, Germany
Hawassa University, Department of Animal and Range Sciences, Box - 05, Awassa, Ethiopia
University of Manitoba, Faculty of Agricultural and Food Sciences, Department of Animal Science, 12 Dafoe Road, Winnipeg, MB, Canada - R3T 2N2

Philippe Chemineau
INRA, DEPE, 147 rue de l'Université, 75007 Paris France

Maria Wurzinger and Johann Sölkner
BOKU - University of Natural Resources and Life Sciences, Department of Sustainable Agricultural System, Division of Livestock Sciences, Gregor - Mendel Strasse 33, A-1180 Vienna, Austria

Luis Escareño
BOKU - University of Natural Resources and Life Sciences, Department of Sustainable Agricultural System, Division of Livestock Sciences, Gregor - Mendel Strasse 33, A-1180 Vienna, Austria
Universidad Autónoma de Zacatecas, Unidad Académica de Medicina Veterinaria y Zootecnia, Carretera Panamericana, Fresnillo Km 31.5 apartado postal 9 y 11 Calera de Víctor Rosales, Zacatecas, Mexico

Francisco Pastor and Homero Salinas
INIFAP- Instituto Nacional de Investigaciones Forestales, Agrícolas y Pecuarias, Campo Experimental La Laguna, Blvd. José Santos Valdez No. 1200 Pte. C.P. 27440; Matamoros, Coahuila; Mexico

Luis Iñiguez
ICARDA-International Center for Agricultural Research in the Dry Areas, Aleppo, Syrian Arab Republic

A. Aslaminejad and M. Nassiri
Dept. Animal Science, Fac. Agriculture, Ferdowsi Univ. of Mashhad, Mashhad, Iran

G. R. Dashab
Dept. Animal Science, Fac. Agriculture, Ferdowsi Univ. of Mashhad, Mashhad, Iran
Dept. Animal Science, Univ. of Zabol, Iran

A. K. Esmailizadeh
Dept. Animal Science, Fac. Agriculture, Shahid Bahonar Univ. of Kerman, Kerman, Iran

D. A. Saghi
Agricultural and Neutral Resource Center, Mashhad, Iran

O. A. Fasae, I. F. Adu and A. B. J. Aina
Department of Animal Production and Health, University of Agriculture, P.M.B. 2240, Abeokuta, Nigeria

Fernando Centurión-Castro, Juan Orihuela-Porcayo, Ricardo J. Aké-López, Juan G. Magaña-Monforte, Rubén C. Montes-Pérez and José C. Segura-Correa
Campus de Ciencias Biológicas y Agropecuarias, Universidad Autónoma de Yucatán, km 15.5 carretera Mérida-Xmatkuil. Apartado postal 4-116, Mérida, Yucatán, México

H. Hernández, P. Poindron, S. Ramírez-Vera, J. A. Flores, J. A. Delgadillo, J. Vielma, G. Duarte, I. G. Fernández and G. Fitz-Rodríguez
Centro de Investigación en Reproducción Caprina, Universidad Autónoma Agraria Antonio Narro, Periférico Raúl López Sánchez y Carretera a Santa Fe, CP 27054, Torreón, Coahuila, México

A. Terrazas
Departamento de Ciencias Pecuarias, Facultad de Estudios Superiores Cuautitlán, Universidad Nacional Autónoma de México. Km 2.5 Cuautitlán-Teoloyucan, San Sebastián Xhala, Cuautitlán Izcalli, México

S. Retana-Márquez and M. Muñoz-Gutierrez
Departamento de Biología de la Reproducción, Universidad Autónoma Metropolitana Iztapalapa. Edificio de Ciencias Biológicas y de la Salud, Av. San Rafael Atlixco No. 186, Colonia Vicentina, 09340 Iztapalapa México D.F.

N. Serafín
Departamento de Neurobiología Conductual y Cognitiva, Instituto de Neurobiología, Universidad Nacional Autónoma de México, Juriquilla, Querétaro, México

L. Sarmiento-Franco and F. Solorio-Sanchez
Faculty of Veterinary Medicine and Animal Science, University of Yucatan (UADY), Mérida, Yucatán, Mexico

F. M. K. Abou-Elezz
Faculty of Veterinary Medicine and Animal Science, University of Yucatan (UADY), Mérida, Yucatán, Mexico
Faculty of Agriculture, Assiut University. Assiut (71526), Egypt

Misba Alewi
Debrezeit Agricultural Research Center, Debre Zeit, Ethiopia

Aberra Melesse
Department of Animal and Range Sciences, Hawassa University, Hawassa, Ethiopia

Juan de Dios Guerrero-Rodríguez and Ángel Bustamante-González
Colegio de Postgraduados, Campus Puebla. Km. 125.5 Carretera Federal México-Puebla, San Pedro Cholula, Puebla, México. CP 72760

Rocío Margarita Flores-Galván, Numa Pompilio Castro-González and Ramiro Escobar-Hernández
Benemérita Universidad Autónoma de Puebla. Unidad Académica de Ingeniería Agrohidráulica-Plantel de Ingeniería Agronómica y Zootecnia. Avenida Reforma No. 165 CP 73900, Tlatlauquitepec, Puebla. México

Rocío Teresa García-Balcázar, Miriam Plata-Rodríguez and Alejandro Villa-Godoy
Universidad Nacional Autónoma de México, Facultad de Medicina Veterinaria y Zootecnia (UNAM-FMVZ), Departamento de Fisiología y Farmacología. Edif. 2, 2313, Ciudad Universitaria, Ciudad de México, CP 04510, México

Antonio Díaz-Cruz
UNAM-FMVZ, Departamento de Nutrición Animal y Bioquímica. Ciudad Universitaria, Ciudad de México, 04510, México

Joel Hernández-Cerón
UNAM-FMVZ, Departamento de Reproducción Animal. Ciudad Universitaria, Ciudad de México, 04510, México

Karla Rodríguez-Hernández
Instituto Nacional de Investigaciones Forestales, Agrícolas y Pecuarias, CIR-Norte Centro. Matamoros, Coahuila, 27440, México

O. J. Makinde
Department of Animal Science, Federal University, Gashua, Nigeria

M. Inuwa
Department of Animal Production Technology, Federal College of Wildlife Management, PMB 268, New Bussa, Niger State, Nigeria

J. Gusha, S. Katsande and P. I. Zvinorova
Department of Para clinical Veterinary studies, Faculty of Veterinary Science, University of Zimbabwe, MP167, Mount Pleasant, Harare

T. Chiuta
Makoholi Research Institute, P. Bag 9182, Masvingo, Zimbabwe

Esther Evaris, Luis A. Sarmiento-Franco, Jose C. Segura-Correa and Concepción M. Capetillo-Leal
Departamento de Nutrición Animal, Facultad de Medicina Veterinaria y Zootecnia, Universidad Autónoma de Yucatán, Carretera Mérida-Xmatkuil km 15.5. Apdo. 4-116 Itzimná, C.P. 97100, Mérida, Yucatán, México

Girma Chalchissa
Adami Tulu Agricultural Research Center, Ziway, Ethiopia

Yoseph Mekasha
International Livestock Research Institute, Addis Ababa, Ethiopia

Mengistu Urge
Haramaya University, School of Animal and Range Sciences, Dire-Dawa, Ethiopia

James O. Daramola, John A. Abiona and Olusiji F. Smith
Department of Animal Physiology

Olubukola A. Isah
Department of Animal Nutrition

Olusiji S. Sowande
Department of Animal Production

Moshood O. Olaniyi
College of Veterinary Medicine

Jimoh A. Olanite
Department of Pasture & Range Management Federal University of Agriculture Abeokuta, PMB 2240, Nigeria

D. E. Oliveira
Santa Catarina State University, Lages, SC, Av. Luiz de Camões, 2090 - Conta Dinheiro, 88520-000, Brazil

S. R. Medeiros
Embrapa Beef Cattle, BR 262 km 4, 79002-970 - Campo Grande, MS - Brazil

L. O. Tedeschi
Texas A&M University, 230 Kleberg Center, College Station, TX 77843-2471, USA

S. C. Silva and D. P. D. Lanna
USP/ESALQ, C.P. 09, Piracicaba, São Paulo, 13418-900, Brazil

J. F. J. Torres-Acosta
Facultad de Medicina Veterinaria y Zootecnia, Universidad Autónoma de Yucatán

P. G. González-Pech
Facultad de Medicina Veterinaria y Zootecnia, Universidad Autónoma de Yucatán
Centro Multidisciplinario de Educación, Ciencia y Cultura. Km 15.5 Carretera Mérida-Xmatkuil, Apdo. 4-116 Itzimná.C.P. 97100, Mérida, Yucatán, México

James O. Daramola, Ezekiel O. Adekunle, Oyegunle E. Oke, Opeoluwa Ogundele, Ebun-Oluwa A. Saanu and Adebisi J. Odeyemi
Department of Animal Physiology, Federal University of Agriculture Abeokuta, P.M.B 2240, Abeokuta, Nigeria

L. T. Ogunniyi, A. Ajiboye and W. A. Sanusi
Department of Agricultural Economics and Extension, Ladoke Akintola University of Technology Ogbomoso

Index

* 9 7 8 1 6 3 9 8 7 3 4 5 6 *